The Welfare of Laboratory Animals

Animal Welfare

VOLUME 2

Series Editor

Clive Phillips, *Professor of Animal Welfare, Centre for Animal Welfare and Ethics, School of Veterinary Science, University of Queensland, Australia*

The Welfare of Laboratory Animals

Edited by

Eila Kaliste

*State Provincial Office of Southern Finland Social and Health Affairs,
Hämeenlinna, Finland*

 Springer

A C.I.P. Catalogue record for this book is available from the Library of Congress.

ISBN 978-1-4020-6136-3 (PB)
ISBN 978-1-4020-2270-8 (HB)
ISBN 978-1-4020-2271-5 (e-book)

Published by Springer,
P.O. Box 17, 3300 AA Dordrecht, The Netherlands.

www.springer.com

Printed on acid-free paper

All Rights Reserved
© 2007 Springer
No part of this work may be reproduced, stored in a retrieval system, or transmitted
in any form or by any means, electronic, mechanical, photocopying, microfilming, recording
or otherwise, without written permission from the Publisher, with the exception
of any material supplied specifically for the purpose of being entered
and executed on a computer system, for exclusive use by the purchaser of the work.

Contents

Series Preface   vii

Preface   ix

Contributing authors   xi

GENERAL PRINCIPLES FOR MAINTENANCE AND USE OF LABORATORY ANIMALS   1

Animal welfare - an introduction   3
    Dorte Bratbo Sørensen

Research, animals and welfare. Regulations, alternatives and guidelines   15
    Timo Nevalainen

Infections in laboratory animals: Importance and control   23
    Werner Nicklas

Housing, care and environmental factors   37
    Axel Kornerup Hansen and Vera Baumans

Nutrition and animal welfare   51
    Merel Ritskes-Hoitinga and Jan H. Strubbe

Experimental procedures: General principles and recommendations   81
    David B. Morton

THE WELFARE OF DIFFERENT SPECIES 117

The welfare of laboratory mice 119
  Vera Baumans

The welfare of laboratory rats 153
  Eila Kaliste and Satu Mering

The welfare of laboratory guinea pigs 181
  Norbert Sachser, Christine Künzl and Sylvia Kaiser

The welfare of laboratory rabbits 211
  Lena Lidfors, Therese Edström and Lennart Lindberg

The welfare of laboratory dogs 245
  Robert Hubrecht and Anthony C Buckwell

The welfare of pigs and minipigs 275
  Peter Bollen and Merel Ritskes-Hoitinga

The welfare of non-human primates 291
  Jann Hau and Steven J. Schapiro

Animal welfare issues under laboratory constraints, an ethological
perspective: rodents and marmosets 315
  Augusto Vitale, Francesca Cirulli, Francesca Capone and
  Enrico Alleva

Index 339

Animal Welfare by Species: Series preface

Animal welfare is attracting increasing interest worldwide, but particularly from those in developed countries, who now have the knowledge and resources to be able to offer the best management systems for their farm animals, as well as potentially being able to offer plentiful resources for companion, zoo and laboratory animals. The increased attention given to animal welfare in the West derives largely from the fact that the relentless pursuit of financial reward and efficiency has lead to the development of intensive animal production systems that offend the conscience of many consumers in those countries. In developing countries, human survival is still a daily uncertainty, so that provision for animal welfare has to be balanced against human welfare. Welfare is usually provided for only if it supports the output of the animal, be it food, work, clothing, sport or companionship. In reality there are resources for all if they are properly husbanded in both developing and developed countries. The inequitable division of the world's riches creates physical and psychological poverty for humans and animals alike in all sectors of the world. Livestock are the world's biggest land user (FAO, 2002) and the population, particularly of monogastric animals, is increasing rapidly to meet the need of an expanding human population. Populations of animals managed by humans are therefore increasing worldwide, so there is the tendency to allocate fewer resources to each one.

The intimate connection between animal, stockman and consumer that was so essential in the past is rare nowadays, having been superseded by technologically efficient production systems where animals on farms and in labs are tended by fewer and fewer humans in the drive to increase labour efficiency. Consumers also rarely have any contact with the animals that produce their food. In this estranged, efficient world man struggles to find the moral imperatives to determine the level of welfare that he should afford to animals within his charge. Some aim for what they believe to be the highest levels of welfare provision, such as the owners of pampered pets, others deliberately or through ignorance keep animals in impoverished conditions or even dangerously close to death. Religious beliefs and directives encouraging us to care for animals have been cast aside in a supreme act of human self-confidence, stemming largely from the accelerating pace of scientific development. Instead, today's moral code derives as much from horrific tales of animal abuse portrayed in the media and the assurances that we receive from supermarkets that animals used for their products were not abused in this way. The young were always exhorted to be kind to animals through exposure to fables whose moral message was the benevolent treatment of animals. Such messages are today enlivened by the powerful images of modern technology, but essentially still alert children to the wrongs associated with animal abuse.

This series has been designed to provide academic texts discussing the provision for the welfare of the major animal species that are managed by humans. They are not detailed blue-prints for the management of animals in each species, rather they describe and consider the major welfare concerns of the species, often in relation to similar species or the wild progenitors of the managed animals. Welfare is considered in relation to the animal's needs, concentrating on nutrition, behaviour, reproduction and the physical and social environment. Economic effects of animal welfare provision are considered, and key areas requiring further research.

With the growing pace of knowledge in this new area of research, it is hoped that this series will provide a timely and much-needed set of texts for researchers, lecturers, practitioners, and students. My thanks are particularly due to the publishers for their support, and to the authors and editors for their hard work in producing the texts on time and in good order.

Clive Phillips, Series Editor
Professor of Animal Welfare and Director, Centre for Animal Welfare and Ethics, School of Veterinary Science, University of Queensland, Australia

Reference: Food and Agriculture Organisation (2002). http://www.fao.org/ag/aga/index_en.htm.

Preface

Laboratory animals are important tools in biomedical research to investigate such vital issues as the ontogeny and ageing of mammals, mechanisms of diseases and their prevention or treatment, or health risks in our living environment. In 1999, 9.7 million animals were used in experiments (including 8.7 million mammals) in the European Union member states. They were mostly mice (5.3 million), rats (2.6 million), guinea pigs (0.29 million) and rabbits (0.23 million). Of the larger animal species, 66000 pigs, 22 000 dogs and 7000 primates were used in the same year in the 15 EU member states.

The welfare of laboratory animals is perhaps one of the most distrusted issue concerning animals under human control. The discussion about rights and ethics of animal use is of paramount importance to scientist, authorities and lay people. The use of laboratory animals is strictly regulated by legislation, and the numbers of animals used in research is the subject of intense scrutiny. Today, the principles of 3 R's (Reduction, Refinement and Replacement) are accepted to be the main guide for the use of laboratory animals. Moreover, a lot of research focuses on the welfare issues concerning the maintenance and use of laboratory animals, searching for better alternatives to husbandry routines, experimental techniques, as well as alternatives to animal research. This has led to several principles, guidelines and recommendations, the goal being to ensure the welfare of animals and the reliability of research.

The welfare of laboratory animals includes two main issues: one is their breeding and general maintenance, the other is their handling during experimental procedures. Breeding includes strict control of the genetics, at least with rodents. In the maintenance of laboratory animals, the standardisation and elimination of confounding factors like pathogens are the main principles by which the scientific reliability of experiments is ensured. This means many restrictions on the environment of laboratory animals. They have to eat only a standardised diet, live on the same bedding material, under a regular light rhythm etc., in the facilities with very high hygienic control. Meanwhile, their welfare is preserved as far as possible by enrichmental tools and appropriate care routines, the main goal being that the species specific ethological needs are fulfilled. In experiments, appropriate methods must be used when procedures such as administration of substances, sampling of tissues, anaesthesia and euthanasia are carried out. These procedures should not confound the experimental results and the welfare of animals must be ensured as far as possible. Training and education of personnel undertaking these procedures are important to ensure a good science.

This book has two main parts: part one focuses on the general principles of laboratory animal maintenance and experimental use, as well as factors which have to be taken into account when good research is done with animals. The second part is species specific, concentrating on the species most used as laboratory animals. This part gives a comprehensive description of the welfare questions considered to be important for each species under laboratory conditions.

The authors of this book are leading European scientists in laboratory animal science. I wish to thank all of them for their valuable contribution of this book. The pervading theme of the book is that animal welfare can be enhanced by giving the animals safe living environment which fulfils the species specific needs. The living environment should be without severe stress though the environment should be variable enough to help animals to cope with different challenges when they are taken into the experiment. Indeed, the welfare of laboratory animals should be under continuous evaluation, and the one goal should be it's improvement as far as possible.

Eila Kaliste

Contributing Authors

Enrico Alleva
Department of Cell Biology and Neurosciences, Istituto Superiore di Sanità, Rome, Italy

Vera Baumans
Department of Laboratory Animal Science, University of Utrecht, Utrecht, The Netherlands, and Karolinska Institutet, Stockholm, Sweden

Peter Bollen, Biomedical Laboratory , University of Southern Denmark, Odense, Denmark

Anthony C Buckwell
Division of Biomedical Services, University of Leicester, Leicester, United Kingdom

Francesca Capone
Department of Cell Biology and Neurosciences, Istituto Superiore di Sanità, Rome, Italy

Francesca Cirulli
Department of Cell Biology and Neurosciences, Istituto Superiore di Sanità, Rome, Italy

Therese Edström
Astra Zeneca R & D, Mölndal, Sweden

Jann Hau
Department of Neuroscience, University of Uppsala, Uppsala, Sweden

Axel Kornerup Hansen
Department of Pharmacology and Pathobiology, The Royal Veterinary and Agricultural University, Frederiksberg, Denmark

Robert Hubrecht
Universities Federation for Animal Welfare, Wheathamstead,, Hertfordshire, United Kingdom

Eila Kaliste
National Laboratory Animal Center, University of Kuopio, Kuopio, Finland

Sylvia Kaiser
Department of Behavioural Biology, University of Münster, Münster, Germany

Christine Künzl
Department of Behavioural Biology, University of Münster, Münster, Germany

Lena Lidfors
Department of Animal Environment and Health, Swedish University of Agricultural Sciences, Skara, Sweden

Lennart Lindberg
National Veterinary Institute, Uppsala, Sweden

Satu Mering
National Laboratory Animal Center, University of Kuopio, Kuopio, Finland

David B Morton,
Centre for Biomedical Ethics and the Biomedical Services Unit, University of Birmingham, Birmingham, United Kingdom

Timo Nevalainen
National Laboratory Animal Center; University of Kuopio, Kuopio, and Faculty of Veterinary Medicine, University of Helsinki, Helsinki, Finland

CONTRIBUTING AUTHORS

Werner Nicklas
Central Animal Laboratories, Microbiological Diagnostics, German Cancer Research Center,
Heidelberg, Germany

Merel Ritskes-Hoitinga
Biomedical Laboratory, University of Southern Denmark, Odense, Denmark and Department of Neuroendocrinology, University of Groningen, Haren, The Netherlands

Norbert Sachser
Department of Behavioural Biology, University of Münster, Münster, Germany

Steven J. Schapiro
Department of Veterinary Sciences, The University of Texas M.D. Anderson Cancer Center, Bastrop, TX, USA

Jann Strubbe
Department of Neuroendocrinology, University of Groningen, Haren, The Netherlands

Dorte Bratbo Sørensen
Department of Pharmacology and Pathobiology, The Royal Veterinary and Agricultural University, Frederiksberg, Denmark

Augusto Vitale
Department of Cell Biology and Neurosciences, Istituto Superiore di Sanità, Rome, Italy

GENERAL PRINCIPLES FOR MAINTENANCE AND USE OF LABORATORY ANIMALS

Chapter 1

ANIMAL WELFARE - AN INTRODUCTION

Dorte Bratbo Sørensen
Division of Laboratory Animal Science and Welfare, The Royal Veterinary and Agricultural University, Frederiksberg, Denmark

1. INTRODUCTION

"Why even bother?" Perhaps this is the first question we should ask, when we begin to address the issue of animal welfare. If we do not have any moral or ethical obligations toward animals as pertains to their welfare, there is no reason to consider whether their welfare is good or bad. On the other hand, if we do have such moral obligations, we need to be able to assess the welfare of the animals we work with.

The fundamental assumption in this chapter is that we indeed have a moral obligation to ensure the welfare of animals, and therefore we need ways to evaluate how the animals we work with are faring. A lot of different views and theories exist on the nature of animal welfare, but despite decades of committed work, no final definition has been agreed upon.

Scientists have often tried to define animal welfare in a way which already contains the answer as to how it can be measured. However, agreeing on the nature of animal welfare does not require us to define it like we would define any technical term such as hyperglycemia or hypoplasia, but rather it requires agreement on the basic values that contribute to the well-being of the individual animal (Tannenbaum 1991, Duncan and Fraser 1997). Any conceptualization of animal welfare inherently involves values because it pertains to what is better or worse for the animals. The different research approaches and interpretations that scientists use in assessing animal welfare often merely reflect such value-laden presumptions (Fraser et al. 1997).

So the philosophical question considering which basic values matter most, must be solved before animal welfare can be assessed. If our assumptions on the nature of animal welfare are implicit or perhaps even unclear, we can't be certain that we are asking the right questions, and we certainly can't be sure that we are asking them the right way.

The question of <u>how</u> to measure animal welfare, however, is not a philosophical one but a scientific one, whereas the overall evaluation and interpretation of the results of these measurements calls for both a philosophical and a scientific approach.

The nature of animal welfare falls into one of two categories. Either the basic values are objective, such as good biological functioning or the possibility to perform natural behaviours. These values can be assessed using measures such as reproduction rate, disease prevalence, cortisol levels and occurrence of stereotyped behaviours.

Alternatively, the basic values may be subjective in nature, relating to the inner mental state of the animal, such as feelings and preferences. As we are not yet able to see what's going on inside the animal's head, these values cannot be directly measured. However, evaluating certain behaviours of the animal may provide an indirect measure of these feelings and preferences.

2. DEFINING ANIMAL WELFARE IN TERMS OF PREFERENCES AND FEELINGS.

2.1 Preference theories (or desire-fulfilment theories)

Preference theory (Success theory) holds the notion that the level of welfare relates directly to having desires or preferences fulfilled. The welfare of an animal depends on the satisfaction of preferences - a more preferred environment results in a higher level of welfare (Jensen and Sandøe 1997). This theory, originally relating to human well-being, raises a central question which becomes even more obvious when working with animals: Does it contribute to an individual's welfare to have its desires met, even if the individual does not experience it? Intuitively it does not – if you do not realise that an important desire has been fulfilled, it will not change your situation – you have to experience the fulfilment of a preference for it to influence your welfare. Sandøe (1996) expresses this criterion of experience in his version of the experienced preference satisfaction theory:
"A subjects welfare at a given point in time (t1) is relative to the degree of agreement between what he/it at t1 prefers (is motivated to do, wants, aspires after, hopes for, does not try to avoid or is not indifferent to getting)

and how he/it at t1 sees his/its situation (past, present and future) - the better agreement the greater welfare".

Sandøe states that it is important that you experience the fulfilment of your preferences and that your preferences are fulfilled, while you still have them. In other words, your preferences must exist in the present if fulfilment should result in increased welfare. But this theory also calls for the individual's ability to judge its own situation – both with regard to the past, the present and the future. It can be argued that animals are not capable of such judgements regarding the past and the future, and hence they have no long-term preferences. As discussed below, preference studies support the idea that animals do not experience or at least do not consider long-term preferences.

2.2 Hedonism

2.2.1 Narrow hedonism

Hedonism deals with feelings and mental states that matter to the animal. According to the hedonistic approach, good welfare consists of a life-long presence of pleasant mental states and, just as important, the absence of unpleasant mental states (Appleby and Sandøe 2002). The more pleasant feelings and the fewer unpleasant ones, the better welfare (Jensen and Sandøe 1997).

Several scientists have agreed that feelings are what matters in animal welfare. For example, Dawkins (1990) writes: *"Let us not mince words: Animal welfare involves the subjective feelings of animals."* And Duncan (1996) concludes that: *"It is feelings that govern welfare and it is feelings that should be measured in order to assess welfare"* as well as *"...sentience, in other words feelings, is what welfare is all about".*

2.2.2 Preference-hedonism

The problem with the above mentioned hedonistic view – the so-called narrow hedonism – is that it is difficult to agree on which feelings are positive and which are negative, and to which extent they count. Moreover, different feelings may not have a basic common inherent quality, which makes it very difficult to compare different feelings. One promising way of dealing with this problem is to accept the claims of preference-hedonism, as expressed by Parfit (1984): *"Narrow hedonism assumes, falsely, that pleasure and pain are two distinctive kinds of experience. Compare the pleasures of satisfying an intense thirst or lust, listening to music, solving an intellectual problem, reading a tragedy and knowing that one's child is happy. These various experiences do not contain any distinctive common quality. What pains and pleasures have in common are their relations to our desires. On the use of "pain" which has a rational and moral significance, all pains are when experienced unwanted, and a pain is worse or greater the more it is unwanted. Similar, all pleasures are when experienced wanted, and they are better or greater the more they are wanted. These are the claims of Preference-Hedonism. On this view, one of two experiences is more pleasant if it is preferred."*

Coming back to the experienced preference satisfaction theory of Sandøe, it should be noted that although this theory is a variant of success theory, it allows room for having preferences for certain feelings, which again relates it to the theory of preference-hedonism.

When working with animal welfare, preference-hedonism seems to be a good starting point. Preference-hedonism appeals only to desires about one's present state of mind (Parfit 1984), and animals most likely do not have long-term preferences or preferences regarding the past.

2.2.3 Measuring feelings

Animals' feelings are very difficult to measure, which make the hedonistic view hard to work with on a practical level. However, if we assume that an animal would prefer situations which are linked to pleasant feelings, and avoid situations arousing negative ones, we can indirectly assess feelings by measuring the preferences of the animal.

2.2.4 Measuring preferences

Measuring an animal's preferences can be done in a strictly scientific way, as demonstrated below. However, it is worth noting that it is difficult – if not impossible – to discern whether an animal acts the way it does due to certain preferences or because preferences are merely a means to reach a certain goal, namely the preferred mental state.

Preferences can be assessed using tests such as choice tests or operant tests. In a choice test, an animal must choose between two or more resources (stimuli) provided to the animal by the researcher. In an operant test, an animal is trained to perform a simple response (e.g. pressing a lever) in order to obtain something good – a positive stimulus. If the animal is motivated to obtain the stimulus, it will work for the stimulus. The more important the stimulus is to the animal, the harder the animal will work to obtain it. In choice tests there is also an element of work, since the animal has to decide on for instance going from A to B, when making a choice, but it is not possible to assess the strength of a certain preference to the same extent as when using operant tests.

In a preference test, the animal is thus presented with a choice of certain environmental factors, and it is assumed that the animal will choose according to its preferences, and that these choices will be made in the best interest of its own welfare (Duncan 1992). However, the answer is not so straightforward. Using preference testing to assess animal welfare calls for caution on several points.

2.2.5   Problems of preference testing

First, we do not know for sure what we are testing. In operant preference testing, it is most likely not just the preferences of the animal, but rather the decisions made by the animal on the basis of both preferences and environmental factors, which may influence the amount of work the animal must perform to reach its goal (Sørensen 2001).

Moreover, a preference test tends to only give an idea of the relative properties of the choices given. If the animal is given the choice between two aversive conditions, it may show strong preference for one of the conditions. Nevertheless, the animal's welfare is reduced even by exposure to the "preferred" condition. In the same manner, exposure to two appealing but not essential resources may indicate a strong preference for one, but actually the lack of both resources will not affect the welfare of the animal severely (Duncan 1992).

2.2.6   Animals and long-term preferences

On the assumption that the nature of welfare corresponds with the theory of experienced preference satisfaction as mentioned previously, one particular point must be carefully considered when evaluating animal welfare. Even though animals show anticipatory behaviour (Ladewig et al. 2002, Van der Harst et al. 2003) and are able to anticipate the consequences of a choice over a short time (i.e. seconds) (Abeyesinghe et al. 2003), it often seems as if animals do not consider long-term preferences (preferences on what might be good in the long run (days, weeks or even years) as opposed to instant gratification). Compared to humans, animals are probably less able to think about past and future situations, and therefore there may be lack of agreement between what the animal prefers in a preference test, and what the researcher knows is best for the animal on a long term basis. Broiler breeding stock has been selected for increased growth, and therefore increased appetite, to such an extent that they will become obese if they are allowed free access to food. Even though the resulting obesity will reduce long-term welfare of the animals, feed restriction will most likely also result in reduced welfare (Duncan 1992). It is also relevant that rodents fed *ad libitum* will have a higher risk of obesity, tumours and a reduced life-span. This corresponds to a situation most people can relate to, namely that fulfilling the preferences of children for eating gummy bears, chips and watching junk cartoons does not improve the child's welfare in the long run, even though the child feels it is all that matters here and now. If children and animals on the other hand are Preference-Hedonists, meaning that it's only fulfilling the present preferences for preferred mental states that counts, then the chickens and the rats as well as the child in the above examples are experiencing good welfare.

However, working with animal welfare, we can not allow ourselves to ignore the importance of long-term effects of the environment on animal welfare. Environmental factors, which we as informed humans, believe or know are important, may prove indifferent to the animal, when tested. For example, to us, locomotion is an important factor in animal welfare - it keeps the animal fit and strengthens the joints and muscles. Pigs in operant preference testing show very little interest in walking for it's own sake (Matthews and Ladewig 1994, Ladewig and Matthews 1996). People who jog often do so, because it makes them feel good, probably because they know it will improve their health on the long term. Pigs do not realise that walking in a treadmill may prevent health problems in the future, but if walking is important to pigs, why don't they have build-in preferences to ensure that they keep themselves fit?

The answer probably lies in the way we house our animals. In wild animals factors such as locomotion are an integral part of normal behaviour displays, including foraging, escaping predators, hunting and searching for shelter. Hence the animal does not need any motivation for locomotion per se. Evaluation of the result of preference testing therefore must contain a careful judgement of apparent lack of preferences for environmental factors or behaviours obviously important for the welfare of the animal. It is important to realise that if we consider behaviours that the animal is highly motivated to perform (assessed by preference testing), the need for factors having a positive impact on long-term welfare will often be fulfilled in the process. For example, gnawing and chewing behaviours with no consummatory function are often considered an integral part of rat behaviour, but assuming that rats will show gnawing behaviour for no reason at all is - from an ethological point of view - counter-intuitive. Gnawing is a behaviour which will ensure the necessary wear and tear of the teeth of the rodent, but this fact does not mean that rats have built-in preferences for gnawing *per se*. If rats live a varied life in a complex environment, gnawing will be performed as an integral part of different behaviours such as eating, exploring, gaining access to desirable environments or escaping from aversive ones and probably nest building. Hence, physically-based needs for gnawing are fulfilled in the process of performing behaviours for which the rats are motivated. It would be reasonable to suggest that also in laboratory rats there is a direct purpose of gnawing such as preparing materials for nest-building or trying to gain access to something desirable. Another obvious reason to gnaw would be to try to escape from an aversive environment (Sørensen et al. in press).

It is essential to realise that a preference test only provides information on the current demands of the animal, and these demands must be important to the animal – otherwise the animal would not have worked to obtain them. But since animals cannot be expected to make rational choices, taking into consideration the long-term consequences on their welfare, it is the scientists job to balance the current desires of the animal with the scientific knowledge that tells us which environmental factors the animal will benefit from in the long run.

3. DEFINING ANIMAL WELFARE IN TERMS OF THE NATURE OF THE ANIMAL

3.1 Perfectionism

The theory of Perfectionism states that in order to have a good life one must realise certain genetically-based, species-specific potentials. In other words, the animal must be able to express its nature in order for the animal to live a good life. Rollin (1989) writes: *"We would expect its (the animal's) behaviour to be appropriate to its telos – the unique, evolutionary determined, genetically encoded, environmentally shaped set of needs and interests which characterise the animal in question – the 'pigness' of the pig, the 'dogness' of the dog, and so on"*.

One of the main objections to this idea is that often an animal in the wild or in a semi-natural environment expresses behaviours which intuitively does not seem to add to the animal's welfare while it is expressing this behaviour (for example, escaping a predator). Such behaviours are adaptations that have evolved to enable animals to cope with aversive environments or situations. Critics of perfectionism will say that it is reasonable to assume that an animal under such aversive, or even life-threatening, circumstances has reduced welfare or even suffers, and hence performing the full behavioural repertoire does not necessarily increase the welfare of the animal.

However, even though an animal possesses a set of conditional behavioural patterns that governs the performance of behaviours, it doesn't necessarily mean that the animal must perform these behaviours to enjoy welfare, but rather that the animal should be able to use these mechanisms for adaptation, if the circumstances should require it to (Fraser et al. 1997). So laboratory rodents do not need to run away from cats from time to time to have good welfare, but if a cat appears, the rodent should be allowed to express relevant flight-related behaviour.

3.2 Evaluating "natural" behaviours

To evaluate animal welfare at the basis of Perfectionism demands that we have thorough knowledge on how the animal would behave in a natural environment. Laboratory rodents have been bred by humans for several generations, and furthermore mice and rats are highly adaptive, both factors that may raise doubts as to what really constitutes the true natural behaviour of these species. But the ancestors of our laboratory rodents are still living in the wild, and the study of their behaviour will provide a very good starting point. To further refine the obtained results, laboratory rodents can be studied in environments of variable complexity, preferable similar to those of their ancestors. Last, the study of feral animals, i.e. animals released or escaped into the wild, having adapted to a life without humans caring for them may provide valuable information. These approaches probably will not provide the entire truth, but hopefully it will bring us much closer.

4. DEFINING ANIMAL WELFARE IN TERMS OF THE BIOLOGICAL FUNCTIONING OF THE ANIMAL

Good biological functioning constitutes the basis of a great deal of welfare evaluating methods. The view that the functioning of the animal is what matters is represented e.g. by McGlone (1993), who suggests that *"an animal is in a state of poor welfare only when physiological systems are disturbed to the point that survival or reproduction are impaired."* Another theory emphasising the need for good biological functioning is the theory of coping: *"The welfare of an individual is its state as regards its attempts to cope with its environment."* (Broom 1986, 1996).

According to Broom, the attempts to cope and the results of failure to cope can be measured using variables such as mortality rates, disease incidence, reproductive success, severity of injury, extent of adrenal activity and so on (Fraser and Broom 1990). In other words, the better the animal is able to adapt using its physiological mechanisms, without these mechanisms being challenged beyond their capability, the better the animal's welfare.

The previous problem of preference-testing and long-term preferences would be solved using this approach. If animals are not showing any preferences for resources that they will benefit from in the long run, then their biological functioning may be jeopardised – just like the child eating nothing but chips and gummy bears. So according to this theory feeding on a diet consisting of gummy bears will in fact result in decreased welfare due to impaired biological functioning.

However, this theory raises another problem. Consider an animal whose biological functioning is in fact impaired, but which doesn't experience any resulting negative mental states at all. This would be the case of a vasectomised male rat, used for mating embryo transfer recipient females. The vasectomised male is allowed to perform courtship and mating behaviour, but he can not produce offspring, It seems contra-intuitive that such a rat should experience bad welfare due to impairment of his biological functioning.

5. SCIENTIFICALLY BASED HYBRID VIEWS ON ANIMAL WELFARE

Scientists working with animal welfare seem to agree that accurate assessment of animal welfare should be based on a blend of these different theories. The theory put forward by Broom (1986, 1996) also holds elements of perfectionism in that it is the ability of the animal to function according to its nature that counts. Moreover, Broom also states that the measuring of welfare should include behavioural and physiological indicators of pleasure – thus making pleasure count which is clearly a hedonistic approach (Broom 1996).

A previously mentioned theory on welfare is that of Dawkins (1990). It is based on hedonism, but relates to several other theories. Dawkins (1990) states that: *"Suffering occurs when unpleasant subjective feelings are acute or continue for a long time, because the animal is unable to carry out the actions that would normally reduce risks to life and reproduction in those circumstances."*

So, suffering originates from not being able to cope by using evolutionary-determined, species-specific behaviours. So far, suffering, and thus poor welfare, relates to hedonism, perfectionism and the biological functioning of the animal. But furthermore, according to Dawkins, not being able to do what you want will result in mental states which negatively impact on the animal's welfare. Hence not having one's preferences fulfilled will lead to a decrease in welfare. Dawkins (1990) provides this example: *"Wild birds may have little chance of surviving, if they do not migrate, so the caged one (of the same species and well cared for) is behaving 'as if' death through failure to migrate were very likely. In other words, the canonical costs (risks to fitness) of not migrating may be very small, but the animal may suffer nevertheless."*

So in fact, Dawkins' view represents a mixture of many philosophical approaches to the nature of welfare. Such a hybrid view is also presented by Simonsen (1996), whose definition shares qualities with most of the previously mentioned theories. This definition states that: *"Animal welfare consists of the animals' positive and negative experiences. Important negative experiences are pain and frustration and important positive experiences are expressed in play, performance of appetitive behaviour and consummatory acts. Assessment of animal welfare must be based on scientific knowledge and practical experience related to behaviour, health and physiology."* (Simonsen 1996).

The first part of the definition is truly hedonic in character. Poor welfare originates in negative experiences or mental states such as pain and frustration. The second part involving positive experiences does not mention the positive experiences in themselves, but rather their expression. It is reasonable to assume that the animal would prefer to have the opportunity to perform behavioural patterns such as play, appetitive behaviour and consummatory behaviour, since these behaviours express good welfare. Not being able to perform these behaviours will lead to frustration and hence to reduced welfare (as frustration is a negative mental state). The second part of the definition put forward by Simonsen is therefore related to preference theories. Moreover, the behavioural patterns mentioned by Simonsen, are species-specific normal behaviours. Performing these behaviours is connected to experiencing good welfare, and therefore elements of perfectionism are present.

The five freedoms, as described by Webster (2001) relate to all of the above mentioned theories on animal welfare:

Freedom from hunger and thirst
Freedom from physical discomfort and pain
Freedom from injury and disease
Freedom from fear and distress
Freedom to conform to essential behaviour patterns

The first four freedoms all pertain to aversive experiences and thus relates to hedonism. However, it is possible for an animal to be sick without perceiving it, and still the disease would affect the animal's natural functioning, one of which is reproduction. For example, if a pregnant rat is infected with Kilham rat virus, the infection will not affect the pregnant rat, but if the virus crosses the placental barrier, it may result in abortion or malformations of the developing foetuses. And if an animal is distressed (the fourth freedom), it is most likely experiencing a situation difficult to cope with. In both cases the natural functioning of the animal has been compromised.

The last of the five freedoms clearly relates to Perfectionism – the animal must be allowed to express natural species-specific behaviour to have good welfare. Unfortunately, the formulation is rather vague, failing to define the term "essential behaviour patterns."

6. CONCLUSION

If the serum level of corticosterone is increased, we conclude that the animal is acutely stressed. The crucial question is then: "Why is an elevated corticosterone level an indicator of poor animal welfare?" Is it because the natural functioning of the animal is jeopardised? Or is it because the animal experiences an aversive situation? Or is it because the animal is not allowed to display its natural behaviour?

Working with animals, scientists have proposed that the assessment of animal welfare should include a mixture of different philosophical theories such as hedonism, perfectionism and preference theories. Intuitively, this holistic approach considering the entire animal is appealing. However, in many cases the conclusion drawn on the basis of a scientific evaluation of animal welfare will depend on how the nature of welfare is defined. Consider a dog having behavioural problems caused by fear of being left alone. The anxiety-related behaviour can be eliminated using psychotherapeutic drugs such as tricyclic antidepressants. The medication enables the dog to be at home alone without showing any signs of fear or anxiety. If the dog is not feeling anxious or frightened then, according to a hedonist, the welfare of this dog is not compromised. And if the dog does not have an unfulfilled preference for company, the preference theoretic does not see any problems, either. However, according to a perfectionist there is a reduction in welfare, since the dogs natural functioning and behaviour is compromised (dogs are pack animal and therefore it is natural for a dog to be anxious when left alone). So there is no simple answer to what constitutes animal welfare. That makes it even more important that, when evaluating animal welfare, the underlying assumptions regarding which values are important for animal welfare are made explicit.

REFERENCES

Abeyesinghe SM, Hartnell SJ, Nicol CJ, Wathes CM. Can domestic fowl show self-control? Proceedings of the 37th international congress of the ISAE, 24th -28th of June. Abano Therme, Italy 2003; 85

Appleby MC, Sandøe P Philosophical debate on the nature of well-being: Implications for animal welfare. Anim Welf 2002; 11:283-294

Broom DM. Indicators of poor welfare. Br Vet J 1986; 142:524-526

Broom DM. Animal welfare defined in terms of attempts to cope with the environment. Acta Agric Scand Section A - Animal Science 1996; 27(Suppl):22-28

Dawkins MS. From an animal's point of view: Motivation, fitness, and animal welfare. Behav Brain Sci 1990; 13:1-61

Duncan IJH. Measuring preferences and the strength of preferences. Poult Sci 1992; 71:658-663

Duncan IJH. Animal welfare defined in terms of feelings. Acta agric Scand Section A - Animal Science 1996; 27(Suppl):29-35

Duncan IJH, Fraser D. Understanding animal welfare. In Animal Welfare. Appleby MC, Hughes BO eds, CAB International, Wallingford 1997

Fraser AF, Broom DM. Farm Animal Behaviour and Welfare, Third edition. Bailliére Tindall, London, UK 1990

Fraser D, Weary DM, Pajor EA, Milligan BN. A scientific conception of animal welfare that reflects ethical concerns. Anim Welf 1997; 6:187-205

Jensen KK, Sandøe P. Animal welfare: Relative or absolute? Appl Anim Behav Sci 1997; 54:33-37

Ladewig J, Matthews LR. The role of operant conditioning in animal welfare research. Acta agric Scand Section A - Animal Science 1996; 27(Suppl):64-68

Matthews LR, Ladewig J. Environmental requirements of pigs measured by behavioural demand functions. Anim Behav 1994; 47:713-719

McGlone JJ. What is animal welfare? J Agric Environ Ethics 1993; 6(Suppl 2):26-36

Parfit D. What makes someone's life go best? In "Reasons and Persons". Oxford University Press, Oxford, UK 1984

Rollin B. Animal pain. In "Animal's Rights and Human Obligations". Regan T, Singer P eds, Prentice-Hall Inc, New Jersey, US 1989

Sandøe P. Animal and human welfare - are they the same kind of thing? Acta Agric Scand Section A - Animal Science 1996; 27 (Suppl):11-15

Simonsen HB. Assessment of animal welfare by a holistic approach: Behaviour, health and measured opinion. Acta Agric Scand Section A - Animal Science 1996; 27(Suppl): 91-96

Sørensen DB. Evaluating animal welfare: Assessing the substitutability of two environmental factors by use of operant conditioning. Ph.D. thesis, The Royal Veterinary and Agricultural University, Copenhagen 2001

Sørensen DB, Ottesen JL, Hansen AK. Consequences of enhancing environmental complexity in laboratory rodents - a review with emphasis on the rat. Anim Welf in press

Tannenbaum J. Ethics and animal welfare: The inextricable connection. J Am Vet Med Assoc 1991; 198:1360-1376

Van der Harst JE, Fermont PCJ, Bilstra AE, Spruijt BM. Access to enriched housing is rewarding to rats as reflected by their anticipatory behaviour. Anim Beh 2003; 66:493-504

Webster AJF. Farm animal welfare: the five freedoms and the free market. Vet J 2001; 161:229-237

Chapter 2

RESEARCH, ANIMALS AND WELFARE
Regulations, alternatives and guidelines

Timo Nevalainen
National Laboratory Animal Center, University of Kuopio, Kuopio and Faculty of Veterinary Medicine, University of Helsinki, Helsinki, Finland

1. INTRODUCTION

High quality of biomedical research and acceptance of animal use overall in science necessitate refined animal welfare. On the legal side, the European Commission (86/609/EEC) states that the EU Member States must actively encourage and support the development, validation and acceptance of methods, which could reduce, refine or replace the use of laboratory animals (3R's).

ESF's (European Science Foundation) 'Use of Animals in Research' statement also strongly endorses the principles of the '3R's'. 'Efforts ought to be taken to replace the use of live animals by non-animal alternatives, to reduce the number of animals used in experiments to the minimum required for obtaining meaningful results and to refine procedures, so that the degree of suffering is minimised. Research aiming at improving the welfare of animals should be encouraged and actively supported (ESF 2001).

Recently started revision of the directive is likely to include cost-benefit analysis for ethical evaluation of animal studies at study level, a development which will undoubtedly emphasize the need for better laboratory animal welfare, both in procedures and in housing (European Parliament 2002).

The answer to all these requirements is obvious: Proper education and training of all involved.

2. EUROPEAN REGULATIONS

Harmonisation of the laws and regulations on use of vertebrate animals in research is the key aim of both the European Directive and the Convention (86/609/EEC, ETS123). These contain articles with almost the same text. The Convention includes Appendix A, housing and care of laboratory animals. This Appendix is experiencing a major revision, which is will definitely improve laboratory animal housing and care, and consequently animal welfare.

Article 5 of the Directive and the Convention states general principles of animal care and housing. 'Any animal...shall be provided with accommodation, an environment, at least a minimum degree of freedom of movement, food, water and care, appropriate to its health and well-being. Any restriction on the extent to which an animal can satisfy its physiological and ethological needs shall be limited as far as practicable' (ETS123).

Appendix A gives much more detailed guidelines for animal housing. It contains minimum space allocations for all laboratory species, facility requirements and routine animal care procedures. Appendix A is under revision, and the new one includes general and species specific parts, some with scientific basis for guidelines. The revision will increase some of the space requirements; emphasize group housing for all gregarious species and implementation of environmental enrichment. As such, the revised document is much larger than the present one, and it is expected to improve animal welfare (The Council's Group of Experts on Rodents and Rabbits 2001). A more detailed description of and the basis for the revised Appendix A is presented in another chapter of this book.

Articles 6-12 deal with the procedure to be carried out for scientific purposes. The key message of these articles can be seen in the text of article 7: 'in a choice between procedures, those should be selected which use the minimum number of animals, cause the least pain, suffering, distress or lasting harm and which are most likely to provide satisfactory results' (ETS123).

What does it mean being harmonized? And how does harmonization relate to other requirements for animal welfare, ethics and science? In this environment harmonization can and should be seen as the minimum standard, below which nobody is allowed to operate. Well above the minimum standard, there should be an area of excellence, where ideals of ethics and science are the driving forces. This relation is illustrated in Figure 2-1.

Improvement of laboratory animal welfare means good science in the vast majority of cases. Yet, there may be conflicts as well. This possibility is acknowledged in passing in articles of both Directive and Convention of animals. Article 7 states that the choice of procedures should be selected on the basis of 'which are most likely to provide satisfactory results' (ETS123).

The revised Appendix A is clearer on this possibility, and says that for instance single housing or raised area in rabbits can be omitted if there is welfare or scientific reason not to use them (The Council's Group of Experts on Rodents and Rabbits 2001).

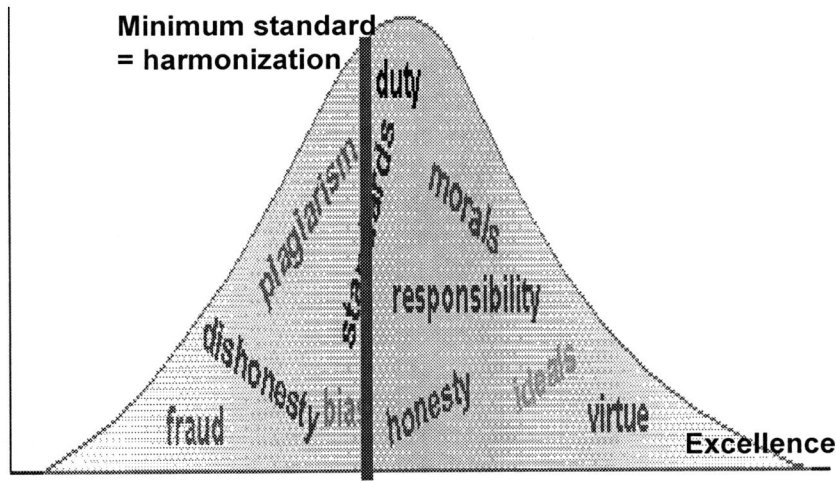

Figure 2-1. Gaussian curve illustrating relationship between harmonization and excellence in all operations, including those aiming at improving laboratory animal welfare.

3. NORTH-AMERICAN REGULATIONS

In the United States, regulations for laboratory animals are included in the Welfare Act, the Public Health Service (PHS) Policy on Humane care and Use of Laboratory Animals, the Guide for the Care and Use of Laboratory Animals, and the U.S. Government Principles for the Utilization and Care of Vertebrate Animals Used in Testing, Research, and Training.

The U.S. Government Principles state that 'the living conditions of animals should be appropriate for their species and contribute to their health and comfort'. Overall, the Policy is a large document and intended to implement and supplement those principles. This Policy is a basic requirement for all PHS-conducted or supported activities involving animals. Compliance must be shown through a written Assurance acceptable to the PHS.

The PHS requires facilities to use the Guide as a basis for an institutional program, and through the Guide that those who care for or use animals in research must assume responsibility for their well-being. Each institution must have an Institutional Animal Care and Use Committee (IACUC) to oversee and evaluate the institution's animal program to assure compliance with all the regulations.

The Guide states that 'proper housing and management of animal facilities are essential to animal well-being' and that 'a good management program provides the environment, housing, and care that permit animals to grow, mature, reproduce, and maintain good health; provides for their well-being; and minimizes variations that can affect research results.' Moreover, the Guide endorses housing which maximises species-specific and minimises stress-induced behaviours.

In Canada detailed guidelines for laboratory animal welfare can be found in CCAC (Canadian Council on Animal Care) Guide to the Care and Use of Experimental Animals and CCAC Guide to the Care and Use of Experimental Animals. These documents state that 'In the past, emphasis has been directed towards providing adequate caging for experimental animals in order to contain them hygienically, to facilitate husbandry, and minimise (husbandry) variables. However, increasing importance is now being placed on reducing the animal's stress, and improving its social and behavioural well-being. Provision of varied environmental enrichment may or may not result in increased cost of operation; however, it is considered that there is often immediate benefit to the animal and ultimately to the researcher and the research.'

The Guide emphasises the social needs of animals to have equal importance as environmental factors such as lighting, heating, ventilation and containment (caging). Singly housed animals must be observed daily to provide social contact for the animal and the animal to become accustomed to the human presence.

Whenever single housing is used the protocols must include measures for meeting the social requirements of the isolated animal. Overall, investigators must justify all deviations from the Guide to an Animal Care Committee (ACC) in order to receive approval. All protocols must be reviewed at least annually by the ACC.

4. ALTERNATIVES AND ETHICS

Refinement, Reduction and Replacement (3R's) are all considered alternatives to laboratory animals (Russell and Burch 1959). When replacement is successful, no welfare problems remain, since no animals are used. Refining procedures or housing has direct consequences for welfare, but reduction has welfare dimension as well. Refinement and reduction (2R's) are not independent of each other; hence understanding of their interplay is crucial.

These 2R's can be considered in a two dimensional model. If a study is evaluated with 2R's, we can see four resulting combinations:

1. Improved welfare, fewer animals
2. Compromised welfare, fewer animals
3. Improved welfare, more animals
4. Compromised welfare, more animals

Of these combinations, option 1 is the preferable direction, and the opposite option 4 is a direction that should be avoided. Option 3 seems to be more acceptable to most people than option 2. This suggests that welfare, or avoidance of pain, suffering and distress, is more important than the number of animals exposed.

In the reduction dimension there is a window of appropriate numbers of animals, below and above which the experiment becomes meaningless and unethical. This is because the first scenario fails to draw conclusions due to too few animals and consequently poor statistical power, and the latter is guilty of unnecessarily large group size. The refinement dimension is more straightforward, the higher the refinement value, the better.

Whenever replacement alternatives cannot be used, as is often the case, the application of the other two alternatives (2Rs) is still necessary and, they are key elements in ethical evaluation. Scientists are concerned with standardization of all factors that may impact on experimental results e.g. a recent Council of Europe expert group emphasized the need for environmental enrichment for all species unless there were scientific or welfare reasons not to do so.

A scientific reason could be potential interference with an experimental protocol, and a welfare reason aggression between incompatible animals. The outcome may simply be a change in the mean, and this may not matter as it should affect all groups, but changes in variance will lead to more animals being used, itself an ethical issue. The opposite could also happen and results could be improved leading to fewer animals being used. Both these option may lead scientists to draw false positive or false negative conclusions and not notice it. It may even be that better science evolves from simultaneous application of the 2Rs.

Any refinement to animal experiments to improve animal welfare requires scientific validation to ensure the refinements are truly beneficial for the animals and do not detract from the scientific integrity of the results.

This is to say that in order to get a comprehensive picture, one should use scientific method to prove both efficacy and 'safety' of a refinement.

A recent EU report called for a cost-benefit analysis as an integral part of any ethical review. The report states that 'Before a license is issued, an ethical and animal-welfare assessment must be carried out setting limits to the level of stress to which the animals may be subjected. Even if it can be shown that certain experiments may be of benefit to animals or humans, they should not be authorized if the stress on the animals used in the experiment exceeds the maximum level' (European Parliament 2002). Since this report started revision of the Directive, these crucial welfare elements will be under a magnifying glass during the revision.

Refinement and reduction, and their breakdown elements, are obvious candidates when assessing costs of a study protocol. Implementation of cost-benefit analysis will identify the weak points of welfare of procedures and housing, and hence inevitably further laboratory animal welfare.

5. EDUCATION AND TRAINING

Modern biomedical research is progressing at an unbelievable pace. Sophistication of methods exposes studies to an increasing number of potentially complicating factors, many totally unknown so far. The animal facilities must be able cope with the change, and all persons, down to the lowest level, should be considered key members of the research teams. In this environment, welfare of the animals used is an essential issue. There is no other way to cope than proper and continuous education and training in laboratory animal science and welfare.

European Convention and Directive required 'competency' of all personnel working with laboratory animals (ETS123, 86/609/EEC). The meaning of the word 'competent' became clear when Federation of European Laboratory Animal Science Associations' (FELASA) working group divided it to four competence categories (FELASA 1995, 1999, 2000):

A. Animal technicians
B. Research technicians
C. Scientist
D. Specialists

Three FELASA working groups have published a curriculum for each of the competence categories. Guidelines for categories A and D are career-type, and those for categories B and C course-type curricula. The 3R's are the guiding principles in all the categories and all curricula contain detailed breakdown on topics taught on the welfare aspects in housing, care and procedures. The newer ones, categories D and B also identify the need for scientific integrity, *i.e.* the 'safety' component of welfare (FELASA 1995, 1999, 2000).

The European Parliament (2002) states that 'Although voluntary guidelines exist on the education and training of persons working with laboratory animals (FELASA), there is currently no EU-wide standard course that can be followed'. FELASA has realized this prior to the report, and established FELASA Accreditation Scheme for Training and Education applicable to all the European categories (FELASA 2002). The scheme is tailored to be a quality assurance system for education, and due to inclusion of welfare, alternative method and ethics elements, well beyond basic harmonization in laboratory animal welfare. When the FELASA Accreditation Scheme is implemented with the FELASA guidelines for all the four categories, it is the single closest thing we have to an EU-wide standard in laboratory animal science and welfare education.

In the United States legal requirement for proper training comes from the Animal Welfare Act, the Health Research Extension Act and the PHS policy. The Animal welfare regulations have more specific topics for training and education than the PHS policy. The Code of Federal Regulations state the responsibility for and goal of training: 'It shall be the responsibility of the research facility to ensure that all scientists, research technicians, animal technicians, and other personnel involved in animal care, treatment, and use are qualified to perform their duties. This responsibility shall be fulfilled, in part, through the provision of training and instructions to those personnel'.

Many research institutes, in United States and elsewhere, are voluntarily accredited by the Association for Assessment and Accreditation of Laboratory Animal Care, International (AAALAC). Since AAALAC scheme is based on the PHS policy and the Guide, training requirements are derived from these federal documents, even for those not otherwise covered by federal laws. In addition to facility and research personnel, proper training of IACUC members is expected by regulatory agencies as well as by AAALAC.

All training in the US is based on performance standard, which emphasizes the outcome of the training process. Training services of the institutes can also be tailored to meet the types of procedures and species used as well as according to experience and competence of the personnel. In reality this means large degree of flexibility.

Attending veterinarians and facility mangers typically have special training in professional programs and many veterinarians have also specialty board certification. Scientists using animals usually have training courses offered by their institutions.

American Association for Laboratory Animal Science (AALAS) offers training and certification for animal care personnel at three levels:

1. Assistant Laboratory Animal Technician
2. Laboratory Animal Technician
3. Laboratory Animal Technologist

The National Institute of Health's Office for Laboratory Animal Welfare strongly endorses 'institutes to offer their staff access to training leading to certification in animal technology, such as available from AALAS'.

REFERENCES

CCAC. Guide to the Care and Use of Experimental Animals. Vol. 2. Canadian Council on Animal Care 1984

CCAC. The CCAC Guide to the Care and Use of Experimental Animals, Vol. 1, 2nd Edition. Canadian Council on Animal Care 1993

Council Directive of 24November1986 on the approximation of laws, regulations and administrative provisions of the Member States regarding the protection of animals used for experimental and other scientific purposes (86/609/EEC) 1986

ETS 123. European Convention for the protection of vertebrate animals used for experimental and other scientific purposes. Strasbourg, 18.III.1986

European Parliament. Report on Directive 86/609 on the protection of animals used for experimental and other scientific purposes (2001/2259(INI) 2002

ESF. European Science Foundation. Use of Animals in Research. ESF 2001

FELASA recommendation on the education and training of persons working with laboratory animals: Categories A and C. Lab Anim 1995; 29:121-131

FELASA recommendations on the education of specialists in laboratory animal science (Category D). Lab Anim 1999; 33:1-15

FELASA guidelines for the education and training of persons carrying out animal experiments (Category B). Lab Anim 2000; 34:229-235

FELASA recommendations for the accreditation of laboratory animal science education and training. Lab Anim 2002; 36(4):373-377

Guide for the Care and Use of Laboratory Animals. Institute of Laboratory Animal Resources. Commission on Life Sciences. National Research Council. National Academy Press. Washington, D.C. 1996

Public Health Service Policy on Humane Care and Use of Laboratory Animals. Office of Laboratory Animal Welfare. Amended August, 2002

Russell WMS, Burch RL. The Principles of Humane Experimental Technique. Special Edition. (1992) Publrs. UFAW. 8 Hamilton Close, South Mimms, Potters Bar Herts, EN6 3QD UK 1959

The Council's Group of Experts on Rodents and Rabbits. Stauffacher M, Peters A, Jennings M, Hubrecht R, Holgate B, Francis R, Elliot H, Baumans V, Hansen AK. Future principles for housing and care of laboratory rodents and rabbits. Report for the revision of the Council of Europe Convention ETS 123 Appendix A for rodents and rabbits, PART B. Background information to GT 123 (2000) 57, 1 February 2001

Chapter 3

INFECTIONS IN LABORATORY ANIMALS: IMPORTANCE AND CONTROL

Werner Nicklas
Central Animal Laboratories, Microbiological Diagnostic, German Cancer Research Center, Heidelberg, Germany

1. INTRODUCTION

The quality of laboratory animals (especially rodents) has constantly improved over the last few decades, however, various agents are still prevalent in numerous colonies of laboratory animals.

Some agents were lost during the domestication process or due to the lack of intermediate hosts for specific agents (e.g. most cestodes). First attempts to eliminate diseases in laboratory animals were made in the 1950s. However, it became obvious that agents with a low potential to induce clinical signs are able to increase variation between individuals, causing an increase in the numbers of animals required for scientific experiments. Many agents, although clinically silent, were shown to influence physiological or immunological functions. Research complications occurred frequently and resulted in the need to eliminate also those agents that cause clinically-silent infections, and to monitor colonies for the presence or absence of such micro-organisms. In fact, most infections do not lead to overt clinical symptoms (disease) and are usually latent. Thus, the absence of clinical manifestations of infection has no or only limited diagnostic value, and subclinical infections are only detected if specific monitoring methods are applied. It is not sufficient that laboratory animals are free of disease, it is essential that they are free of agents that may have impact on the health of animals (and humans) or on the results of animal experiments.

During the last few years health monitoring procedures became increasingly important with the extensive use and exchange of genetically modified animals between research institutions. Such animals are often subclinically infected with various unwanted agents which can be detected only by comprehensive health monitoring. Health monitoring can therefore be an essential prophylactic measure and helps to avoid animal losses or other complications. Some of the micro-organisms that may be present in laboratory animals (e.g. lymphocytic choriomeningitis virus, *Streptobacillus moniliformis*) can also infect humans (zoonoses).

Appropriate health monitoring is therefore an important prerequisite for the use of microbiologically-standardised animals in scientific experiments and for experiments undisturbed by infectious agents. The use of animals which are monitored and free from unwanted agents is the basis to achieve reliable and reproducible results with a minimum of animals and is therefore an important contribution to animal welfare.

2. IMPACT OF INFECTIONS ON EXPERIMENTAL RESULTS

There is no doubt that research complications due to overt infectious disease are significant and that animals with clinical signs of disease should not be used for scientific experiments. But also clinically-silent infections may have severe effects on physiological parameters and on animal experiments. They often remain undetected, and therefore modified results may be obtained and published. There are numerous examples of the influences of micro-organisms on the physiology of laboratory animals and hence of the interference of latent infections on the results of animal experiments. All infections, apparent or inapparent, are likely to increase biological variability and hence result in an increase in animal use.

Research complications may occur in various ways. Whenever pathogens infect laboratory animals, the immune system is activated regardless of the level of pathogenicity, and many agents also have the potential to induce suppression of the immune system. Sometimes, only B cells or T cells, or specific subpopulations are influenced. It is also obvious that infections lead to an increased inter-individual variation within a population resulting in increased numbers of animals necessary to obtain reliable results. Various effects have been observed on the function or the morphology of organs or cell systems. Infected animals may show altered behaviour, suppressed body weight, or reduced life expectancy which may, for example, influence the tumour rate in a population. Micro-organisms present in an animal may lead to contamination of tissue specimens or samples such as sera, cells, or tumours (Nicklas et al. 1993b). This may interfere also with *in vitro* experiments conducted with cells or isolated organs. Finally, latent infections may be activated by environmental factors (e.g. increased humidity or temperature), by experimental procedures such as irradiation, anaesthesia, or by the combination and interaction between various microorganisms.

It may be difficult - especially in genetically modified animals - to distinguish pathogen effects from genetic effects. For example, opportunistic pathogens (e.g. *Pneumocystis carinii*), environmental agents or agents of human origin (e.g. *Staphylococcus aureus*) are increasingly important in immunodeficient animals. Phenotypes are sometimes attributed to genetic modifications, but examples are known where an effect is observed only when specific agents are present.

Several conferences have been held on microbial complications on research, and the knowledge has been summarised almost 20 years ago (Bhatt et al. 1986, Hamm 1986). The literature has later repeatedly been reviewed (e.g. National Research Council 1991, Baker 1998, Nicklas et al. 1999), and hundreds of articles in scientific journals describing research complications by infectious agents stress the importance of using animals for research that are free of infections.

3. PREVENTION OF AGENT INTRODUCTION

Housing of animals is best conducted behind physical barriers. The term "barrier" rather describes a programme aiming at the prevention of pathogen introduction than just a certain design of a facility. Additionally, appropriate management practices are necessary to reduce the risk of introducing agents. These have to take into consideration all relevant factors by which agents might be transmitted (for details, see Nicklas 1993a). The most important sources for an animal population are other animals of the same or closely related species. Most animal facilities are multipurpose and must therefore house a variety of strains coming from various sources. Animals from commercial breeders are usually monitored carefully, and sufficient information on their health status is in most cases available. However, a greater risk arises from animals which have been bred or housed in experimental units where less attention may be paid to preventive hygienic measures compared with breeding units. These animals are much more frequently infected, and especially genetically-modified animals that are usually obtained from numerous facilities, may increase the risk of receiving infected animals. Health reports for these animals are for various reasons often not sufficiently reliable. To avoid the introduction of agents, such animals must be housed under strict quarantine conditions, or the status of such animals must be defined upon arrival to choose appropriate housing conditions (isolator, individually ventilated cages, conventional unit).

The importance of animals from experimental colonies as sources of infections becomes obvious from a survey of biomedical research institutions funded by the National Institutes of Health (Jacoby and Lindsey 1997). In this survey, mouse hepatitis virus (MHV), parvoviruses, ecto- and endoparasites were reported in 10% to more than 30% of the "specific pathogen free (SPF)" mouse colonies. Among non-"SPF" mice, MHV was present in more than 70% of the reporting institutions, pinworms in about 70%, parvoviruses and ectoparasites in about 40%, and Theiler's murine encephalomyelitis virus in more than 30%.

Other important risk factors are biological materials like e.g. cells, tissues, sera, tumours, but also ES (embryonic stem) -cells, sperm, etc. that originate from animals (Nicklas et al. 1993b). Furthermore, xenotransplantats may be contaminated and be sources of an infection (Criley et al. 2001). Such specimens cannot be sterilised, and contamination of these materials by bacteria, viruses, or parasites (e.g. *Encephalitozoon cuniculi*) can therefore only be detected prior to introduction into a unit by careful microbiological monitoring. Two recent outbreaks of ectromelia in the USA both resulted from use of contaminated serum samples (Dick et al. 1996; Lipman et al. 2000). Monitoring of biological materials for viral contamination is usually done by the mouse/rat antibody production (MAP/RAP) test. This test is based on the serum antibody response to microorganisms which is stimulated in pathogen- and antibody-free animals if the material injected is contaminated. Meanwhile, polymerase chain reaction (PCR) assays have been established to replace the MAP test as the preferred method for detecting viral contaminants in biological materials (Compton and Riley 2001, Bootz and Sieber 2002).

Despite thorough testing and appropriate management practices, the introduction of agents into a facility or parts of it cannot always be avoided. It is therefore important that animals in breeding units as well as in experimental units are regularly monitored to confirm their status.

4. DEFINITIONS OF MICROBIOLOGICAL QUALITIES OF LABORATORY ANIMALS

Although attempts have been made long ago to classify different microbiological qualities of laboratory animals, a universal reporting terminology for clear and consistent definition of pathogen status in mouse populations does not exist. Three categories (gnotobiotic, specified pathogen-free and conventional) have definitions that are generally accepted and understood. **Gnotobiotic** animals play a minor role and are at present seldom used for animal experiments. **Conventional** animals are generally housed without special precautions to prevent entry of infectious agents. The risk to get infected is higher than for barrier-housed animals, but they are not necessarily infected with many pathogens. Furthermore, it is frequently considered less important to monitor for the presence of unwanted agents in conventional animals than in their rederived, barrier-maintained counterparts.

The term most frequently used to describe the microbiological quality is **'specified pathogen-free'**. This term is often abused and requires explicit definition every time it is used. It means that the absence of individually listed micro-organisms has been demonstrated for a population by regular monitoring of a sufficient number of animals at appropriate ages by appropriate and accepted methods (Kunstyr and Nicklas 2000). "SPF" animals originate from gnotobiotic animals and subsequently lose their gnotobiotic status by contact with environmental and human micro-organisms. These animals are bred and housed under barrier conditions that aim at preventing the introduction of unwanted micro-organisms. "SPF" animals are morphologically and physiologically normal, well suited for modelling the situation of a human population. Due to the very low risk of introducing agents, influences of micro-organisms on the health of the animals or on the results of experiments are less likely compared to conventionally housed animals. Additional terms (Lindsey 1998) such as 'barrier-reared', 'virus-antibody-free (VAF)', 'clean conventional', 'pathogen-free' or 'murine pathogen-free (MPF)', 'optimal hygienic conditions (OHC)' and 'health-monitored' are used; however, they describe concepts rather than the microbiological quality of animals.

5. HEALTH MONITORING

It is obvious that proper health surveillance is of vital importance for the evaluation of the microbiological status of laboratory animals and the conduction of standardised experiments with a minimum of animals. Sufficient and reliable information on the health status has become more important since new genetically modified rodents are rapidly developed and exchanged world-wide between numerous places. Results of the survey mentioned earlier (Jacoby and Lindsey 1997) show that health monitoring and the quality of rodents are still inadequate even among major research centres in the USA.

A universal testing strategy for the definition of pathogen status in rodent populations does not exist. Proper evaluation of the quality of animals could best be achieved if internationally recognised standards and definitions of their quality were available. Global standards should encompass animals and animal products and should also include testing laboratories. Also, standards for reporting results of microbiological monitoring and additional quality controls should be created. Common language and format of health reports are critical for clarity, accuracy and completeness (Jacoby and Homberger 1999).

Lists of pathogens for which rodents should be monitored have repeatedly been suggested. A comprehensive list of pathogens which should be considered when specifying the "SPF" status in mice and rats has been published by GV-SOLAS (Kunstyr 1988). This list contains over 20 viruses, 25 bacteria, and about 30 parasites and fungi. Waggie et al. (1994) also list about 100 agents that are known or suspected to be pathogenic for mice and rats or may interfere with biomedical research. They selected about twenty agents for regular monitoring on the basis of their prevalence, importance, and availability of tests. Similar viruses are listed in the FELASA (Federation of European Laboratory Animal Science Associations) recommendations (Nicklas et al. 2002), but it is recommended that frequent monitoring is performed only for the most relevant agents, and rarely occurring agents be tested annually. Few bacterial species are named, but additional organisms should be declared as present if they are associated with lesions, clinical signs, or when there is evidence of perturbation of physiological parameters. Few parasites are listed by their species name, but principally all that were detected by the methods recommended should be mentioned in the health report.

Lists of agents to be monitored are never complete as new agents are found from time to time (Wan et al. 2002), others might re-emerge unexpectedly. Other agents are relevant only for specific strains (e.g., for immunodeficient animals) and it is therefore not reasonable to include them in general lists. Too strict adherence to existing lists may therefore bear the risk that monitoring does not include important (e.g. recently detected) agents. Strict adherence to existing lists is wrong, especially in cases of clinical disease, and might not result in detection of a causative agent of a disease. The whole spectrum of micro-organisms as a concept is not a permanent list for all time, it rather represents a moving boundary in which old agents are eradicated and new agents are added. Such regulations should therefore be used for guidance, but the definite decision which agents are unacceptable in a colony or parts of it has to be made by the responsible veterinarian.

The need for health surveillance programmes is generally accepted, but there is a great diversity of opinions about their design. Usually, an individual programme has to be tailored to the conditions it is to serve. Such a programme is dependent on research objectives, but numerous additional factors must be considered, such as physical conditions and layout of the animal house, husbandry methods, sources of animals, number and quality of personnel, availability of tests, local and national conditions, and finances. Some aspects to consider when establishing a health surveillance programme have recently been published (National Research Council 1991, Nicklas 1996, Weisbroth et al. 1998).

6. FELASA RECOMMENDATIONS FOR HEALTH MONITORING

It is to the credit of the Federation of European Laboratory Animal Science Associations (FELASA) that a first working group on health monitoring of rodent and rabbit breeding colonies was established in 1989. It became obvious at that time that exchange of animals would increase in Europe for political reasons (the Schengen agreement was already in preparation). Therefore, the main task of the group was to establish recommendations aiming at getting similar standards of testing and comparable health monitoring reports by harmonisation of health monitoring procedures within Europe. The first FELASA health monitoring recommendation was published in 1994 for breeding colonies of rodents and rabbits (Kraft et al. 1994). After initial hesitations, most parts of this recommendation were accepted by commercial breeders. The first working group already completed a format for a health certificate, but in the printed version only guidelines were given on the lay-out and the information to be included in a health report. As an important step, the uniform format was later printed in the "FELASA recommendation for the health monitoring of mouse, rat, hamster, gerbil, guineapig and rabbit experimental units" (Rehbinder et al. 1996).

It was the intention of the second working group (Rehbinder et al. 1996) not only to recommend health monitoring for experimental units, but the recommendation was also considered to be the first update as hitherto unknown agents (e.g. *Helicobacter* sp., "orphan" parvoviruses, "*Corynebacterium bovis*") had been described in the meantime. However, this addition was not applied to breeding colonies because the title obviously restricted the recommendation to experimental colonies. As another important item, the recommendation for health monitoring of rodent and rabbit experimental units (Rehbinder et al. 1996) states that different monitoring programmes might be necessary for the same facility due to compartmentation of facilities (e.g. barrier units for different purposes, isolators, etc.). In contrast to breeding colonies, conditions are much more variable in experimental units and it was therefore recommended that *"the design of monitoring programmes is dependent on research objectives and numerous other factors"*. Additionally, it was clearly stated that unexpected serological results should be confirmed and previously unnoticed infections be verified.

This publication was followed by a recommendation for the health monitoring of breeding colonies and experimental units of cats, dogs, and pigs (Rehbinder et al. 1998). These recommendation had to take into consideration that the number of agents monitored varies from country to country, because infections declared to be absent in a country or a region by a national authority need not to be monitored. In contrast to rodents, these animal species are usually vaccinated or receive drugs (e.g. for deworming), and individual treatment is easier due to the usually smaller numbers of individuals and their higher economic value. These factors have also impact on the monitoring practices which include both clinical inspection of a colony and laboratory investigations.

A recommendation for health monitoring of non-human primates (Weber et al. 1999) aims at setting standards for a wide group of different species. These animals have been bred in captivity for very few generations, and in many cases parents of existing animals have been captured in the wild. These species are therefore potential carriers of micro-organisms endemic in wild populations. Besides, they are often susceptible to the same agents as humans and may therefore acquire agents that are not present in the natural environment through close contact with humans.

Another FELASA recommendation is dealing with the health monitoring of calves, sheep and goats used for experimental purposes (Rehbinder et al. 2000). Although the use of small ruminants in research has a long tradition, specialised breeding is rare and they are usually purchased from farms.

6.1 Additional FELASA recommendations dealing with health monitoring

FELASA did not only prepare recommendations for the health monitoring of different animal species. A guidance paper was prepared dealing with quality procedures for diagnostic laboratories (Homberger et al. 1999), and it is planned that an additional recommendation will be prepared to describe how an accreditation board for health monitoring programmes can be established. It will be the goal of this board to evaluate health monitoring programmes after voluntary request and to monitor if a programme fulfils the quality standards as defined by FELASA.

7. THE REVISED FELASA HEALTH MONITORING RECOMMENDATIONS: RODENTS AND RABBITS

The most recent report of a FELASA working group recommends health monitoring of rodent and rabbit colonies in breeding and experimental units. The working group was established in 1999 to revise the existing health monitoring recommendations for rodents and rabbits (Kraft et al. 1994, Rehbinder et al. 1996) and to attempt to combine them into a single document. The revision also addresses the following issues:
- updating the list of pathogens
- the changes in animal husbandry (increased compartmentation due to the use of filter-top cages)
- the increasing use of transgenic animals (quarantine issues, emerging new pathogens in immunodeficient animals)
- making the document easy to use and giving guidance on the use of the FELASA recommendations for commercial purposes.

Furthermore, the group felt that more educational information was needed. Also, a number of references were included to give the reader easy access to information on various aspects of health monitoring and on agents that are rarely mentioned in the literature. Another aim of the group was to emphasise that sufficient flexibility is necessary so that health monitoring programmes can be adapted to specific needs. It is stated in the 'General considerations' that *"these recommendations constitute a common approach for the health monitoring of laboratory rodents and the reporting of results. Actual practice may differ from these recommendations ... Health monitoring schemes must be tailored to individual and local needs"* (Nicklas et al. 2002). Increased flexibility and adaptation to individual and local needs may lead to a reduced uniformity of monitoring programmes and health reports. For this reason *"it is recommended that a person with sufficient understanding of the principles of health monitoring be identified as the individual responsible for devising and maintaining a health monitoring policy"*.

7.1 Sample size and frequency of monitoring

As in previous recommendations, the suggested sample size for health monitoring is based on a formula published by Institute for Laboratory Animal Research (ILAR) (1976). Also, the formula and its limitations are explained. Under special circumstances, fewer animals and more frequent monitoring may be reasonable. It is recommended that colonies should be monitored at least quarterly, but more frequent monitoring may be necessary.

The detection rate for a given infection depends also on the test method employed and additional factors, and some general aspects are given on advantages and limitations of test methods (virology, bacteriology, parasitology, pathology). Positive results should be confirmed by other methods, by another laboratory, and/or by repeated testing.

Due to the increasing importance of sentinels in experimental and in breeding colonies (e.g. transgenic colonies, immunodeficient animals), suggestions are given for housing and handling of sentinels and their microbiological quality.

7.2 Agents to be monitored

One list of agents for which animals should be monitored is given for each species regardless of their immune status. Only agents are listed from which natural infections have been reported in a given host species and which have been reported in recent decades to be prevalent in colonies of laboratory animals. Compared to previous recommendations, some additional agents are included (e.g. mouse or rat parvovirus), whereas monitoring for other agents is not generally recommended any longer (e.g. K virus, mouse thymic virus, *Leptospira* sp.). For other agents the frequency of monitoring was adapted to their present prevalence rate in populations of laboratory rodents (e.g. mouse rotavirus). It is clearly stated that monitoring for additional agents and their declaration in health reports is necessary under specific circumstances, e.g., when associated with lesions or clinical signs of disease or when using immunodeficient animals.

7.3 Reporting test results

Guidelines are given on the information to be included in a health report (e.g. vaccination or anthelmintic therapy). As a novel item, an agent may be considered to be eradicated if all results of monitoring done in accordance with FELASA recommendations during 18 months after the last positive results are negative.

As in previous recommendations, uniform monitoring reports for different species have been included in the revision (Appendix 3 in Nicklas et al. 2002). To be in accordance with the FELASA recommendations a separate health report must be prepared for each animal species and location (e.g. barrier unit, room). It contains information on the housing of the animals (e.g. barrier, non-barrier, isolator). Agents (viruses; bacteria; mycoplasma and fungi; parasites) which are considered relevant for certain animal species and must therefore be included in a health report are listed in the same appendix 3. For each agent the testing frequency, latest test date and test results, as well as the testing laboratory and the test method must be listed. It must also be stated if monitoring has not been conducted for an agent listed ("not tested"). Finally, cumulated historical results (≥ 18 months) must be given for each agent.

7.4 Appendices

The first appendix lists some points to consider when monitoring animals from experimental units or various housing systems. It aims at explaining why monitoring programmes have to be tailored to individual needs especially in experimental units. For example it is recommended that the frequency of monitoring depends on the risk of introducing agents. Appendix 2 gives very basic comments on most relevant agents to help to understand better why monitoring for specific agents is recommended or why changes were made compared to previous recommendations.

Forms for health monitoring reports for mice, rats, hamsters, guinea pigs, and rabbits are given in Appendix 3.

8. CONCLUSIONS

Infectious agents do not only cause losses of animals and time due to diseases but may also lead to research complications by influencing the outcome of experiments. This often results in poor reproducibility of research data and the need to conduct additional experiments. Infections frequently lead to an increase in the numbers of animals necessary to obtain reliable research results and are unacceptable not only for scientific but also for animal welfare reasons. Efforts must therefore be undertaken to eliminate infectious agents from colonies of laboratory animals and to prevent their introduction, and indeed most agents have successfully been eradicated from colonies of commercial vendors and from many experimental units. However, the use of genetically modified animals (primarily mice) led to an increased exchange of animals between research institutes during the last decade and to re-emergence of agents (viruses, bacteria, parasites). Health monitoring is therefore of crucial importance, together with appropriate quarantine and isolation procedures as well as derivation methods, to prevent the introduction of agents into experimental colonies by animals that are obtained from experimental units and other sources. Global standards for sampling and testing strategies to define the health status of laboratory animals do not exist, neither has a uniform form been developed that is worldwide accepted for reporting health monitoring results. However, FELASA has established recommendations that aim at creating standards for sampling and testing and also suggest a uniform format for health monitoring reports to give the reader sufficient information for the evaluation of the microbiological status of an animal population.

REFERENCES

Baker DG. Natural pathogens of laboratory mice, rats, and rabbits and their effect on research. Clin Microbiol Rev 1998; 11:231-266

Bhatt PN, Jacoby RO, Morse HC, New A eds. Viral and Mycoplasma Infections of Laboratory Rodents: Effects on Biomedical Research. Academic Press, New York 1986

Bootz F, Sieber I. Replacement of mouse and rat antibody production test. Comparison of sensitivity between the in vitro and in vivo methods. Altex 2002; 19(Suppl 1):76-86

Compton, SR, Riley LK. Detection of infectious agents in laboratory rodents: traditional and molecular techniques. Comp Med 2001; 51:113-119

Criley JM, Carty AJ, Besch-Williford CL, Franklin CL. Coxiella burnetii infection in C.B-17 Scid-bg mice xenotransplanted with fetal bovine tissue. Comp Med 2001; 51:357-360

Dick EJ, Kittell CL, Meyer H, Farrar PL, Ropp SL, Esposito JJ, Buller RML, Neubauer H, Kang YH, McKee AE. Mousepox outbreak in a laboratory mouse colony. Lab Anim Sci 1996; 46:602-611

Hamm TE ed. Complications of Viral and Mycoplasmal Infections in Rodents to Toxicology Research and Testing. Hemisphere Publ. Co, Washington DC 1986

Homberger FR, Boot R, Feinstein R, Hansen AK, van der Logt J. FELASA guidance paper for the accreditation of laboratory animal diagnostic laboratories. Lab Anim 1999; 33(Suppl 1):19-38

ILAR (1976). Long term holding of laboratory rodents. ILAR News 19, L1–L25

Jacoby RO, Homberger FR. International standards for rodent quality. Lab Anim Sci 1999; 49:230

Jacoby RO, Lindsey JR. Health care for research animals is essential and affordable. FASEB J 1997; 11:609-614

Kraft V, Blanchet HM, Boot R, Deeny A, Hansen AK, Hem A, van Herck H, Kunstyr I, Needham JR, Nicklas W, Perrot A, Rehbinder C, Richard Y, de Vroey G. Recommendations for health monitoring of mouse, rat, hamster, guineapig and rabbit breeding colonies. Lab Anim 1994; 28:1-12

Kunstyr I ed. List of pathogens for specification in SPF laboratory animals. Society for Laboratory Animal Science Publ. No. 2, Biberach 1988

Kunstyr I, Nicklas W. Rat pathogens: Control of SPF conditions, FELASA standards. In The Laboratory Rat. Handbook of Experimental Animals Series. Krinke GJ ed, Academic Press, San Diego 2000; Chapter 8, 133-142

Lindsey JR. Pathogen status in the 1990s: abused terminology and compromised principles. Lab Anim Sci 1998; 48:557-558

Lipman NS, Perkins S, Nguyen H, Pfeffer M, Meyer H. Mousepox resulting from use of ectromelia virus-contaminated, imported mouse serum. Comp Med 2000; 50:425-435

National Research Council, Committee on Infectious Diseases of Mice and Rats. Infectious Diseases of Mice and Rats. National Academy Press, Washington DC 1991

Nicklas W. Possible routes of contamination of laboratory rodents kept in research facilities. Scand J Lab Anim Sci 1993a; 20:53-60

Nicklas W. Health monitoring of experimental rodent colonies: an overview. Scand J Lab Anim Sci 1996; 23:69-75

Nicklas W, Kraft V, Meyer B. Contamination of transplantable tumors, cell lines and monoclonal antibodies with rodent viruses. Lab Anim Sci 1993b; 43:296–300

Nicklas W, Homberger FR, Illgen-Wilcke B, Jacobi K, Kraft V, Kunstyr I, Maehler M, Meyer H, Pohlmeyer-Esch G. Implication of infectious agents on results of animal experiments. Lab Anim 1999; 33(Suppl 1):39-87

Nicklas W, Baneux P, Boot R, Decelle T, Deeny AA, Fumanelli M, Illgen-Wilcke B. FELASA recommendations for the health monitoring of rodent and rabbit colonies in breeding and experimental units. Lab Anim 2002; 36:20-42

Rehbinder C, Baneux P, Forbes D, van Herck H, Nicklas W, Rugaya Z, Winkler G. FELASA recommendations for the health monitoring of mouse, rat, hamster, gerbil, guineapig and rabbit experimental units. Lab Anim 1996; 30:193-208

Rehbinder C, Baneux P, Forbes D, van Herck H, Nicklas W, Rugaya Z, Winkler G. FELASA recommendations for the health monitoring of breeding colonies and experimental units of cats, dogs and pigs. Lab Anim 1998; 32:1-17

Rehbinder C, Alenius S, Bures J, de las Heras M, Greco C, Kroon PS, Gutzwiller A. FELASA recommendations for the health monitoring of experimental units of calves, sheep and goats. Lab Anim 2000; 34:329-350

Waggie K, Kagiyama N, Allen AM, Nomura T eds. Manual of Microbiologic Monitoring of Laboratory Animals. 2nd edn. U.S. Department of Human Health and Human Services, NIH Publication No. 94-2498, 1994

Wan C-H, Söderlund-Venermo M, Pintel DJ, Riley LK. Molecular characterization of three newly recognized rat parvoviruses. J Gen Virol 2002; 83:2075-2083

Weber H, Berge E, Finch J, Heidt P, Kaup F-J, Perretta G, Verschuere B, Wolfensohn S. Health monitoring of non-human primate colonies. Lab Anim 1999; 33 Suppl 1:3-18

Weisbroth SH, Peters R, Riley LK, Shek W. Microbiological assessment of laboratory mice and rats. ILAR J 1998; 39:272-290

Chapter 4

HOUSING, CARE AND ENVIRONMENTAL FACTORS

Axel Kornerup Hansen[1] and Vera Baumans[2]
[1]*Department of Pharmacology and Pathobiology, The Royal Veterinary and Agricultural University, Frederiksberg, Denmark and* [2]*Department of Laboratory Animal Science, Utrecht University, Utrecht, The Netherlands/Karolinska Institute, Stockholm, Sweden*

1. PRINCIPLES OF HOUSING AND CARE

1.1 The need for standardisation

Over the last forty years the 3 R´s, reduction, refinement and replacement, have been the basis for laboratory animal science. Refinement among other things means that we should aim at reducing the impact that the experiment may have on the well-being of the animal, e.g. by housing them as close to their needs as possible. Reduction means that we should make our experiments as precise as possible, and we should reduce the number of animals as much as possible. One of the main ways of achieving these goals is to reduce the variation between the animals (Figure 4-1). Therefore, animals in experiments need to be housed in a standardised way.

Conflicts may occur between fulfilling the needs of the animals and the need for standardisation. However, animal needs may also be fulfilled in a standardised way, giving all animals in experiments the same possibilities and stimulating environment thereby not ruining standardisation. Furthermore, the need for reduction does not overrule the need for refinement, i.e. it may be considered morally acceptable to use more animals if the increase leads to less suffering in an individual animal (Hansen et al. 1999).

Environmental causes of variation in animal experiments include cage material, structure and size, cage mates, lighting, noise, NH_3 and CO_2 levels, climate and bedding. Some other important factors, e.g. microbiology and nutrition, will not be dealt with in this chapter, as these are covered elsewhere in this book.

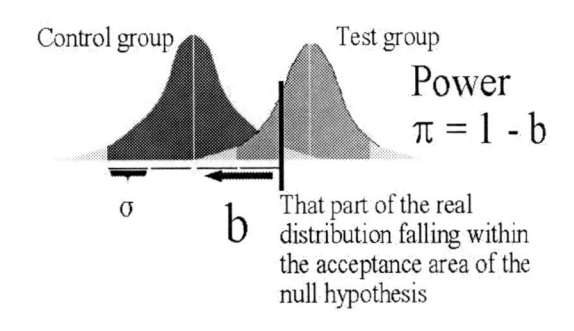

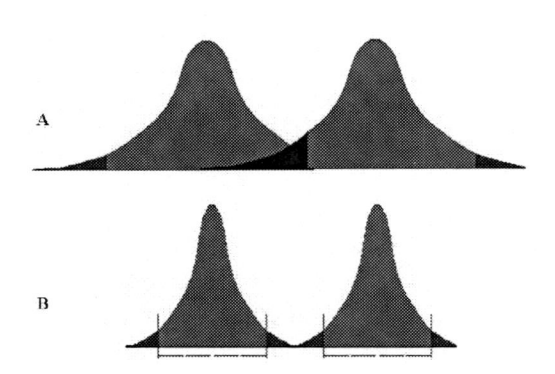

Figure 4-1. **Variation:** Most animal experiments aim at showing a difference between a control group not under influence of a test factor and a test group under influence of a test factor. If there really is a difference between the two distributions, there will be a distribution covering all animals in the control group and another, different distribution covering all animals in the test group. However, animal experiments only use a very limited sample of animals, and if the overlap between the two distributions (b) is too large the chances of showing the difference between control and test group is reduced accordingly. The larger the natural variation (σ) the larger the overlap (b). **A** shows an experiment where the variation is large, b is large and power is low. **B** shows an experiment where variation is small, b is close to zero and power is close to 100 %. Therefore, to reduce the number of animals used, it is necessary to reduce the variation by e.g. standardisation.

The animal aims at being in balance with its environment (homeostasis). In order to stabilise important parameters, the organism must necessarily compensate by variation in other parameters as a response to changes in the environment. Figure 4-2 shows how the organism attempts to keep a constant body temperature as a response to changes in room temperature. As animals are often not housed at temperatures within the homeothermic zone, changes in room temperature may have serious impact on e.g. pharmacokinetics (Clough 1982). Furthermore, if heat production is increased, food consumption needs to be increased as well, which is also an obvious source of variation, e.g. if a drug is dosed in the diet. It is generally accepted that if the variation in the surrounding temperature of the animal is kept within ± 1°C dramatic fluctuations in metabolism are not observed. However, in this context it should be remembered, that the macroclimate, i.e. the climate in the animal room, might differ significantly from the microclimate, i.e. the climate in each individual cage and within different areas of the cage rack (Figure 4-3). More than 5°C difference in temperature may be shown between cages in the top and the bottom of a rack (Clough 1984). One way to minimise such variation may be to house the animals in individually ventilated cage (IVC) racks (Figure 4-4), although variation may also be considerable in these systems, depending on other factors such as lighting. Moreover, to protect the animals from draught due to the forced ventilation, nesting material or a hiding place should be provided (Baumans et al. 2002).

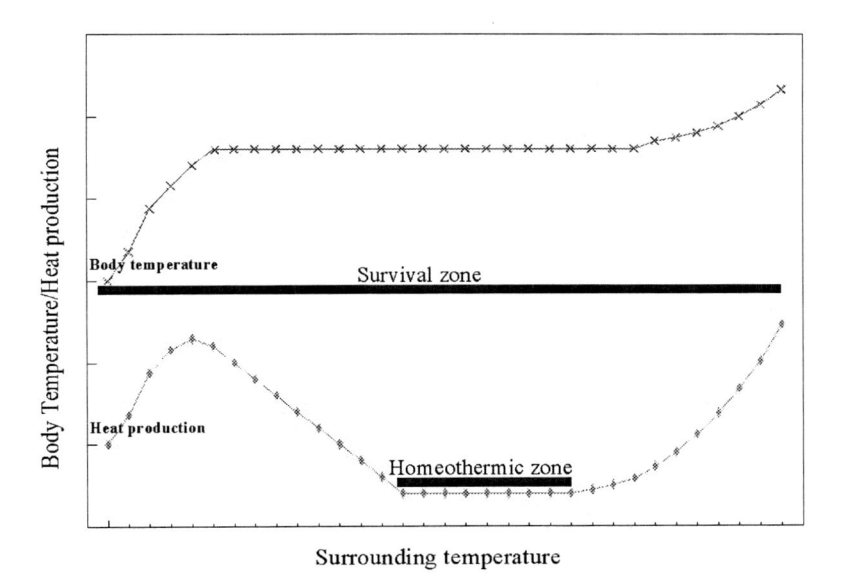

Figure 4-2. **Heat production versus environmental temperature:** Variation in heat production caused by fluctuations in the surrounding temperature.

Figure 4-3. **Standardised housing of laboratory rodents:** Even though animals are housed in similar cages with the same stocking density, variation may occur between different locations in a rack.

Gregarious animals such as mice, rats, gerbils, guinea pigs and female rabbits, may become stressed by single housing, showing both behavioural and physiological abnormalities usually referred to as 'isolation stress' or 'isolation syndrome' (Brain 1975, Haseman et al. 1994, Ader and Friedman 1964). Stress influences metabolism, causing differences in enzymatic activity and pharmacokinesis between single housed and group housed animals (Dairman and Balazs 1970), although their corticosterone levels will only differ within the first two weeks (Goldsmith et al. 1978). Inadequate housing may therefore be hazardous to both the quality of the experiment and the welfare of the animals.

Differences in cage cleaning frequency (Vesell et al. 1973) and the act of cage cleaning itself lead to an arousal in the animals, frequently causing severe aggression in male mice. Transfer of familiar odours from the nest reduces aggression, whereas odours from soiled bedding lead to an increase in fighting (Van Loo et al. 2000). The type of bedding (Cunliffe-Beamer et al. 1981) may affect the animal, probably due to induction of liver cytochrome oxidase activity induced by trace gases such as ammonia in the dirty bedding or by resin in pine bedding. In bedding and other environmental enrichment materials the risk of subjecting the animals to different types of chemical residues should always be considered. Some materials for bedding or environmental enrichment, such as wood products, e.g. cedar wood, have been suspected of containing carcinogens (Vlahakis 1977, Jacobs and Dieter 1978, Tennekes et al. 1981). Therefore, all such materials should be batch controlled, setting the same standards on these materials as for diets (Fox 1979).

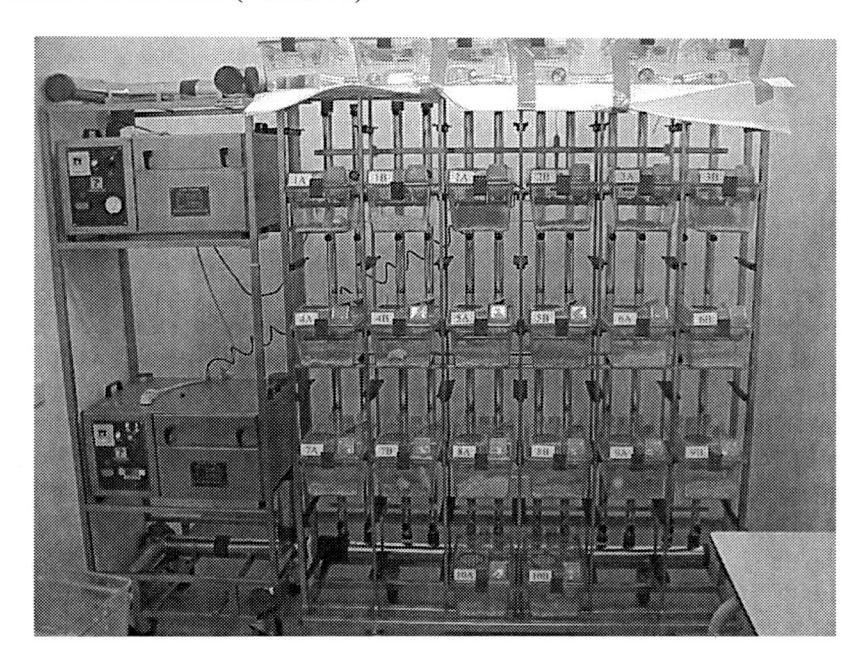

Figure 4-4. **Individually ventilated cage (IVC) racks:** In IVC racks each cage is closed and independently ventilated.

There seems to be some doubt about, whether environmental enrichment conflicts with the standardisation of experiments. Some researchers fear that 'enriched' animals show more variability in their response to experimental procedures because they show more diverse behaviour. It has been shown that in many cases enrichment does not increase variability or might even decrease it and a decrease reduce the number of animals needed. The effects of enrichment seem to be dependent of the parameter studied, the type of enrichment and the species/strain of the animal. The possible conflicting effects of enrichment should not be generalised, potentially at the cost of improved well-being of the animals and even when enrichment increases variation it should not be overstated, but balanced against the improved well-being of the animals in successful enrichment programmes (Augustsson et al. in press, van de Weerd et al. 2002).

1.2 The impact of housing on animal welfare

Most laboratory animals spend their entire life in their homecages. Their ability to cope with experimental procedures depends for a great part on coping with the environment in the home cage.

Housing is important regardless of whether welfare is defined from a hedonistic, a coping or a preference point of view, as poor quality of housing may prevent the animal from coping with the environment or have its preferences fulfilled, causing stress in the animal and thereby jeopardizing the validity of the experimental results.

Single housing of gregarious animal species is considered detrimental to welfare. Whereas group-housing of male mice may be difficult in terms of aggression, depending on strain, previous experience, and type of cage enrichment, housing of single sex groups of male and female rats does not pose problems. However, subordinate male mice prefer company to being housed individually, even if that companion is dominant (Van Loo et al. 2001) and also solitary living animals, such as hamsters, may benefit from social housing as shown by increased growth (Borer et al. 1988), although this may also be linked with fighting and enlarged adrenals (Arnold and Estep 1994). Mature male rabbits, on the contrary, are in general difficult to house socially, as they may fight severely and even castrate one another. For group housed animals the optimal group size is determined by sex and age of the animals, cage size and complexity and experimental design. It is important to form harmonious groups and to keep group size and composition stable to avoid stress by altering the established hierarchy.

Space allocation for animals should be considered from two points of view. The animal should - together with its cage mates - have access to a certain space, i.e. a minimum acceptable floor area has to be defined. For a group of animals it should also be considered how much space should be allocated to each animal, i.e. how densely should the animals be stocked. It is a common understanding - and to a certain extent misunderstanding - that more space correlates directly to increased welfare. It is quite obvious that in a too limited space the animal will have severe problems with a range of normal behaviours. For example, some of the smallest cages, such as the 180 cm^2 accepted for mice in the 1986 European Convention (Council of Europe 1986), may be too limited, as it has been shown that, when space up to 1600 cm^2 is provided for mice, they make good use of it (Sherwin and Nicol 1997). Moreover, it will be difficult to provide enrichment in such small cages. However, it seems to be more the degree of structuring of the environment rather than space which is important for the preferences and absence of behavioural disorders in mice (Baumans et al. 1987, Würbel et al. 1996), and increasing amounts of empty space as well as inappropriate structuring may stimulate territorial aggression among male mice (Haemisch et al. 1994). Therefore, welfare of animals should never be calculated only by the quantity of space they are allocated but also by the quality. Generally, a too high stocking density is considered bad for group-housed animals, but it is difficult to calculate exactly how much space would be acceptable per animal from a welfare point of view. Although several such attempts have been made for laboratory rodents (Weihe 1978, Hackbarth et al. 1999, Bruhin 1988, Gärtner et al. 1976), these are seldom based upon scientific investigations, and it is not realistic to allow animals the size of a natural territory. Stocking densities are often defined in grams body weight per square cm, for many animal species the larger animals are often the older and less active animals, while the smaller animals are the younger ones who tend to be more active. Therefore, it may be reasonable to define stocking densities according to the final adult weight of the animals, allowing young playful animals more space.

Animals can be housed on a solid floor with bedding or on some kind of perforated floor allowing the excrements to be collected underneath the cage or pen. There are many types of perforated floors, from wire mesh over grids to floors with small holes. Rodents such as rats prefer solid floors (Blom 1993, Manser et al. 1995) and get stressed by housing on grid floors (Krohn et al. 2003). Today, it is generally accepted that most animal species should be housed on bedding, but there is also a growing understanding that further structuring of the environment, so-called environmental enrichment, is necessary to stimulate exploratory behaviour and reduce boredom (Wemelsfelder 1997), e.g. by providing climbing accessories, shelters, exercise devices or nesting material. Such structuring of the environment may be more beneficial than simply providing a larger floor area, although a minimum floor area is needed to provide such a structured space. Scattering food materials in the bedding may allow the animal a semi-natural way of finding food, as in nature a large part of the time-budget is spent on foraging. In primates much time may be spent getting a sticky food material out of the holes of a wooden block (puzzle feeders). Different research results concerning the need for gnawing objects in rodents are more difficult to interpret and it can be argued that excessive gnawing may be indicative for primary frustration and hence reduced welfare (Sørensen et al. in press), e.g. rats gnaw on aspen blocks especially when housed without bedding (Kaliste-Korhonen et al. 1995). However, mice seem to gnaw less on their nesting material when provided with gnawing sticks.

Gentle and competent handling, training and other types of human contact often benefit the animals (Weininger 1956). To be able to observe and handle the animals properly appropriate lighting must be provided, but excessive light or exposure to continuous high level light may cause retinal damage, which is especially important in albino animals (Semplerowland and Dawson 1987). Giving animals the opportunity to hide from excessive light by providing a shelter or nest is important.

2. INTERNATIONAL RULES AND GUIDELINES FOR HOUSING AND CARE OF LABORATORY ANIMALS

Guidelines and legislation on housing of laboratory animals should balance the needs of science and the animals. From a legislative point of view two different ways of regulating housing and care of laboratory animals exist; the European and North American systems. The European system enforces minimum standards through legislation, whereas the North American system employs a self justice regulation, where projects and institutions must be evaluated by local committees not necessarily working according to a law code.

Most countries inside Europe as well as other countries like Australia and New Zealand regulate the European way, while U.S.A. and Canada regulates the North American way. Laboratory rodents and birds are not protected by US law.

2.1 Appendix A of the Council of Europe's Convention ETS 123

The Council of Europe's convention ETS 123 (Council of Europe 1986) is presented in Chapter 2 in this book. Appendix A of this convention provides guidelines for housing and care of vertebrate laboratory animals.

Conventions of the Council of Europe are in principle not mandatory for the member states. First, they can reject participation in the negotiations, secondly they can decide not to sign the convention, and thirdly they can decide not to ratify the convention. However when member states sign and ratify the Convention, the guidelines have to be implemented in national legislation. Today the Convention ETS 123 has been signed and ratified by 14 member states. For the European Union (EU) members the situation is somewhat different. In 1986 the EU issued a Directive (86/609/EEC) (European Economic Community 1986) with guidelines similar to the Convention ETS 123. From the year 2000 EU has decided that all member states must ratify the Convention. Consequently, guidelines of housing and care in appendix A are to be considered mandatory parts of the legislation within EU member states.

Currently Appendix A is under revision. This is due to the fact that in 1997 a multilateral consultation of the Council of Europe in relation to the Convention ETS 123 agreed upon a resolution on accommodation and care (Council of Europe 1997). This determined which housing and care procedures should be improved in order to meet species specific needs of these animals. It was decided that the environment should be enriched allowing social interaction, activity-related use of the space and appropriate stimuli and materials. Therefore, housing in stable and harmonious groups or pairs is preferred for all gregarious species, and the cage should contain a structured environment without negative effects on welfare. Animals should be regularly handled, be in social contact with humans, with particular attention to the socialisation period in species such as cats and dogs. Rodents, rabbits and dogs should be provided with solid floors with bedding instead of grid floors. Although high hygienic standards should be maintained, it is advised to maintain odour patterns left by the animals, preferably from the nest area.

In 1998, several expert groups for the different species were established to propose changes in the appendix A to the multilateral consultation. So far all expert groups have presented their proposals. These proposals are divided into two parts. Part A gives the proposal for the revision of the appendix, while part B provides the background information, based on scientific results or best practice. The animal species covered by the revised appendix A are mice, rats, Syrian and Chinese hamsters, gerbils, guinea pigs, rabbits, dogs, cats, ferrets, minipigs, farm animals, non-human primates, fish, birds, reptiles and amphibians.

The current Appendix A (Council of Europe 1986) consists of a general introduction and species-specific tables and figures. The latter gave recommendations of minimum floor space and cage heights as well as stocking densities. In the revised Appendix these figures and tables will be supplemented with some legends for each species giving some basic information on their needs. Moreover, recommendations for enriching and structuring the animal's environment are given. In order to fulfil the requirement for environmental enrichment the minimum floor area is generally increased. For example, it will no longer be allowed to house mice on a floor area of 180 cm^2 (Macrolon type I cage), as this does not allow a properly structured environment. An example of differences between the 1986 Appendix A and the revised Appendix A is given for mice in Table 4-1.

If the housing guidelines of the Appendix have to be disregarded due to veterinary or scientific reasons, this should be regarded as a procedure to be licensed by the regulatory authority. This could for example be the case when housing animals in metabolic cages.

Table 4-1. Comparison of space allowances for mice during procedures of the Council of Europe Convention ETS 123 Appendix A from 1986, the revision of the Appendix A expected in 2004, and the U.S. National Research Council's Guide for the Care and Use of Laboratory Animals (Institute of Laboratory Animal Resources 1996) (converted from inch to cm) to show differences in principles.

	1986 Appendix A	Revised Appendix A	Guide for the Care and Use of Laboratory Animals
Cage height	12 cm	12 cm	13 cm
Minimum floor area	180 cm^2	330 cm^2	Not recommended
Floor area / animal			
<10 g	40 cm^2	60 cm^2	39 cm^2
Up to 15 g	50 cm^2	60 cm^2	52 cm^2
Up to 20 g	60 cm^2	60 cm^2	77 cm^2
Up to 25 g	70 cm^2	70 cm^2	77 cm^2
Up to 30 g	80 cm^2	80 cm^2	97 cm^2
Up to 35 g	90 cm^2	100 cm^2	97 cm^2
Up to 40 g	100 cm^2	100 cm^2	97 cm^2
> 40 g	100 cm^2	100 cm^2	97 cm^2

The cage height is essentially the same for all sizes of mice, as 12 or 13 cm probably does not interfere dramatically with the well-being of mice of any sizes. The 1986 Appendix A follows a straight linear curve and is actually shown as a graph in the Appendix. The revised Appendix favours the smaller mice, being the younger ones and thereby more active. For other sizes no real differences exist between the old and the revised Appendix A. The Guide for the Care and Use of Laboratory Animals follows a close to linear curve, but in contrast to the 1986 Appendix A it only indicates four different body weights of mice. The most dramatic differences between the three are the larger space allowances for small mice and the large increase in minimum floor area in the revised Appendix A.

2.2 US National Research Council's Guide for the Care and Use of Laboratory Animals

Whereas in Europe principles for housing and care of laboratory animals are included in the national legislation, in North America, the protection of laboratory animals is based on a system of self justice among institutions working with laboratory animals. The U.S. Animal Welfare Act applies to warm-blooded animals, but the U.S. Department of Agriculture has exempted rats, mice, birds and farm species. However, each individual institution or company must have an Institutional Animal Care and Use Committee (IACUC), which has to approve research projects using animals. Facilities must either be inspected every six months by the IACUC or be approved by the Association for Assessment and Accreditation of Laboratory Animal Care International (AAALAC). Basis for project and institution evaluation in relation to housing and care is the Guide for the

Care and Use of Laboratory Animals (Institute of Laboratory Animal Resources 1996).

The guide provides some guidelines on cage size and describes regulations, policies, and principles as well as elucidating some more general aspects such as veterinary care, qualifications and occupational health. Furthermore, the guide also deals with aspects related more to performing good research rather than to the welfare of the animals, such as genetics and nomenclature.

The guide applies to any vertebrate animal, either traditional laboratory animals, farm animals, wildlife, or aquatic animals used in research, teaching, or testing.

2.3 Differences and similarities between the Appendix A and the Guide for the Care and Use of Laboratory Animals

Similarities and differences between the Appendix A and the Guide for the Care and Use of Laboratory Animals can be shown in a very simple way by e.g. comparing space allocations for mice given in these two papers (Table 4-1).

The Guide for the Care and Use of Laboratory Animals is a more educational paper than Appendix A, due to differences in their status, the Guide being a guide for animal users, IACUC and AAALAC, while Appendix A is an official document to be implemented in national laws. Some of the more general items, such as veterinary care, qualifications and occupational health are covered both by the Council of Europe Convention ETS 123 and its resolutions, and the Guide for the Care and Use of Laboratory Animals.

3. CONCLUSIONS

Housing of laboratory animals should both secure an acceptable level of welfare for the sake of the animals as well as an acceptable level of standardisation for the sake of science and to avoid increased numbers of animals used due to excessive variation. When housing laboratory animals the quality of space seems more important than the quantity of space, although too little space may not allow the expression of natural behaviour. The most common laboratory rodents, i.e. rats and mice, are social animals which generally should be group housed. Furthermore, the housing environment should be enriched by various types of structurisation, such as gnawing blocks, hiding tubes and nesting materials. It is unclear whether and to which extent such enrichment has an impact on the variation between the animals. In Europe very exact measures for housing are given in an appendix to a Council of Europe Convention, which is mandatory for EU member states, while in North America the Institutional Animal Care and Use Committees judge housing according to the Guide for the Care and Use of Laboratory Animals.

REFERENCES

Ader R, Friedman SB. Social-factors affecting emotionality and resistance to disease in animals 4. Differential housing, emotionality, and Walker 256 carcinosarcoma in the rat. Psychol Rep 1964; 15:535-541

Arnold CE, Estep DQ. Laboratory Caging Preferences in Golden-Hamsters (Mesocricetus-Auratus). Lab Anim 1994; 28:232-238

Augustsson H, van de Weerd HA, Kruitwagen CLJJ, Baumans V. Within and between group variation in the light/dark test in mice: The importance of strain and environmental enrichment. Lab Anim in press

Baumans V, Stafleu FR, Bouw J. Testing housing system for mice - the value of a preference test. Z Versuchstierkd 1987; 29:9-14

Baumans V, Schlingmann F, Vonck M, van Lith HA. Individually ventilated cages: beneficial for mice and men? Contemp Top Lab Anim Sci 2002; 41:13-19

Blom HJM. Evaluation of housing conditions for laboratory mice and rats - The use of preference tests for studying choice behaviour. Ph.D. Thesis, University of Utrecht 1993

Borer KT, Pryor A, Conn CA, Bonna R, Kielb M. Group housing accelerates growth and induces obesity in adult hamsters. Am J Physiol 1988; 255:R128-R133

Brain P. What does individual housing mean to a mouse. Life Sci 1975; 16:187-200

Bruhin H. Planung und Struktur von Versuchstierbereichen tierexperimentell tätiger Institutionen. GV-SOLAS, Biberach 1988

Clough G. Environmental effects on animals used in biomedical research. Biol Rev 1982; 57:487-523

Clough G. Environmental factors in relation to the comfort and well-being of laboratory rats and mice. In: Standards in Laboratory Animal Management. 1. Anonymous, UFAW, London 1984; 7-24

Council of Europe. European Convention for the Protection of Vertebrate Animals used for Experimental and other Scientific Purposes (ETS 123). Council of Europe, Strassbourg 1986

Council of Europe. Resolution on the accommodation and care of laboratory animals: Council of Europe 1997

Cunliffe-Beamer TL, Freeman LC, Myers DD. Barbiturate sleeptime in mice exposed to autoclaved or unautoclaved wood beddings. Lab Anim Sci 1981; 31:672-675

Dairman W, Balazs T. Comparison of liver microsome enzyme systems and barbiturate sleep times in rats caged individually or communally. Biochem Pharmacol 1970; 19:951

European Economic Community. Council Directive 86/609/EEC of 24 November 1986 on the approximation of laws, regulations and administrative provisions of the Member States regarding the protection of animals used for experimental and other scientific purposes. EEC Official Journal L358 1986; 1-28

Fox J. Selected aspects of animal husbandry and good laboratory practices. Clin Toxicol 1979; 15:539-553

Gärtner K, Küpper W, Maess J. Zum artgemässen Bewegungsbedürfnis der Versuchstiere (Maus, Ratte, Meerschweinchen, Kaninchen, Hund, Katze). Fortschr Vet med 1976; 25:130-138

Goldsmith JF, Brain PF, Benton D. Effects of the duration of individual or group housing on behavioural and adrenocortical reactivity in male mice. Physiol Behav 1978; 21:757-760

Hackbarth H, Bohnet W, Tsai PP. Allometric comparison of recommendations of minimum floor areas for laboratory animals. Lab Anim 1999; 33:351-355

Haemisch A, Voss T, Gärtner K. Effects of environmental enrichment on aggressive behavior, dominance hierarchies, and endocrine states in male DBA/2J mice. Physiol Behav 1994; 56: 1041-1048

Hansen AK, Sandøe P, Svendsen O, Forsman B, Thomsen P. The need to refine our notion on reduction. In Proceedings of International Conference on the Use of Humane Endpoints in Animal Experiments for Biomedical Research, Zeist, The Netherlands. Hendriksen CFM, Morton DB eds, The Royal Society of Medicine Press Limited, London 1999; 139-144

Haseman JK, Bourbina J, Eustis SL. Effect of individual housing and other experimental-design factors on tumor-incidence in B6C3F1 mice. Fund Appl Toxicol 1994; 23:44-52

Institute of Laboratory Animal Resources. Guide for the Care and Use of Laboratory Animals. Commission on Life Sciences. National Research Council. National Academy Press, Washington DC 1996

Jacobs BB, Dieter DK. Spontaneous hepatomas in mice inbred from Ha-Icr swiss stock - effects of sex, cedar shavings in bedding, and immunization with fetal liver or hepatoma-cells. J Natl Cancer Inst 1978; 61:1531-1534

Kaliste-Korhonen E, Eskola S, Rekilä T, Nevalainen T. Effects of gnawing material, group size and cage level in rack on Wistar rats. Scand J Lab Anim Sci 1995; 22:291-299

Krohn TC, Hansen AK, Dragsted N. Telemetry as a method for measuring impacts of housing conditions on rats. Anim Welf 2003; 12:53-62

Manser CE, Morris TH, Broom DM. An investigation into the effects of solid or grid cage flooring on the welfare of laboratory rats. Lab Anim 1995; 29:353-363

Semplerowland SL, Dawson WW. Cyclic light-intensity threshold for retinal damage in albino- rats raised under 6-lx. Exp Eye Res 1987; 44:643-661

Sherwin CM, Nicol CJ. Behavioural demand functions of caged laboratory mice for additional space. Anim Behav 1997; 53:67-74

Sørensen DB, Ottesen JL, Hansen AK. Consequences of enhancing environmental complexity in rats and mice - a review. Anim Welf in press

Tennekes HA, Wright AS, Dix KM, Koeman JH. Effects of dieldrin, diet, and bedding on enzyme function and tumor-incidence in livers of male Cf-1 mice. Cancer Res 1981; 41:3615-3620

van de Weerd HA, Aarsen EL, Mulder A, Kruitwagen CLJJ, Hendriksen CFM, Baumans V. Effects of environmental enrichment for mice: variation in experimental results. J Appl Anim Welf Sci 2002; 5:87-109

Van Loo PLP, Kruitwagen CLJJ, van Zutphen B, Koolhaas JM, Baumans V. Modulation of aggression in male mice: influence of cage cleaning regime and scent marks. Anim Welf 2000; 9:281-295

Van Loo PLP, de Groot AC, van Zutphen LFM, Baumans V. Do male mice prefer or avoid each other's company? Influence of hierarchy, kinship and familiarity. J Appl Anim Welf Sci 2001; 4:91-103

Vesell ES, Lang CM, White WJ, Passanan GT, Tripp SL. Hepatic drug-metabolism in rats - impairment in a dirty environment. Science 1973; 179:896-897

Vlahakis G. Possible carcinogenic effects of cedar shavings in bedding of C3H-Avy Fb mice. J Natl Cancer Inst 1977; 58:149-150

Weihe WH. Recommendation for caging space of small laboratory-animals. Z Versuchstierkd 1978; 20:305-309

Weininger O. The effects of early experience on behavior and growth characteristics. J Comp Physiolog Psychol 1956; 49:1-9

Wemelsfelder F. Life in captivity: its lack of opportunities for variable behaviour. Appl Anim Behav Sci 1997; 54:67-70

Würbel H, Stauffacher M, VonHolst D. Stereotypies in laboratory mice - Quantitative and qualitative description of the ontogeny of 'wire-gnawing' and 'jumping' in Zur:ICR and Zur:ICR nu. Ethology 1996; 102:371-385

Chapter 5

NUTRITION AND ANIMAL WELFARE

Merel Ritskes-Hoitinga[1,2] and Jan H. Strubbe[2]
[1]*Biomedical Laboratory, University of Southern Denmark, Odense, Denmark and*
[2]*Department of Neuroendocrinology, University of Groningen, Haren, The Netherlands*

1. INTRODUCTION

One of the concepts of good animal welfare focuses on the fact that animals should function and feel well (Carter et al. 2001). There are two categories of basic values (see the text by Sørensen in this book). The first category is more or less objective, in that an animal should function well biologically and/or has the possibility to perform natural behaviours. This can be assessed by e.g. the absence of pathology and behavioural abnormalities such as stereotyped behaviour. The second category refers to the subjective approach, which relates to the inner mental state of the animals. An increased welfare is then related to an increased positive subjective state and reduced negative mental state (Fraser et al. 1997). In scientific research attempts are made to measure these subjective states indirectly by e.g. preference testing.

Each organism strives to maintain homeostasis, both physiologically and mentally. Homeostasis refers to a regulated state of internal stability or balance (Strubbe 2003). Such a state can never be a stable permanent situation, as there will always be fluctuations, such as activity versus resting, eating versus not eating, social interaction, etc., and it is essential that the homeostatic state is reached again. Eating a meal will automatically lead to a certain disturbance of the homeostasis (e.g. the thermogenic effect), but as long as this stays within certain limits, this offers no real threat. It is probably even stimulating animal welfare when fluctuations arise, as biological rhythms are at the basis of virtually all natural processes. Many functions show circadian rhythms, i.e. rhythms of approximately 24 hours.

But these fluctuations should probably remain within an upper and lower boundary, otherwise the organism has more difficulties to return to the homeostatic state. In order to maintain welfare, it is possible that a certain amount of species-specific natural fluctuations in the feeding process have to be introduced in the laboratory setting in order to improve the welfare of laboratory animals.

Our goal in this chapter is to evaluate several aspects of nutrition in relation to the welfare of animals (focused on rodents) in experimental conditions. The aims are further to present essential knowledge and data that are known to prevent the occurrence of pathological and behavioural disorders. This knowledge can also contribute to a better standardisation of experiments. When doing animal experimentation, it is important to strive for standardisation within and between experiments, between institutes, nationally as well as internationally, in order to make data comparable (see also Haseman 1984, Roe 1994). In the more subjective approach, examples from results of different types of tests like e.g. preference testing will be provided, with a discussion of the possible welfare implications for the animal species involved. It is hypothesized that the more the environmental factors fit into the species-specific adaptive capacities (for being able to return to homeostasis) the better the welfare (Crok 2003).

By presenting the already published knowledge from animal experiments on nutrition and related animal behaviour, the insight into the laboratory animal in an experimental situation can be improved. By comparing the guiding principles of the concept of standardisation versus the natural rhythmicity, tools are given to researchers to make conscious and well-based decisions about the experimental design. Directly and indirectly this improved understanding can contribute to optimising the experimental conditions, leading to simultaneous improvement of animal welfare and experimental results.

2. GOOD BIOLOGICAL FUNCTIONING

Since this chapter focuses on nutrition and animal welfare, we first pose the question why eating is necessary. Food needs to be ingested, as many processes in the body depend on the (continuous) supply of energy and nutrients. Moreover, to maintain the body temperature at a certain level, which is critical for many vital functions in mammals, it is necessary to supply energy to the system. Nutrients like proteins and amino acids are necessary to build muscle tissue and keep it in good health, and minerals need to be a part of the diet e.g. for a proper bone mineralization as well as good muscle function. Certain trace elements and vitamins are essential factors in maintaining enzymatic functions in the body. By overconsumption during certain periods, energy deposits can be formed. This occurs for instance in hibernating animals. This is a basic mechanism applied by animals in nature, in order to prepare the organism for more difficult periods, in which no or hardly any food is available. On the other hand, the continuous supply of food to rodents in the laboratory, where the animals never have to make use of these reserves, may therefore not be the best way of feeding.

2.1 Nutritional requirements

From a general perspective, the nutritional requirements of farm animals and laboratory animals are similar. All require energy, protein, carbohydrate, lipid, macrominerals, vitamins and trace elements supplied in diets that should be palatable and free from chemical and biological contaminations (Ritskes-Hoitinga and Chwalibog 2003). Much information regarding farm animals may be applicable to laboratory animals, however, the aim of laboratory animal nutrition is not the efficient or maximum production, in contrast to the situation in farm animals. The documents published by the National Research Council provide currently the best documented basis for the nutrient needs of each animal species These guidelines are based on studies supporting maximum/optimum growth and production, which is not necessarily the same as goals in animal experimentation, where optimum performance and nutritionally unbiased results in biomedical experiments are the aims (Ritskes-Hoitinga and Chwalibog 2003). Fiber is not included in most nutrient requirements, but depending on the species, there can be a need to include fiber and/or add it to the diet for maintaining and/or improving good health. Rabbits and guinea pigs are examples of species where extra fiber will typically contribute to good biological functioning.

As regards energy requirements, animals will eat an amount of food determined by their energy need of that moment, mainly based upon the stage of life they are in (growth, maintenance, reproduction, lactation) (Beynen and Coates 2001). So, based on the energy density of the diet and the energy need, it can be calculated what the expected food intake will be under *ad libitum* conditions. It is beyond the scope of this chapter to give the exact calculations, and the reader is referred to Beynen and Coates (2001) and Ritskes-Hoitinga and Chwalibog (2003) for details. Upon changing the energy density of the diet, e.g. by adding fat, the animal will eat less food mass since fat contains more calories per gram. When changing the amount of fat in the diet, isocaloric exchange is advised in order to achieve optimal standardisation (see also Beynen and Coates 2001). Moreover, on a high fat diet animals (may) compensate for the higher caloric density and will therefore eat less of the other compounds. On the other hand, when diluting the diets, e.g. by adding fiber, one has also to make sure that the animals will still be able to ingest the necessary minimum amount of all essential nutrients. In other words, is the necessary ingested volume of diet too big in relation to the capacity of the stomach. Also the type of fiber needs to be chosen carefully: a certain short-type cellulose (Arbocel R B-00) is known to cause intestinal obstruction and death in rats (Ritskes-Hoitinga and Chwalibog 2003, Speijers 1987).

2.1.1 Fats

Essential nutrients need to be part of the diet in order to maintain good biological functioning. This will secure animal welfare and sound scientific results. De Wille et al. (1993) studied the influence of dietary linoleic acid levels on mammary tumour development. The control diet did not contain any linoleic acid (De Wille et al. 1993). This resulted in premature deaths of a substantial number of control animals as well as in biased results, as linoleic acid is essential for building cell membranes (National Research Council 1995). By leaving out dietary linoleic acid completely, it is possible that tumours cannot develop.

2.1.2 Proteins

Ewen and Pusztai (1999) showed that genetically modified potatoes could lead to crypt hyperplasia in the jejunum coinciding with an increased T lymphocyte infiltration. However, the diet which was fed to young growing rats contained only 6% protein. According to the guidelines of the National Research Council (1995) a young growing rat needs 15% high quality protein in the diet. Such a low protein diet as used by Ewen and Pusztai is therefore considered deficient for young growing rats, exerting a negative influence on the biological functioning, thereby compromising welfare as well as biasing results.

2.1.3 Vitamins

Dietary composition needs to fulfil species specific needs. For example, if a rat diet is fed to guinea pigs, disease and death will arise within a 14 days period due to vitamin C deficiency. Thus, vitamin C is an essential nutrient for the guinea pig, and not for rats. Another complicating factor arises from the shape of the pellet: the diameter of the pellets for rats is larger than for guinea pigs, as rats use their incisivi for gnawing. Guinea pigs will typically take the entire pellets into their mouth and grind it between their molars. So, besides knowledge on nutrient needs, species specific components of ingestive behaviour must be considered as well.

Several animal species perform coprophagy. This process is not prevented when animals are housed on grid floors, as they eat the faeces directly from the anus. Due to bacterial synthesis, animals supply themselves with adequate amounts of vitamin K and B12. As germfree animals lack this bacterial synthesis, they need to be fed with diets containing extra vitamins (Beynen and Coates 2001).

The goals of the experiment determine whether one wishes to leave out an essential nutrient. In case one examines the biological functioning of an essential nutrient, it may be worthwhile to leave it out entirely. However, in some cases it may be more worthwhile to use subdeficient levels instead; deficient levels will probably cause premature deaths, whereas the use of subdeficient concentrations are expected to lead to a more subtle pathology, providing more refined and detailed information on what the function of the nutrient is. This can be seen in line with the development of alternative tests in toxicological studies. Instead of using the LD50 (searching for a dose that kills 50% of the animals), the alternative fixed-dose-procedure provides more detailed and refined information on the working mechanism of a chemical that is tested.

2.2 Types of diets

The use of different diets between experiments and institutes can be a source of variation in experimental results. Therefore, major care should be given to the composition of diets, even when they do not affect health or lead to certain deficiencies. In laboratory animal units, generally two types of diets are used: 1) diets based on natural-ingredients (chow diets) and 2) purified diets.

2.2.1 Natural-ingredient diets

Natural ingredient diets, containing e.g. maize gluten, wheat, lard, corn oil and soya bean meal, are frequently used, as their usual composition makes it relatively easy to produce pellets, which are easy to handle. Also, these diets are generally much cheaper than purified diets. Many "standard" diets for growth/maintenance and reproduction are commercially available for each species. These diets are composed according to the species-specific needs in such a way that the concentration of all essential nutrients for each species and each condition (growth, reproduction) are met more than sufficiently. Depending on where the natural ingredients used in chow diets originate from (soil composition, use of pesticides, etc.) and the type of weather conditions they have been subjected to, the natural-ingredient diets can show large variation, which may be a cause of variation in experimental results (Table 5-1; Ritskes-Hoitinga et al. 1991). There is also variation between diets from different manufacturers (between-brand variation) and between batches from the same firm (between-batch variation) (Beynen and Coates 2001). Because a variable dietary composition can sometimes induce certain pathologies, variably compromised welfare in the form of reduced biological functioning can be the case. For instance, nephrocalcinosis is a condition which is induced in rats and rabbits as a result of higher dietary phosphorus levels (Ritskes-Hoitinga and Beynen 1992, Ritskes-Hoitinga et al. in print). Nephrocalcinosis has been found to be associated with reduced kidney function (Ritskes-Hoitinga et al. 1989, Al-Modhefer et al. 1986), indicating reduced biological functioning.

When producing natural-ingredient diets it is of importance that the minimum essential nutrient needs of each species are met, but at the same time it is essential to refrain from too high nutrient levels that can cause pathological conditions. A specific analysis certificate for each batch of diet is recommended, in order to make it possible for the buyer/researcher to judge the nutrient (and contaminant) levels before the start of each experiment.

Table 5-1. Batches of rodent maintenance diets from 10 different commercial firms fed to young female Wistar rats during a 4 week period (Ritskes-Hoitinga et al. 1991). The range of a different number of parameters is presented here.

Parameter	Group mean range
Growth (g/d)	2.5 – 3.5
Food intake (g/d)	12.1 – 15.6
Water intake (ml/d)	18.2 – 24.8
Incidence of nephrocalcinosis	0/6 – 6/6
Urinary pH	6.2 – 7.9
Caecal weight (g/100 g body weight)	1.8 – 3.9

2.2.2 Purified diets

Purified diets are much more expensive than natural-ingredient diets and are formulated from a combination of more purified macronutrients, pure chemicals, and ingredients of varying degrees of refinement. Purified diets are mainly used when the investigator wants to use a diet of quite a different composition than can be made from natural ingredients, e.g. when investigating (sub)deficient levels of nutrients (Zhou et al. 2003). By using purified diets there is more control of the ingredients and final dietary levels than in natural-ingredient diets, leading to more reproducible results within and between institutes (see also Ritskes-Hoitinga and Chwalibog 2003, Ritskes-Hoitinga et al. 1996). The nutrient levels are mostly close to the desired levels. By following the guidelines of the National Research Council (1995) and the American Institute of Nutrition (AIN93 diet, Reeves et al. 1993) guidelines, a standardised purified diet can be produced that fulfils the minimum required levels of essential nutrients of rodents. The nutrient levels are such that no pathology is induced due to uncontrolled variation or (unknown) superfluousness of certain nutrients.

One of the disadvantages of using purified diets is the fact that the composition is often such that it is unpelletable. Feeding the diet in a powder form to rodents may not meet their (essential?) need of gnawing. Gnawing makes sure that the continuously growing incisor teeth are shortened. If gnawing does not occur, incisor teeth can become too long which makes regular cutting necessary. Therefore, when feeding these soft diets, cage enrichment in the form of wood blocks/sticks may help to prevent these adverse effects, so that welfare becomes positively influenced. One of the disadvantages of using purified diets may be that they lack so far undiscovered nutrients, which can lead to deficiencies, thereby compromising biologically healthy functioning. One example for instance is that in the long-term, purified diets can cause degenerative problems and a decline in reproductive success (Ritskes-Hoitinga et al. 1993). Why this happens is unclear. Caution is warranted when using refined sources of protein (casein), as was demonstrated by Sanders et al. (1984). Five different types of casein were included in purified diets, and it appeared that not all casein sources could promote growth. When offering a choice situation, rats showed a reduced preference for the purified diet that they had experienced earlier which did not promote growth. What applies to different sources of protein, could also apply to other sources of nutrients. By allowing animals to select and compose their diets themselves, more insight in the necessary dietary elements can be made visible. Animals will learn and experience what is good for their well-being and can "tell" us their experiences in preference tests (see also 3.1 in this chapter).

2.2.3 Palatability

Palatability (defined as the relative acceptability of the feed, as determined by taste and texture) is another relevant factor that has to be taken into account, as the minimum level of essential nutrients must be ingested. When an unknown substance has to be mixed into the diet, palatability can be changed. Therefore it is wise to do a pilot study in a few animals, in order to judge the effect, before the real experiment is executed. The palatability of dietary fish oil for herbivorous rabbits was such, that a habituation period of about 6 months was necessary in order to make the rabbits ingest it in the proper amounts, i.e. as is required for a good health (Ritskes-Hoitinga et al. 1998).

2.2.4 Avoiding toxic levels

In order to make sure that animals do not become diseased as a result of toxic levels of nutrients and/or contaminants it is important to keep nutrient and contaminant concentrations similar to or below the "recommended" levels, unless the goal of the experiment is to examine this toxicity. The National Research Council documents provide background data on known toxic levels of each nutrient. In the case of contaminants, various guidelines provide maximum limits (British Association of Research Quality Assurance 1992, Environmental Protection Agency 1979, GV-Solas 2002, guidelines from production firms). These different guidelines do not necessarily provide the same maximum allowable concentrations. One of the most popular guidelines used by toxicologists worldwide is issued by the Environmental Protection Agency (1979). Remaining under these levels for all contaminants, will in general prevent animals and results from becoming negatively influenced. Which guidelines are used, will depend on the purpose of the experiment. If it is known that a certain contaminant will interfere with the particular purpose of an experiment, then the maximum tolerated level will be determined by the concentration known to cause interference. So in this case a specific maximum allowable level can be determined for the particular purpose of an experiment. For obvious reasons, by using purified ingredients, the level of toxic contaminants can be kept much lower than that found in natural-ingredient diets.

2.3 Choice of dietary composition in relation to optimal results and welfare

When designing the experimental dietary composition, important and relevant choices need to be made. These will partly depend on the goal of the study and the animal species. A thorough literature study should precede animal studies in order to avoid unnecessary duplication, suffering and biasing of experimental results. Pilot studies in a few animals under close observation are recommended in case the palatability and/or toxic effects of a certain diet/nutrient concentration are unknown. By using the information from pilot studies, experiments can be designed in a better way, so that an accurate number of animals can be used. Also post mortem examinations are important to reveal unwanted side-effects, as they may be relevant for the interpretation of the study as well as for the welfare. A few examples are outlined below.

Rabbits have often been used as a model in atherosclerosis research, since atherosclerotic plaques arise relatively easy (Jayo et al. 1994). When the effects of dietary fish oil on atherosclerosis in the rabbit were tested, it appeared that the amount of atherosclerosis in the aorta rose with increasing levels of fish oil in the diet (Ritskes-Hoitinga et al. 1998). This was quite unexpected, as dietary fish oil is associated with a reduced level of atherosclerosis in epidemiological studies in the human population (eskimo's) (Ritskes-Hoitinga et al. 1998). Detailed post mortem macro- and microscopic examinations revealed clear adverse effects of fish oil on the liver and a significant positive relationship was detected between the group levels of liver pathology scores and the quantity of atherosclerosis in the aorta. As the herbivorous rabbit usually does not eat fish, it is concluded that the rabbit is not a suitable animal model for testing these long-type unsaturated fatty acids. This is supported by anecdotical evidence about rabbits in the wild: in case of food scarcity, rabbits do eat dead fish found on the beach. As a result of eating the highly unsaturated long-chain fatty acids, the rabbits develop vitamin E deficiency symptoms (yellow fat disease). A similar picture of liver pathology like that resulting from fish oil feeding, arises when cholesterol is added to the rabbit diet in order to induce atherosclerosis. Cholesterol is not part of the normal rabbit diet, and relatively high concentrations of dietary cholesterol will induce liver pathology. It is therefore advisable to use dietary cholesterol levels that do not exceed 0.5 %, at least in cases where the diet contains added fat (Jayo et al. 1994), as this will induce atherosclerotic plaques, but no irreversible liver pathology. Testing fish oil in pigs does not lead to liver pathology: as the pig is omnivorous it can utilise these long type unsaturated fatty acids as well as dietary cholesterol.

Special diets used to induce atherosclerosis in mice are "western-type" diets containing about 21% fat and 0.15% cholesterol, and "atherogenic" diets, which contain 15% fat, 1.25% cholesterol and 0.5% cholic acid (Moazed 1998). This last diet is also referred to as "Paigen's diet". Historically, this diet was used to induce gallstones. This diet is known to be hepatotoxic and to induce a proinflammatory state (Moazed 1998). When using the Paigens diet, atherosclerotic plaques can be induced, however, hepatotoxicity and gallstones are induced simultaneously. Hepatotoxicity may interfere with the development of atherosclerotic lesions, as was seen in rabbits (Ritskes-Hoitinga et al. 1998). Therefore it is considered necessary to (at least) evaluate the condition of the liver. As gallstones are very painful in humans, it may be the case in mice as well.

The use of transgenic mouse models instead of feeding "extreme" diets to wild type mice models may be a good alternative solution, as it then becomes possible to leave out e.g. the cholate from the diet. On a normal chow diet, apoE-deficient mice develop plasma cholesterol levels that are at least 10 times as high as in wild type mice and most of the cholesterol is in the form of the highly atherogenic VLDL (Very Low Density Lipoproteins) form (Moazed 1998). Apo-E deficient mice develop atherosclerotic lesions on a normal chow and on a Western type diet, however, lesions develop more rapidly and at an earlier age on the Western type diet. The atherosclerotic lesions in apoE-deficient mice have strikingly similar pathological characteristics and anatomical distributions as compared to humans (Moazed 1998). The LDL-receptor-deficient mouse (LDL = Low Density Lipoproteins) does not develop atherosclerosis on a normal chow diet, but this can be induced by feeding a high fat diet. Lesion characteristics are the same as in the apoE-deficient mouse, but lesion formation is better controllable by dietary changes (Moazed 1998). Plasma cholesterol levels are lower than in the apoE-deficient mice and thereby more human-like. Although the use of transgenic animal models is preferable in certain studies, there are also negative welfare aspects surrounding their use. In a survey of the information on transgenic mouse strains submitted by researchers to the Danish Inspectorate, 30% of the strains were reported to have considerable welfare problems (Thon et al. 2002). A reduced breeding performance is one of the reported problems associated with breeding transgenic animals. This has led to the special production of high fat diets (9% fat) on the market, in order to attain higher reproduction levels in transgenic strains (Tobin 2003). An increased fat content also decreases the hardness of the food pellets, likely to have a very significant effect on food intake of weaker animals (Tobin 2003). From the point of view that welfare encompasses good biological functioning, the development of transgenic strains exerting poor breeding performance can be considered a negative development. By modifying dietary composition, this welfare disturbance is tackled succesfully. Whether the development of these strains is acceptable is an ethical question, which falls outside the scope of this chapter.

2.4 Methods of food administration

2.4.1 Feeding schedules

Certain feeding schedules can be a possible tool in the experimental design to achieve increased standardisation. Depending on the species and the scientific question, *ad libitum* or restricted feeding schedules are applied. In food restriction, one restricts the energy supply, while still ensuring nutritional adequacy, i.e. that all essential nutrients are supplied in the required minimum amounts (Hart et al. 1995). For dogs, pigs, cats, monkeys, etc. it is considered bad veterinary practice to feed *ad libitum*, as the animals will become obese (Hart et al. 1995). Therefore these species are fed restricted rations. Rabbits are sometimes fed *ad libitum*, at other times restricted, dependent on the laboratory policy. Rodents are fed *ad libitum* in the majority of all experimental studies. *Ad libitum* means that food is available at all times.

Feeding rodents *ad libitum* may be attractive from a practical point of view, however, on the basis of results from long-term toxicological studies, it can be questioned whether this is a sound scientific or welfare approach. *Ad libitum* feeding leads to clearly negative long-term health effects as compared to restricted feeding (75% of *ad libitum* intake): more obesity, shorter survival time, more degenerative kidney and heart disease, more cancer at an earlier age (Hart et al. 1995). *Ad libitum* feeding schedules in Wistar rats can induce severe kidney degenerative disease at an age of 6-9 months. Food restriction prevents this kidney degeneration. Also in the short-term, food restriction can have favourable effects. The relative body weight reduction 48 hours after surgery (jugular cannulation) is smaller in food restricted animals as compared to *ad libitum* fed rats (both at 3-4 and 17-18 months of age) (Hart et al. 1995). It is possible however, that this effect is mainly caused by the reduced size of the body fat compartments in the food restricted animals. It is claimed that animals become more "robust" when fed restrictedly, i.e. they can cope better with experimental stressors/procedures (Keenan 1999). Diluting the diet, e.g. by including a higher fiber content under *ad libitum* conditions, does not give the same positive health effects as compared to dietary restriction. It may be that a certain fasting period each day has a more positive health effect than feeding an energy diluted diet *ad libitum*. Thus, it is mainly the energy intake which is associated with body weight and the degree of obesity that is causal in these health affecting processes.

Keenan (1999) has demonstrated that *ad libitum* feeding is our least controlled experimental factor in the laboratory. Considerable variation in experimental results from rodents on *ad libitum* feeding schedules arose during the 1980's to 1990's. One of the possible explanations could be the continuous selection of faster growing individuals in outbred strains (Keenan 1999). Feeding rodents at around 75% of the *ad libitum* food intake is recommendable for long-term toxicity studies in order to make a sufficient number of animals survive the requested 2 year period (Hart et al. 1995). Under *ad libitum* feeding conditions, the body fat content can be more than 25% (Toates and Rowland 1987). By restricting food intake to 85% of *ad libitum* intake, the body fat content will be less than 10%, which is similar to that found in wild-captured animals (Toates and Rowland 1987). In order to make sure that all individual animals eat the same amount of food, it is advised to individually house the animals (Keenan 1999). However, the consequence may be a reduced welfare due to the level of food restriction and isolation, as rodents are social animals. In order to improve welfare, one of the labour intensive possibilities is to feed animals individually, whereafter they come back into the group. In line with agricultural practice, we are currently examining the use of microchip and computer technologies to develop automated feeding devices for feeding individual rats restrictedly under permanent group housing conditions. This is expected to improve health (cause less pathologies) and at the same time fulfil the need for social housing.

2.4.2 Pair-feeding

If voluntary food intake of the test group differs from the controls, it may be necessary to equalize the intake between the groups. This can occur when the palatability of the test diet is negatively influenced, or in cases where the substance influences health negatively, causing a reduced food intake in an indirect manner. From the point of view of standardisation, the test and control animals must ingest a similar amount of food at a similar time of day, otherwise it will become impossible to solely judge the effect of the test substance. Whether a reduced food intake has a negative influence on biological functioning, depends on to what level the food intake becomes lowered. As the NRC requirements are based on obtaining maximum growth, a level of 75% of the NRC recommendations is still considered sufficient to fulfil all needs. Four possible methods for achieving pair-feeding are: (1) weighing; (2) coupling of food dispensers; (3) gavage/permanent stomach cannula; (4) feeding machine. A feature of method (1) is that the amount of food eaten by the test group is weighed, and the control group is fed the same amount of food the next day. A disadvantage of this method can be that the total amount of food is provided at once to the control animals, perhaps at an unnatural time point of the day (the light resting phase in rodents). This may induce quite a different eating pattern in the controls as compared to the test animals, preventing proper comparison between test and control animals. In method (2), by using food dispensers, rats can be trained to obtain a pellet by pressing a lever. By coupling food dispensers of one test and one control animal, the dispenser of the control rat will provide a pellet at exactly the same time as the test animal "asks for" and eats a pellet. This system creates an optimal standardisation of amounts eaten and ingesting patterns. In method (3) permanent stomach cannula's and gavage make it possible to provide all animals with exactly the same amount of food at the same time points (Balkan et al. 1991). In method (4), feeding machines with regulated opening of valves make it possible to open and close food hoppers at the same time points of the day in test and control animals.

Meal training can be another possible solution as a pair-feeding schedule. Rats can easily be trained to ingest the required amount of food in a limited amount of time. Within a week they can learn to ingest the necessary food intake in e.g. 2 meals of 0.5 hr during the light phase (Ritskes-Hoitinga et al. 1995). Also one meal of 2 hours per day can be sufficient for a rat to obtain the necessary energy and nutrients. This will depend on the physiological stage of the animal. The exact time of day when the food is offered is critical in relation to the welfare, and is discussed under the section on natural behaviours. There are clear species differences in the rhythmic patterns and training/learning abilities. Although rats can be trained within one week to eat one or two meals per day during the light phase, hamsters could not cope with a schedule of one meal of 2 hours per day. A very rapid drop in body weight (20% in one week) occurred in these hamsters, making it necessary to stop this experiment (Ritskes-Hoitinga, unpublished observations). A restricted feeding schedule will usually lead to a more efficient handling of the diet, which leads to e.g. a slower gastrointestinal passage time and a hypertrophy of the gastrointestinal wall. This ensures that the animal can cope with the situation, despite the fact that the feeding schedule can be unphysiological and uncomfortable.

2.4.3   Gavage / stomach-tubing

When compounds such as pharmaca or nutrients are to be tested, it is important to avoid the possible negative influence of a bad taste on food intake. Therefore, gavage is used in order to apply meals/substances directly into the stomach. Upon comparison of the application of a meal directly in the stomach with the voluntary eating of a similar meal, a difference in experimental results was obtained (Vachon 1988). After the voluntary consumption, the results were more comparable to results from human studies. Disadvantages of stomach tubing are that it omits the oropharyngeal process, thereby omitting the physical effects of chewing and the addition of salivary enzymes that initiate the digestive process. Gavage can also cause stress to the animals, thereby suppressing gastrointestinal activity. By training the animals to become accustomed to the procedure and giving them positive rewards, the use of the stomach tubes becomes gradually less stressful. Another possibility is to insert permanent cannulas in the stomach (or small intestine), which makes it possible to apply substances without stress directly into the stomach or small intestine (Strubbe et al. 1986a).

These permanent cannulas are e.g. used to measure the (satiety) effects of filling the gastrointestinal system with purified nutrients such as glucose at different anatomical locations. These cannulas can also be used to make sure that chemicals and radioactive substances are applied directly into the gastrointestinal system without risk of contamination of the environment and the person executing it.

3. POSSIBILITY TO PERFORM NATURAL BEHAVIOURS

3.1 Behaviour in the laboratory

Although restricted feeding schemes and solitary housing have advantages for standardisation and execution of experiments, there are also opinions that restricted feeding schedules and solitary housing may affect welfare adversely, because it is unnatural. However, it is questionable whether natural conditions always induce better welfare, since nature provides the free-living animals with predation and various kinds of infectious threats. Although predation and health risks like infections are usually excluded, there are still many disadvantages, even when natural condition-like situations in the laboratory are introduced. The introduction of various natural behavioural repertoires for the rat is discussed. Under *ad libitum* feeding conditions, rats will eat most of their meals during the dark phase and only a few meals during the light phase (Spiteri 1982). This is considered a remainder from natural behaviour where the light period is more dangerous due to a heavy predation risk and is probably genetically determined. There are approximately 2-3 peaks in food intake during the dark phase. The first peak arises in the first few hours after the start of the dark phase and is probably needed to compensate for the energy deficit which has arisen during the light (resting) phase. In the middle of the night a relatively small peak is seen. During dawn another peak occurs, even though the stomach is still filled. The dawn peak is considered necessary to build reserves, in order to successfully overcome energetically the dangerous light phase where predation pressure is the highest. This dawn peak will occur typically in the 2 hours before it gets light. Upon changing the dark-light schedule, the dawn peak will always shift to the 2 hours before the start of the light phase. When food is removed in the 2 hours before the start of the light phase, the rat will not learn to consume extra food at an earlier time point, but compensates during the day. By eating in advance of the period of sleep which is during the light phase, the rat does not need to come into the open when the predation pressure is high. Thus, this feeding behaviour is probably the result of natural selection and genetically determined in the body clocks. The circadian clock directs/controls the daily feeding activity, and is located in the suprachiasmatic nucleus (SCN), which is also called light entrainable oscillator (LEO). If the SCN is lesioned, the typical day-night rhythm in food intake behaviour immediately disappears (Figure 5-1, Strubbe et al. 1987).

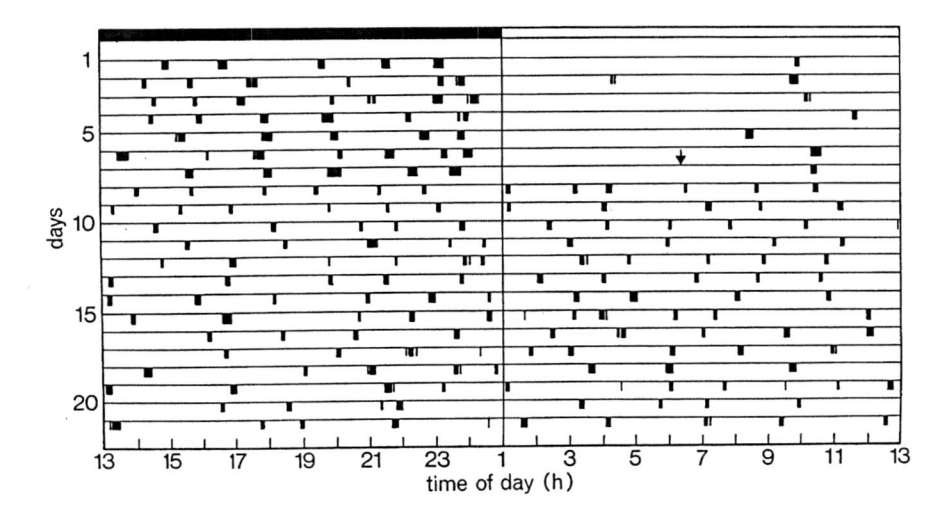

Figure 5-1. Daily pattern of food intake of a rat on consecutive days. The black squares represent bouts of feeding activity. On day 7, the arrow marks the time of electrolytic lesion of the suprachiasmatic nucleus. The horizontal black bar indicates the dark phase. (From: Strubbe et al. 1987)

Rats can be trained to eat one or two meals per day, but the internal rhythm of the circadian clock is maintained. Therefore, in these scheduled feeding designs, there is a strong interference with the circadian feeding behaviour and concomitant gastrointestinal physiology. As soon as the *ad libitum* food intake is reinstalled, rats will immediately revert to their original pattern of food intake (Spiteri 1982). Well-known effects are gastrointestinal problems associated with jetlag and shiftwork in humans, when subjects are "forced" to eat at a different time in the daily cycle to that dictated by the internal clock.

Although food intake on a regular chow diet can be used for numerous experiments, nature provides a greater variety of ingredients. The question can be posed whether a greater variety and the opportunity to choose and select is better for welfare? When rats were given the choice between various diets providing energy from different sources, the rats chose carbohydrate rich diets in the evening hours, and fat rich diets just before the start of the light period (see Figure 5-2, from Strubbe 1994a). The function may be that carbohydrate rich diets will quickly provide the animal with an easy accessible energy source, necessary to stop the deficits that have arisen during the light phase. On the other hand, eating fat rich diets (with their higher energy content per gram) creates energy reserves for covering the next light phase, with the initial fasting/resting phase (Strubbe 1994a). Another example is that monkeys which had access to a herb garden in the Zoo Apenheul, made a selection of specific herbs which was in accordance to their health state.

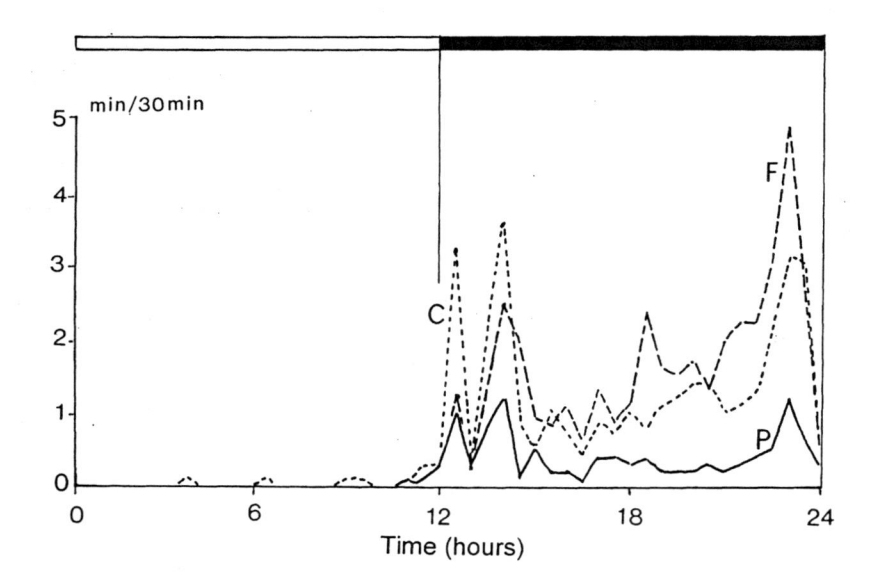

Figure 5-2. Daily rhythm of time spent consuming macronutrients in the rat. C = Carbohydrate, P = Protein, F = Fat. (From Strubbe 1994a)

Food intake is adjusted in accordance with energy expenditure. At lower ambient temperatures, the extra heat loss will be compensated by increased food intake. During increased levels of energy expenditure, the rat will in first instance compensate by increasing meal size. Thereafter, the meal frequency will change as well, when an increased meal size is not sufficient. During lactation, especially the third week postpartum, females will even eat meals during the initial stages of the light phase as well, in order to meet the high energy demands (Strubbe and Gorissen 1980). In diabetic animals the loss of glucose via the urine is immediately compensated by increasing meal size, whereas frequency and timing of meals are unaffected (Strubbe 1994b).

Usually, ingestive behaviour is associated with food-anticipatory behaviour, which in the rat is expressed as increased locomotor behaviour. This is particulary clear when rats are kept on a restricted feeding schedule, e.g. a one hour meal per day. This food-anticipatory behaviour gradually disappears when *ad libitum* conditions are reinstated (Stephan 1984, Brinkhof et al. 1998). If the SCN is lesioned, the rats will still display the food anticipatory activity (Stephan 1984). This suggests that another rhythmic clock determines meal-associated rhythms (the Food Entrainable Oscillator = FEO). Food anticipatory activity may be a functional reaction, in that e.g. the body temperature rises, probably by increasing the efficacy of enzymatic processes involved in nutrient handling (Strubbe and van Dijk 2002).

3.2 Experimental design and circadian rhythms

Experimental animals can be diurnal or nocturnal, i.e. being active during the light or dark phase, respectively. Some species do not have a clear circadian rhythm, e.g. the guinea pig is active during the entire 24 hours per day and has short periods of rest in between activity periods. Many experiments in the laboratory are performed during daytime, when nocturnal animals like rats are normally asleep (Toates and Rowland 1987). If food is presented during the light phase only, rats quickly learn that food has to be ingested at that time as well. However, the circadian pacemaker rhythm remains and does not adapt: as soon as ad lib conditions are reinstated, the rats will immediately return to the original natural ingestive pattern of food intake (Spiteri 1982, Strubbe et al. 1986b, Strubbe 1994b, Brinkhof et al. 1998). By feeding only during the light phase instead of the dark phase, substantial differences in physiology, neurochemistry and behaviour can result. Also the response to pharmacological agents can be different (Toates and Rowland 1987, Claassen 1994). The circadian pacemaker rhythm does not change, but the feeding schedule dictated by the experimental design demands adaptative behaviour. Feeding rats during the light phase can be compared to the effects of shift work in humans. Shift work is associated with disturbed physiology, pathological stomach bleeding, disturbed sleeping behaviour and disturbed food processing in humans. This is because the physiological and behavioural responses are not in accordance with the internal biological clocks. This can result in reduced health and welfare and negative interference with experimental results, unless of course shift work is the focus of study. The following experiment will illustrate how this "shiftwork" induces a less optimal digestive processing of food. The bile flow in rats provided with a permanently implanted bile fistula, shows a clear circadian rhythm under *ad libitum* feeding conditions, with the highest levels during the dark phase, where most food intake takes place (Vonk et al. 1978). Feeding six meals of 0.5 hr each throughout the dark phase gives a virtually identical bile flow response as that under "real" *ad libitum* feeding conditions (Figure 5-3, Strubbe unpublished observations). When rats are offered six meals during the light period only, the bile flow shows a distinctly different pattern: there is a reduced bile flow at night time, and there is a peak in the bile flow, each time a meal is eaten. As the absolute level of the bile flow (ml/hr) is less when being fed during the light phase as compared to the feeding of six meals during the dark phase, the digestive process is expected to be less efficient. Therefore, if it is the goal to study rats that eat e.g. for a few hours per day, feeding the rats during the dark phase instead of the light phase is in better accordance with their gastrointestinal physiological activity patterns. In order to make this practically possible, the dark phase can be shifted to a time point of the day which is more convenient for the experimenter/personnel.

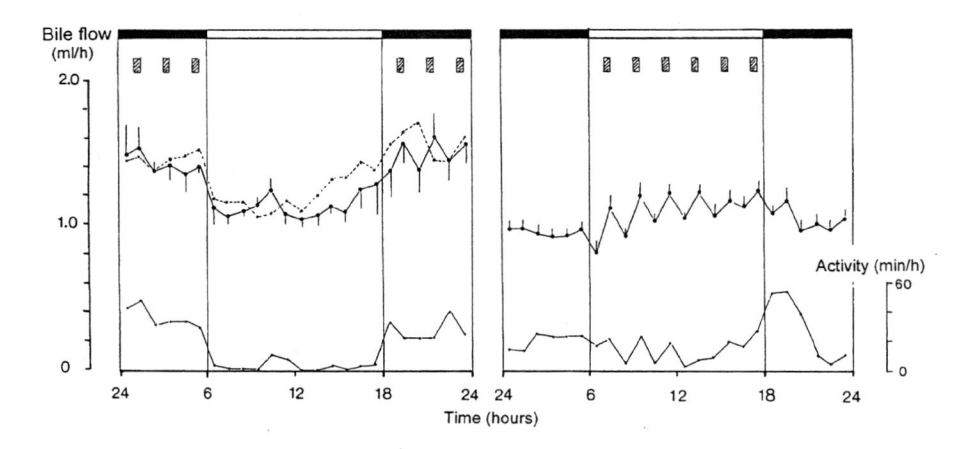

Figure 5-3. Bile flow and physical activity in rats during the daily cycle. Bile flow was measured with permanently implanted bile fistulas. In the lower graphs a measure for general locomotor activity is depicted. The left side of the figure shows the bile flow in a normal *ad libitum* condition (dotted line) and when a meal schedule is offered in 6 meals (dashed bars) spread over the dark phase (black bar on top). The right part of the figure shows the bile flow and concomitant activity when meals are spread over the light phase. (Strubbe unpublished observations)

Rabbits fed a restricted amount of food just before the dark phase, when in nature most of the food is eaten (Hornicke et al. 1984), have a low frequency of stereotyped behaviour (Krohn et al. 1999), compared to rabbits fed a restricted amount of food in the morning or rabbits fed *ad libitum*. As stereotyped behaviour is thought to reduce well-being, these results indicate that by feeding animals a restricted amount of food at a natural time point of the day, welfare can be increased. Not only the timing of providing the food is important, but also the schedule. For instance, when providing rabbits with a restricted amount of food (60% of *ad libitum*) at the beginning of the light phase (7.30 am), a significantly increased food conversion and water intake occurred, compared to rabbits that had food freely available from 7.30 am to 2.30 pm, even though the food intake in grams per day was similar (Table 5-2; Ritskes-Hoitinga and Schledermann 1999). These examples illustrate that it matters how a restricted feeding schedule for a certain species is executed.

Table 5-2. The influence of various dietary restriction schedules in rabbits during a 6 week period. (Ritskes-Hoitinga and Schlederman 1999)

Parameter	AL	FR	NFR	DFR	Probability (P)
Growth (g/d)	30.1±1.6 [a]	13.5±0.3 [b]	18.2±1.5 [b]	30.1±3.0 [a]	<0.05
Food intake (g/d)	206±18 [a]	125±0 [b]	126±3 [b]	213±7 [a]	<0.05
Food conversion (g food/g growth)	6.8±0.3 [a]	9.3±0.2 [b]	6.9±0.5 [a]	7.1±0.7 [a]	<0.05
Water intake (g/d)	255±34 [a]	381±46 [b]	244±44 [a]	284±6 [a]	<0.05

AL = *Ad libitum* feeding, fresh food supplied at 7.30 am each day
FR = 60% of the AL amounts eaten was provided at 7.30 am each day
NFR = Food was freely available from 7.30 am to 2.30 pm.
DFR = Food was freely available from 2.30 pm to 7.30 am.
Means and SD are given for 3 rabbits per group.
Results in the same row not bearing the same superscript are significantly different.

Rats in the wild show a neophobic reaction towards novel objects in their environments and will often avoid them for days. Novel food evokes a similar careful behavioural pattern. After carefully sampling a small amount of unknown food, the rat learns to distinguish safe from poisonous foods, and nutritionally poor foods from those of a higher value (Nott and Sibly 1993). When food has proven to be "safe", larger amounts will be eaten by the rats. Knowledge of these basic behaviours is relevant, in order to design experiments effectively. Upon changing dietary composition in meal-trained rats (cross-over design), a typical neophobic response results in an immediately reduced food intake (Ritskes-Hoitinga et al. 1995). If one wants to perform a cross-over design, animals must have been accustomed to this typical dietary composition in an earlier phase. So, when they are exposed to this diet a second time, they will not reduce their food intake. Diets varying in macronutrient composition can influence behavioural and physiological stress responses, as was found by Buwalda et al. (2001). A high dietary fat level appeared to have an ameliorating effect on behavioural and physiological parameters as compared to a diet with a high carbohydrate level, after rats were exposed to psychosocial or physiological stressors. So it has to be established whether rats wish to eat "functional foods" in order to cope better with stressful situations.

3.3   Group housing

Rats in the wild live in colonies and have social interactions. The question can be posed whether group housing is always favourable for welfare or not? The way we house the animals in the laboratory and the husbandry conditions can influence both the amount of food they ingest as well as their meal patterns. Whether an animal is group or individually housed has implications for the food intake and the results. Individually-housed mice of both sexes had a higher food intake than group-housed mice (2, 4 or 8 mice per cage) (Beynen 2001, Chevdoff et al. 1980) The mice housed individually or at 2 per cage had a higher body weight and body weight variability than the other groups. In case the mice were housed 8 per cage, an increased gastritis frequency occurred compared with individually housed mice (Chevdoff et al. 1980). These results indicate that 4 mice per cage was the optimum group size in this particular study. Peters and Festing (1990) showed that the maximum body weight gain at different cage densities depended on the cage size and the mouse strain chosen: the inbred Balb/c thrived best in a high density housing, whereas the outbred MF1 had a higher body weight gain during low density housing. If mice are housed in a group, they usually lie together, thereby reducing the total surface area as compared to individual housing. A reduced heat loss per animal in the group is the result, and due to this "behavioral thermoregulation" the food intake becomes reduced. It has been hypothesized that individuals from an inbred strain may produce identical pheromones, which may make them more tolerant to high density (Peters and Festing 1990).

Another aspect of group housing concerns the possible interference of the social status of the individuals. Behavioural and ingestive patterns can be influenced by the social structure in the group. A dominant animal may prevent others from eating at certain times. On the left side of Figure 5-4 food intake patterns of a socially compatible group of S3 rats have been given: no perceivable dominant rat is present in this colony. The normal feeding pattern with a clear dawn peak can be seen for all individuals. On the right side, the dominant animal (D) prevented all the other rats lower in hierarchy from having a dawn peak at the natural time point. Some had to shift to an earlier time point during the dark phase, and some even shifted to the beginning of the light phase (Strubbe and Koolhaas, unpublished observations). It is so far unclear whether the phase of circadian rhythm had shifted permanently or not.

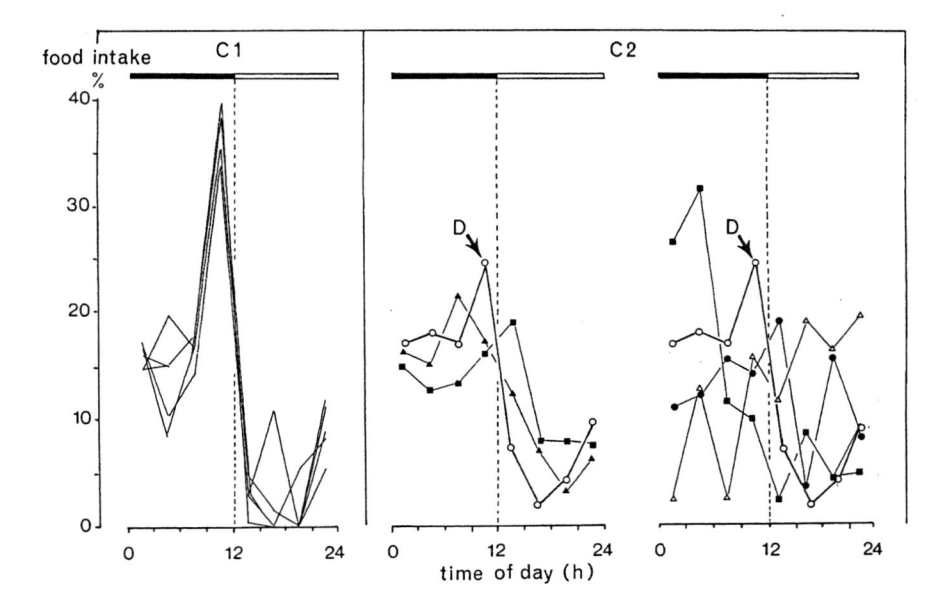

Figure 5-4. Feeding activity of members of two rat colonies. The left part of the figure shows a relatively calm colony (C1) without a dominant rat. Rats are feeding together at dawn at the feeding place. In the right part, the feeding rhythm of a colony (C2) is depicted where a dominant rat (indicated by D) is present. On the left side of C2, the D and subdominant rats are depicted. Notice that D eats at the same time (dawn peak) as the members of colony C1 and that the subdominants eat in the proximity of the dawn peak. In the right part of C2, the feeding activity of the other more submissive members of the colony are depicted in comparison with D. The "dawn peak" of the submissive animals shifts more than those of the subdominant rats. The dark phase is indicated by the black bars on the top of the figures. (Strubbe and Koolhaas, unpublished observations)

Thus, group housing can also differentially influence the experimental outcome, either by influencing food intake/meal patterns and/or due to other unknown factors. For instance, by feeding a high phosphorus diet (0.5% P) to females of the RP and LE strains, nephrocalcinosis was induced in the RP strain only. The severity of nephrocalcinosis in the group-housed RP rats was significantly higher than in individually-housed animals (Ritskes-

Hoitinga et al. 1992). This illustrates the interaction that can occur between genetic and environmental (dietary composition and housing) factors. Thus although group housing is more natural than individual housing, no clear-cut general guidelines on the optimum group size can be given. It depends on the purpose of the experiment, the species, the strain of animals used, the sex, the age, the experimental design, etc., which optimum group size needs to be chosen. Social interactions in group-housed animals can also lead to (chronic) stress responses, thereby compromising the welfare of some individuals and reducing the standardisation of results. We observed in a colony of female group-housed Beagle dogs, that even though the animals had been separated for the duration of the meal (half an hour), they could vomit up to 4-6 hours later (Ritskes-Hoitinga et al. unpublished observations). It is clear that this caused a large variation in results, as the stomach contents were eaten by one other individual. In the case of dogs it is very important to socialize and train the animals accurately in order to establish social compatible groups. A good socialisation and training program will make dogs feel secure and will create a stable hierarchy in the group. In the case of male animals it may be unwise and impossible to group-house them under laboratory conditions, due to aggressive attacks without possibilities for escaping or hiding. Therefore it is likely that the formation of compatible social groups will add to achieving standardisation of results as well as optimal welfare.

3.4 The nutritional state of fasting

Fasting (food deprivation) is commonly used in pharmacological research. It can be used to standardise experiments: an overnight fast assures that the last meal has been taken at least 12 hours before. This guarantees that there cannot be any interference with this last ingested meal. However, as a result of fasting, the animals are in a quite different metabolic state (Strubbe and Prins 1986). Sometimes it is necessary to have an empty stomach to judge the absorption and bioavailability of medicine after oral application, without interference of dietary components. In other experiments animals are made hungry, so that they will immediately eat a meal as soon as it is offered, or use food as a reward to quickly train animals. The time which is used for fasting varies considerably: rats are sometimes fasted up to 72 hours (Claassen 1994). When fasted blood samples need to be collected, usually rats are fasted overnight. Vermeulen et al. (1997) examined how quickly the rat stomach is emptied. They found that the stomach is empty after just 6 hours of food deprivation. If one wishes the intestines to be empty as well, a 22 hour period of food deprivation is necessary (SGV newsletter 2001). When the food deprivation period in rats exceeded a 6 hour period during the dark phase, a clearly increased locomotor and grooming behaviour and increased hair contents in the stomach was detected (Vermeulen et al 1997). No stereotyped behaviour was seen in the rat, in contrast to food-deprived mice (Schlingman et al. 1993). Because of their small body size and high metabolic rate, mice are extremely sensitive to fasting and may die when it lasts too long. The increased locomotor behaviour in rats fasted for longer than 6 hours was thought to be the result from an increased food searching behaviour. For nutritional studies, fasting is not recommended, as the effects of the diet/nutrients are the goal and not the effects of fasting. In some cases fasting cannot be avoided, as high levels of blood lipids can increase turbidity in the blood, which makes the analysis of certain nutrients impossible. From the point of view of the rat and sound results, a fasting period exceeding 6 hours during the dark phase is not necessary when the goal is to obtain an empty stomach. Thus, dependent on the scientific question, one can apply the fasting condition. However, one has to be realize that a different metabolic rate is present at the time of testing due to the fasting condition. Thus, although, long fasting periods can reduce welfare, short-term fasting can be used to standardise experiments.

3.5   Essential needs in the environment

The execution of certain natural behaviours in the laboratory setting are an important prerequisite for good animal welfare, as they represent basic essential needs. It is a well-known fact that zoo-animals develop stereotyped behaviour, because some essential needs, especially in the field of feeding and locomotion are not met. Also caged laboratory mice develop stereotyped behaviour. In the agricultural setting pigs can develop e.g. sham-chewing and tail-biting as a stereotyped behaviour, as their need for rooting for food is not satisfied. Chickens normally use a large part of their awake hours searching, scratching and pecking for food. When they are housed in batteries, many develop feather pecking behaviour, which may be the result from not fulfilling their food searching needs. Which essential needs have to be fulfilled, depends on the species and circumstances. Each species has its own specific needs, which need to be taken into account. An attempt has been made to prioritise categories of needs at the Dahlem workshop nr. 87 in Berlin (Broom 2001). Providing rats and mice with food, water, social contact and the possibility to sleep are presuppositions for basal welfare. Deficiencies of these factors will lead to poor welfare. By providing the possibility of mating, climbing, nest building, rearing and additional space (curiosity satisfaction), conditions are created for good welfare (Gartner 2002).

3.6 Working for food

Rats in nature need to find their food in the neighbourhood of their nest. Therefore, they have to perform action. In the laboratory situation the food is always close or delivered at fixed times, i.e. there is no need for seeking for food. Rats with a permanent stomach cannula, had learned to supply food directly into the stomach by pressing a lever (Strubbe et al. unpublished observations). The rats were able to maintain their normal body weight, however, "sham chewing" near the food hopper was observed. This may indicate that the satisfaction of oropharyngeal factors are an essential part/need of the ingestive process.

Rats have been made to put more effort for obtaining food by shortening the distance between the food hopper bars (Figure 5-5, Strubbe, unpublished observations). The first time they had to work their way through, they spent a considerable amount of extra time gnawing their way through the decreased spaces to gnaw off the food. But, after a certain period they worked through the narrowed distance between the bars much more efficiently. The final result was that they used about the same amount of time on the ingestive process as compared to the situation before. This indicates that working for food may quickly loose its value as a possible enrichment factor.

Another example of working for food is when rats choose between working for food or eating freely available components (Kaufman and Collier 1981). Sunflower seeds with and without hulls were offered. After about 5 days, rats developed a strong stable preference for seeds without hulls. The range of preferences depended on the individual rat: 68.6 – 94.9% of total sun flower seed intake consisted of seeds without hulls. This indicates that rats preferred the food item with the lowest handling cost after a habituation period of about 3-7 days, but that the preference was not absolute (Kaufman and Collier 1981). The fact that the rats did not develop an absolute preference could be a failure of the rat to discriminate, or could be an adaptive strategy that makes it possible for the foraging animals to monitor a changing environment, or may indicate a partial preference, i.e. part of an optimal diet needs to be without hulls. Rats are expected to monitor the consequences of consumption indirectly, as the consequences of ingestion are usually not immediate. They will judge the caloric and nutrient content, but are also sensitive to the cost it takes in time and energy in order to obtain a certain food (Kaufman and Collier 1981).

NUTRITION AND ANIMAL WELFARE 75

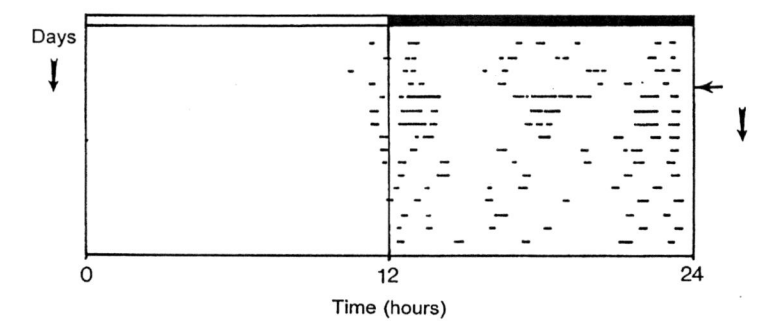

Figure 5-5. Daily pattern of food intake of a rat on consecutive days. The black lines represent bouts of feeding activity. On day 5, the distance between the foodhopper bars was shortened (arrows), requiring increased effort to obtain food through the bars. Notice the large meals of extremely long duration in the beginning. After a few days, the animals could work their way through the narrowed food hopper bars just as fast as before (Strubbe unpublished observations).

It is difficult to determine how much time/energy ought to be spent on working for food. The choice the animals make is not necessarily leading to an improved welfare. Sometimes the animals are inclined to eat a large amount of easily available food fast, in order to make energy deposits. This in turn, may have a negative impact on the animals health and welfare in the long-term. As a result of domestication and selection, animals may change. This may also change their needs for optimal welfare. Upon comparison of the behaviour of the white leghorn and its ancestor the wild jungle fowl, a remarkable difference occurred in the food searching behaviour. The chickens got the choice between freely available food and food that was hidden in a semi-natural environment. The white leghorn chose to eat 70% of the total amount of food from the freely available food. In contrast, the wild jungle fowl ingested 70% of the total diet from the hidden food (Jensen 2003). This illustrates that there is a behavioural adaptation to an increased production level.

4. CONCLUDING REMARKS

In this chapter many experimental conditions in relation to feeding have been discussed and advantages/disadvantages for animal welfare have been evaluated. What implications do they have for research on animals with regard to the animals welfare and the standardisation of experimental results?

Providing animals with food that fulfils their species-specific nutrient needs is a necessary item for securing basal welfare and reliable experimental results. However, in what way we need to provide the food in order to maintain or increase animal welfare is an important question. Is it enough to fulfil essential nutrient needs, or do we need to do more? Is it important for the animal to have the possibility to select from various food items, i.e. have more influence on composing the diets themselves? When rats were offered various diets differing in macronutrient composition, their preference varied with the time of the night. This is an indication that the needs may vary during the circadian cycle. More investigations are needed to solve whether self-selection of food is important for animal health and welfare. Whether it is important for the animals' well-being to work for obtaining food, remains a subject of debate. When given the choice, domesticated chickens clearly show a lower preference for obtaining food by working for it as compared to their wild ancestors. However, the percentage of food obtained by working is still 30%. Results from preference tests with rats give the same result. This indicates that animals in captivity still like to work for food, at least to a certain extent. The method that can be used to make animals work for food will depend on the species-specific behaviour and needs. In case a method is chosen, it is not certain that it will remain satisfactory as it can quickly loose its novelty value as was shown in the example where rats initially had to increase their efforts for obtaining food through narrowed distance between food hopper bars.

As *ad libitum* feeding in long-term toxicity studies have been clearly associated with negative health effects, restricted feeding is advised. How and when to feed is important for the animals' welfare, as e.g. the frequency of stereotyped behaviour in rabbits is increased when they are fed *ad libitum* or restrictedly at an "unnatural" time point of day as compared to restricted feeding at a "natural" time point (just before the dark phase). Stereotyped behaviour will induce increased variation in results, as the degree of behaviour will vary for each individual and thereby the effects on energy utilisation. We believe that an adaptation of feeding schedules to the normal circadian patterns will contribute positively to welfare as well as standardisation of results. Feeding at "unnatural" times will cause wider fluctuation and perturbed physiology and will bring animals out of homeostasis, thereby decreasing welfare. Although group housing of social species is preferred over individual housing, it may be unpractical (e.g. catheterized animals, aggressive males) and a source of unwanted extra variance. Dependent on the species, sex, experimental goals, etc. groups should be composed and monitored carefully, as group composition and behaviour can compromise standardisation and welfare.

Whether variation in food items provided is an essential factor to increase welfare is a matter of debate. Providing this dietary variation may be in conflict with the striving for standardisation of results, as the choice of individuals may be different. Also the choice of individuals during the circadian cycle may vary, contributing to a higher variation in experimental results. An indirect indication of how animals subjectively value the provision of food variation is obtained in preference testing, also involving the level of effort animals are willing to invest for obtaining this goal. By evaluating the level of investment an animal is willing to give, an indication of the motivation to reach a certain goal can be obtained.

Trying to improve welfare and fulfilling the animals' needs for a good welfare may give a potential conflict with trying to standardise experiments and reducing variation in results. On the other hand, an improved welfare may also lead to an animal that is more in balance physiologically and psychologically, thereby resulting in reduced variation as it can cope better with environmental challenges. This will depend on the circumstances and parameters measured. By carefully monitoring the animals' species-specific physiology and behaviour in experimental studies exploring the relation between nutritional factors, welfare and variation in results, more insight into these factors can be obtained.

REFERENCES

Al-Modhefer AKJ, Atherton JC, Garland HO, Singh HJ, Walker J. Kidney function in rats with corticomedullary nephrocalcinosis: effects of alterations in dietary calcium and magnesium. J Physiol 1986; 380:405-414

Balkan B, Steffens AB, Bruggink JE, Strubbe JH. Hyperinsulinemia and glucose tolerance in VMH-lesioned obese rats: Dependence on food intake and route of administration. Metabolism 1991; 49:1092-1100

Beynen AC, Coates ME. Nutrition and experimental results. In Principles of Laboratory Animal Science. Bert FM, van Zutphen LFM, Baumans V, Beynen AC eds, Elsevier Scientific Publishers, Amsterdam 2001; 111-128

Brinkhof MWG, Daan S, Strubbe JH. Forced dissociation of food and light-entrainable circadian rhythms of rats in a skeleton photoperiod. Physiol Behav 1998; 65:225-231

British Association of Research Quality Assurance (BARQA). Guidelines for the manufacture and supply of GLP animal diets. 1992

Broom DM (ed.). Coping with Challenge. Dahlem University Press, Berlin 2001

Buwalda B, Blom WAM, Koolhaas JM, van Dijk GJ. Behavioral and physiological responses to stress are affected by high-fat feeding in male rats. Physiol Behav 2001; 73:371-377

Carter CS, Fraser D, Gartner K, Lutgendorf SK, Mineka S, Panksepp J, Sachser N. "Group report: Good welfare. Improving quality of life". In Coping with Challenge. Broom DM ed, Dahlem University Press, Berlin 2001; 79-100

Chevdoff M, Clarke MR, Faccini JM, Irisarri E, Monro AM. Effects on mice of number of animals per cage: an 18-month study (preliminary results). Arch Toxicol 1980; 4(suppl):435-438

Claassen V. Neglected Factors in Pharmacology and Neuroscience Research. Elsevier, Amsterdam 1994

Crok M. Bij de beesten af. Natuur en Techniek 2003; 4:46-48

De Wille JW, Waddell K, Steinmeyer C, Farmer ST. Dietary fat promotes mammary tumorigenesis in MMTV/v-Ha-*ras* transgenic mice. Cancer Lett 1993; 69:59-66

Environmental Protection Agency. Proposed health effects test standards for toxic substances control act test rules, Good laboratory standards for health effects in Federal Register, 1979; vol 44, no. 91

Ewen SWB, Pusztai A. Effect of diets containing genetically modified potatoes expressing Galanthus nivalis lectin on rat small intestine. Lancet 1999; 354:1353-1354

Fraser D, Weary DM, Pajor EA, Milligan BN. A scientific conception of animal welfare that reflects ethical concerns. Anim Welf 1997; 6:187-205

Gartner K. Good, basal and poor welfare of laboratory animals. Different contributions of behavioural patterns (poster). 8th FELASA symposium, Aachen, June 2002

GV-Solas. Guidelines for the Quality-assured production of laboratory animal diets. Gesellschaft für Versuchstierkunde – German Society for Laboratory Animal Science 2002

Hart RW, Neumann DA, Robertson RT eds. Dietary Restriction: Implications for the Design and Interpretation of Toxicity and Carcinogenicity Studies. ILSI Press, Washington 1995

Haseman JK, Huff J, Boorman GA. Use of historical control data in carcinogenicity studies in rodents. Toxicol Pathol 1984; 12:126-135

Hornicke H, Ruoff G, Vogt B, Clauss W, Ehrlein H-J. Phase relationship of the circadian rhythms of feed intake, caecal motility and production of soft and hard faeces in domestic rabbits. Lab Anim 1984; 18:169-172

Jayo JM, Schwenke DC, Clarkson TB. Atherosclerosis research. In The Biology of the Laboratory Rabbit, 2nd Edition. Manning PJ, Ringler DH, Newcomer CE eds, Academic Press, san Diego 1994; 367-380

Jensen P. Behaviour, stress and welfare – genetic and phenotypic side-effects of selection for production traits (presentation). Symposium "Grenzen aan welzijn & dierlijke productie". Utrecht, April 10, 2003

Kaufman LW, Collier G. The economics of seed handling. Am Nat 1981; 118:46-60

Keenan KP, Ballam GC, Soper KA, Laroque P, Coleman J.B, Dixit R. Diet, caloric restriction, and the rodent bioassay. Toxicol Sci 1999; 52(2 Suppl):24-34

Krohn TC, Ritskes-Hoitinga J, Svendsen P. The effects of feeding and housing on the behaviour of the laboratory rabbit. Lab Anim 1999; 33:101-107

Moazed TC. Continuing Education Seminar, The American Society of Laboratory Animal Practitioners. October 1998, Cincinnati, USA

National Research Council (NRC). Nutrient Requirements of Sheep 1985; of Dogs 1985; of Beef Cattle 1984; of Mink and Foxes, 1982; of Laboratory Animals (Rat, Mouse, Guinea pig, Hamster, Gerbil, Vole), 1995; of Poultry 1994; of Fish 1993; of Horses 1989; of Dairy Cattle 1989; of Swine 1998; of Cats 1986; of Goats 1981; of Nonhuman Primates 1978; of Rabbits 1977. National Academy Press, Washington DC

Nott HMR, Sibly RM. Responses to novel food by rats: the effect of social rank. Crop Prot 1993; 12:89-94

Peters A, Festing M. Population density and growth rate in laboratory mice. Lab Anim 1990; 24:273-279

Reeves PhG, Nielsen FH, Fahey GCJr. AIN-93 purified diets for laboratory rodents: final report of the American Institute of Nutrition Ad Hoc writing committee on the reformulation of the AIN-76A rodent diet. J Nutr 1993; 123:1939-1951

Ritskes-Hoitinga J, Beynen AC. Nephrocalcinosis in the rat: a literature review. Prog Food Nutr Sci 1992; 16:85-124

Ritskes-Hoitinga J, Schledermann C. A pilot study into the effects of various dietary restriction schedules in rabbits. Scand J Lab Anim Sci 1999; 26:66-74

Ritskes-Hoitinga J, Chwalibog A. "Nutrient Requirements, experimental design and feeding schedules in animal experimentation." In Handbook of Laboratory Animal Science. Hau J, Hoosier GL eds, CRC Press, Boca Raton 2003; 281-310

Ritskes-Hoitinga J, Lemmens AG, Danse LHJC, Beynen AC. Phosphorus-induced nephrocalcinosis and kidney function in female rats. J Nutr 1989; 119:1423-1431

Ritskes-Hoitinga J, Mathot JNJJ, Danse LHJC, Beynen AC. Commercial rodent diets and nephrocalcinosis in weanling female rats. Lab Anim 1991; 25:126-132

Ritskes-Hoitinga J, Mathot JNJJ, Van Zutphen LFM, Beynen AC. Inbred strains of rats have differential sensitivity to dietary phosphorus-induced nephrocalcinosis. J Nutr 1992; 122:1682-1692

Ritskes-Hoitinga J, Mathot JNJJ, Lemmens AG, Danse LHJC, Meijer GW, Van Tintelen G, Beynen AC. Long-term phosphorus restriction prevents corticomedullary nephrocalcinosis and sustains reproductive performance but delays bone mineralization in rats. J Nutr 1993; 123:754-763

Ritskes-Hoitinga J, van het Hof KH, Kloots WJ, de Deckere EAM, van Amelsvoort JMM, Weststrate JA. Rat as a model to study postprandial effects in man. In Proceedings of The World Congress on Alternatives and Animal Use in the Life Sciences: Education, Research, Testing. Goldberg AM, van Zutphen LFM eds. Alternative methods in Toxicology and the Life Sciences, Series volume 11, 1995; 403-410

Ritskes-Hoitinga J, Meijers M, Timmer WG, Wiersma A, Meijer GW, Weststrate JA. Effects of two dietary fat levels and four dietary linoleic acid levels on mammary tumor development in Balb/c-MMTV mice under ad libitum feeding conditions. Nutr Cancer 1996; 25:161-172

Ritskes-Hoitinga J, Verschuren PM, Meijer GW, Wiersma A, van de Kooij AJ, Timmer WG, Blonk CG, Weststrate JA. The association of increasing dietary concentrations of fish oil with hepatotoxic effects and a higher degree of aorta atherosclerosis in the ad lib.-fed rabbit. Food Chem Toxicol 1998; 36:663-672

Ritskes-Hoitinga J, Grooten HNA, Wienk K, Peters M, Lemmens AG, Beynen AC. Lowering dietary phosphorus concentrations reduces kidney calcification, but does not adversely affect growth, mineral metabolism, and bone development in growing rabbits. Brit J Nutr in print

Roe FCJ. Historical histopathological control data for laboratory rodents: valuable treasure or worthless trash? Lab Anim 1994; 28:148-154

Sanders S, Ackroff K, Collier GH, Squibb R. Purified diets: some cautions about casein. Physiol Behav 1984; 33:457-463

Schlingmann F, van de Weerd HA, Blom HJM, Baumans V, van Zutphen LFM. Behavioural differentation of mice housed on different cage floors. Proceedings of the fifth FELASA symposium; 1993 June 8-11; Brighton UK, Royal Society of Medicine Press 1994; 335-357

SGV Newsletter nr. 24 (spring 2001): http://www.sgv.org/Newsletter/news-24.htm

Speijers GJA. Voedingsvezel en haarballen. Proceedings of the NVP symposium "voeding en kwaliteit van proef en dier", 1987; 79-86.

Spiteri NJ. Circadian patterning of feeding, drinking and activity during diurnal food access in rats. Physiol Behav 1982; 28:139-147

Stephan FK. Phase shifts of circadian rhythms in activity entrained to food access. Physiol Behav 1984; 32: 663-771

Strubbe JH, Gorissen J. Meal patterning in the lactating rat. Physiol Behav 1980; 25:775-777

Strubbe JH. Neuro-endocrine factors. In Food Intake and Energy Expenditure. Westerterp-Plantenga MS, Fredrix EWHM, Steffens AB eds, CRC Press, London 1994a, 175-182

Strubbe JH. Circadian rhythms of food intake. In Food Intake and Energy Expenditure. Westerterp-Plantenga MS, E.W.H.M. Fredrix EWHM, Steffens AB eds, CRC Press, London 1994b, 155-174

Strubbe JH. "Hunger, meals and obesity." In Encyclopedia of Cognitive Science. Nadel L ed, Nature Publishing Group, London 2003

Strubbe JH, Alingh Prins AJ. Reduced insulin secretion after short-term food deprivation in rats plays a key role in the adaptive interaction of glucose and free fatty acid utilization. Physiol Behav 1986; 37:441-445

Strubbe JH, Van Dijk G. Temporal organisation of feeding behavior and its interaction with regulation of energy balance. Neurosci Behav Rev 2002; 26:485-498

Strubbe JH, Keyser J, Dijkstra T, Alingh Prins AJ. Interaction between circadian and caloric control of feeding behavior in the rat. Physiol Behav 1986a; 36:489-493

Strubbe JH, Spiteri NJ, Alingh Prins A. Effects of skeleton photoperiod and food availability on the circadian pattern of feeding and drinking in rats. Physiol Behav 1986b; 36: 647-651

Strubbe JH, Alingh Prins AJ, Bruggink J, Steffens AB. Daily variation in of food-induced changes in blood glucose and insulin in the rat and the control by the suprachiasmatic nucleus and the vagus nerve. J Auton Nerv Syst 1987; 20:113-119

Thon R, Lassen J, Kornerup Hansen A, Jegstrup I.M, Ritskes-Hoitinga J. Welfare evaluation of genetically modified mice in Denmark. An inventory study of the reports from 1998 to the Animal Experiments Inspectorate. Scand J Lab Anim Sci 2002; 29:45-55

Toates FM, Rowland NE eds. Feeding and Drinking. Amsterdam: Elsevier, Amsterdam 1987

Tobin G. Current concepts – Natural ingredient diets. Harlan Continuing Education Conference, 2003 March 12 – 14; Vaals, The Netherlands.

Vachon C, Jones JD, Nadeau A, Savoie L. A rat model to study postprandial glucose and insulin responses to dietary fibers. Nutr Rep Int 1988; 37:1339-1348

Vermeulen JK, de Vries A, Schlingmann F, Remie R. Food deprivation: common sense or nonsense? Anim Technol 1997; 48:45-54

Vonk RJ, Van Doorn ABD, Strubbe JH. Bile secretion and bile composition in the freely moving unanaesthetized rat with a permanent biliary drainage: influence of food intake on bile flow. Clin Sci Mol Med 1978; 55:253-259

Zhou J, Moeller J, Ritskes-Hoitinga J, Larsen ML, Austin RC, Falk E. Effects of vitamin supplementation and hyperhomocysteinemia on atherosclerosis in apoE-deficient mice. Atherosclerosis 2003; 168(2):255-262

Chapter 6

EXPERIMENTAL PROCEDURES: GENERAL PRINCIPLES AND RECOMMENDATIONS

David B. Morton
Centre for Biomedical Ethics and the Biomedical Services Unit, University of Birmingham, Birmingham, UK

1. HANDLING, DOSING AND SAMPLING

The use of animals in research presents a moral conflict for all those involved. The application of the Three Rs helps determine when animals have to be used but it is only a starting point for both the animals and humans who are subsequently involved. One of the Rs, Refinement, I have defined as a redirection of research towards those methods which aim to alleviate or minimise the potential pain, distress or other adverse effects suffered by the animals involved, or which enhance their wellbeing (Reese 1991, Dean 1999). I have emphasised the promotion of animal welfare because not only do animals potentially suffer during an experiment, but they are also able to suffer through ill health and during their husbandry. Poor health control and husbandry can be stressful and lead to mental and physiological distress to animals (see earlier chapters in this book). A significant reduction in the level of animal suffering can be achieved through improved human animal interactions (a considerate and empathetic attitude to animals is very important: see Hemsworth and Coleman 1998), improved experimental techniques, and the use of humane endpoints. Moreover, the contribution of these factors is essential to the practice of good science. In this chapter I will focus on general principles to do with handling, the administration of substances, the removal of body fluids, anaesthesia, analgesia and euthanasia.

1.1 Handling

All phyla used in research (fish, amphibians, reptiles, birds, mammals) have different needs and the relevant texts should be consulted (e.g. UFAW 1999, JWGR 2001, AWI 2002). Animals vary in their tractability for handling depending on species, strain, age, sex, acclimation and training (i.e. habituation), prior experience, 'personality', mood and, perhaps, any injury that may cause pain and suffering to the animal at the time of handling. Wild caught animals raise additional considerations as often such animals would normally flee from humans but are unable to do so in the confinement of a laboratory, and so they may become very distressed and even aggressive. While domestication will reduce the stress of handling through the selection of docile animals for future breeding stock, this effect is not always predictable. Difficult animals to handle might include large animals like pigs, young ferrets, certain strains of mice (e.g. C57Bl6), rats (GAERS, female Wistars) and primates. Other animals are relative docile, such as guinea-pigs, sheep, some breeds of dogs in particular the beagle, and certain strains and sex of rats (e.g. male Wistars). No matter what the circumstances, whenever animals are stressed by being caught, handled and restrained, the impact on the science has to be considered. The collection of data remotely through the use of telemeters and data loggers may therefore provide significant improvements for both the science and the animals (JWGR 2003).

Signs of stress are attempting to escape by wriggling, biting, scratching, vocalising, running and, paradoxically, staying very still. This 'tonic immobility' is seen in chickens and rabbits and some wild animals and is one of the strategies they use to escape predators. Animals respond to being handled firmly but not tightly, and wrapping animals in a towel for example (rabbits), or letting them enter a black bag (wild rats) seem to provide some animals with a sense of security and are less stressful than a loose grip (see Grandin 1993). Handlers need to be confident, as lack of confidence can be transferred to the animals and they may then become nervous and even aggressive. It is worth playing a thought game and putting oneself in the animal's position (a notion of empathy). Take, for example, the commonly used small research animals, holding a mouse upside down by its tail or high off the floor is probably frightening, as might also be being held and moved rapidly at human waist high level e.g. to a set of scales for weighing. On the other hand, being tucked into the crook of an arm and held gently may be a more secure position akin to 'hiding'. Training animals to cooperate in experimental and husbandry procedures, for example primates (Reinhardt 1992, 1997) can minimise the stress associated with being squeezed in a crush cage or being chased and caught with a net. The effects of handling on science are numerous (Claassen 1994 is very helpful).

Good practices

1. Ease of handling will be promoted by the following: acclimatising the animal to the surroundings where they are going to be handled; training animals such as dogs and primates to be handled at an early age; habituation to the procedures to be carried out; and training and competence assessment of staff involved. Very often speaking to animals or making some recognisable noise can help the animal gain confidence in the handler.
2. Animals should always be handled quietly, gently, firmly, even petted. However, some species may find this threatening, such as primates and mice. Rewarding animals in some way during or after the procedure e.g. giving corn flakes to rodent, treats to dogs and primates, is a positive reinforcement for the desired procedure.

1.2 Administering substances

Two comprehensive reviews of this topic have recently been published (JWGR 2000, Diehl et al. 2001). The JWGR paper draws attention to important considerations such as the preparation of the substance and solutes, the choice of animal, the routes and volumes to be injected, the training of staff and animals, and troubleshooting when things go wrong, whereas the Diehl paper is more orientated to techniques commonly used in the development of safety testing of new medicines. For the impact on science see the review by Claassen (1994).

Good practices

1. Substances to be injected should be biocompatible e.g. in terms of pH, osmolarity, and irritancy. Care should be also taken to sterilise the injectable and its formulation should be suitable for injection e.g. viscosity. Normally the substance should be pure and stable.
2. The commonest solvents are saline and water, but organic ones are also used. The solvent should lack pharmacological effects and be stable, non-toxic, non-irritant and non-sensitising.
3. The chosen route should be biologically suitable for the substance for the desired effect, be the least painful (e.g. oral or subcutaneous rather than intramuscular) and, if necessary, be suitable for repeated dosing.

4. The volume should not have any physiological effects (e.g. fluid overload) and should be the minimum consistent with accuracy and minimal adverse effects on the animal.
5. If repeat dosing is needed, consideration should be given to the use of slow release formulations and to the use of osmotic pumps.
6. If the substance is novel and there is little or no background data and particularly if it has not been used at the institution where the work is being undertaken, then pilot studies are recommended. It has been repeatedly shown that a recommended dose will vary between institutions for a variety of reasons, including diet, environmental conditions, health status, varying basal stress levels in the colony, strain and even sub-strain of animal etc (see e.g. van den Heuvel et al. 1990).
7. As far as the animals are concerned, they may take time to habituate to the procedure and, indeed, some may not do so and be severely stressed, in which case they should be withdrawn from the study.
8. It may be possible to train some species of animals to cooperate in the dosing procedures, such as dogs and primates (see Reinhardt 1997).
9. The training of personnel that carry out the procedures is of vital importance – a simple procedure badly done not only has welfare implications but also scientific ones.
10. Finally, animals that have been dosed should be checked approximately within the next 15 minutes to pick up any adverse effects such as inadvertent puncture of a blood vessel, or an unexpected adverse reaction to the substance, or a misplacement as in gavaging.

1.3 Removal of body fluids

The common body fluids taken from animals include blood, urine, faeces, saliva and ascites (although this should be rarely done now for harvesting monoclonal antibodies as there are good *in vitro* techniques available).

1.3.1 Blood

Two good publications include the JWGR (1993) and Diehl et al. (2001), and the Norina website has good links that give information on methodology. As a rule of thumb it has been found that 10% of the circulating blood volume can be removed without harm, and up to 15-20% over a month. If in doubt haematological studies should be carried out to look for anaemia and the degree of deviation from normal measured and interpreted. Usually a superficial vein is used in mammals such as the cephalic vein in the larger species like dogs, cats and even sheep, the marginal ear vein in guinea-pigs, rabbits and pigs, the jugular in ferrets, rabbits, dogs and cats. In rodents, such as mice, rats and guinea-pigs, the jugular can also be used (direct the needle just in front of the pectoral muscle into the brachiocephalic vein on the right hand side), or the recurrent tarsal vein just above the hock, or the tail vein. With rodents the tip of the tail may be removed but this should be avoided if possible and a needle-stab at the base of the tail may yield sufficient blood. Retro-orbital bleeding should always be avoided because of the occasional side effects such as damage to the eye and orbit. Whatever method is chosen, equipment should be sterile and the skin cleaned of all debris and fur or feather (on some occasion it may be sufficient to simply part the pelage) and a disinfectant such as 70% alcohol applied which also is helpful in highlighting the contour of the vein. The venous return should be blocked through the application of pressure proximal to the vein (i.e. heart side).

Good practices

1. The penetration of the skin to access the blood vessel causes transient pain and discomfort and so the use of a local or even a general anaesthetic may be indicated for some species (rodents and rabbits). EMLA (Eutectic Mixture Long Acting) cream is a local anaesthetic cream but it has to be applied some 30 minutes or so before blood is taken.
2. Any puncture of a blood vessel may lead to subsequent leakage of blood. It may become severe enough to require attention to stop the flow. This is more likely if an artery has been used to obtain the blood (as with the central ear artery in rabbits, or the femoral in other species). It is always good practice to maintain pressure on the site for at least 30 seconds for a vein and 60 seconds for an artery, and then to come back 5 minutes later to check all is well.
3. For repeated sampling consider using a plastic catheter and taping it in position for the period of sampling, where this is in less than a day. If repeated samples are required over a period of weeks then a permanent indwelling catheter may be the best method, with an access port at the base of the neck (see JWGR 1993; Healing and Smith 2000).
4. In small rodents it may be necessary to warm them gently at 30 °C for 10-15 minutes to promote vasodilation of the tail veins particularly, but animals should be under constant supervision to avoid causing them distress, indicated by panting. In some species, such as rabbits, the use of a suitable chemical restraint can lead to vasodilatation (e.g. Hypnorm™ and midazolam).

1.3.2 Urine, faeces and saliva

Some species such as mice and rats may urinate or defecate on handling because of fear, and this is also true of larger animals that may be held in stalls such as cattle and sheep. This can be an easy way to collect samples non-invasively. Alternatively, animals may be housed in a metabolism cage or a closed environment with a grid floor if necessary, at times when they are most likely to eliminate, i.e. in the first hour or so after feeding. Dogs can be trained to eliminate on command, which is preferable to catheterising them but it will take time to train them.

Animals salivate when they are anticipating food, nauseous, and given a secretogogue. Depending on the volume required it is not difficult to obtain a sample of saliva for most animals, but smaller animals may require some substance to induce secretions.

2. ANAESTHESIA, ANALGESIA AND EUTHANASIA

Some experimental procedures in research require the use of anaesthesia and analgesia in order to practise good science. In many countries anaesthesia is legally required for all experiments unless the scientific objective could not be achieved with its use, or the stress of giving the anaesthetic or analgesic is likely to be greater than carrying out the procedure without it. Effective and humane anaesthesia, analgesia and euthanasia are also a scientific necessity to avoid the results being skewed by the stress responses of the animals concerned. This chapter deals with the basic principles involved and identifies good and bad practices. More specific information can be found in published literature (e.g. Green 1979, Flecknell 1996, ACLAM 1997, Close et al. 1996, 1997, AVMA 2000).

2.1 Anaesthesia

Anaesthesia is the state of being unaware of the surroundings and also of any internal and external feelings. In common terms it is about being unconscious (see Table 6-1 for definitions). An anaesthetic may be given in order to carry out a surgical procedure, or sometimes simply to restrain an animal so that it is easier to carry out a minor procedure or examination. This state of a reduced awareness of the surroundings is caused by a depression of the CNS, and the deeper that depression, so consciousness is increasingly lost. Unconsciousness is indicated by an animal's response to a stimulus and so the depth of loss of consciousness can vary according to what reflexes are being tested. Hearing is one of the last to go, responses to the paw pressure pinches go earlier, and coordination and righting reflexes (e.g. ability to stand up) earlier still.

Table 6-1. Definitions of common terms (G=Greek, L=Latin)

Anaesthesia:	(G *an* negative + G *aesthesis* feeling and sensation); drugs that remove feelings and sensations
Neuroleptics:	(G *neuron* nerve + G *lepsis* taking hold); drugs that change behaviours e.g. anti-psychotics
Sedatives:	(L *sedativus* to calm); cause sedation by a general depression of the CNS
Narcotic:	(G *narcosis* to numb); cause non-specific and reversible depression of the CNS
Tranquillizers:	(L *tranquillus* = quiet, calm); a general state of depression of all organs; minor ones are anxiolytics

An anaesthetic has to produce an appropriate depth of anaesthesia for an appropriate time in the experimental animal irrespective of its age, sex, strain, diet and, sometimes, even its disease status. It should produce no adverse effects on the animal e.g. tissue reaction, pain on injection, and ideally have 0% mortality. It should also be appropriate for the type of surgery and experimental requirements. The mortality will depend very much on the type of work being carried out, the depth and duration of anaesthesia and so on; Table 6-2 provides a guide for benchmark standards for mortality.

Table 6-2. Approximate benchmark standards for mortality rates.

Human	1/10,000 - 1/25,000 (elective)
Dog	1/700 (healthy)
Cat	1/700 (healthy)
Horse	1/250
Rat	1/200
Mouse	1/500
Rabbit	1/50
Mouse 3-4 weeks	<1/5000*

* This low level is due to the very short duration – just a few minutes – and was done for tail tipping for genotyping.

There are three basic elements that comprise effective anaesthesia: (1) preventing the animal moving around i.e. restraining it; (2) providing analgesia so it feels no pain; and (3) providing good muscle relaxation for good surgical access. These can all be achieved by use of a single agent causing depression of the CNS. Or sometimes, combinations of agents are used to provide each of the three elements – so-called balanced anaesthesia. The advantage of using combinations (such as neuroleptanalgesics) is that they may be safer than single agents as the animal need not be taken to such a deep level of CNS depression. Depression of the CNS causes a reduction in homeostatic mechanisms and so anaesthesia is often accompanied by a depression of the cardiovascular and respiratory systems, and an inability to thermoregulate. If this depression is not controlled in terms of depth or compensated for in some way, it may lead to death.

2.1.1 Monitoring anaesthetic depth

There are many demands made on animal anaesthetists: keeping animals deep enough not to feel pain but providing the right degree of depth for good muscle relaxation and analgesia, in a predictable and stable manner. Deepening and lightening of anaesthesia follow the same pathways and reflexes and homeostatic mechanisms are lost and gained in roughly the same order, i.e. recovery from an anaesthetic follows the reverse of induction. The depth of anaesthesia (see Table 6-3) should be assessed before starting surgery by giving the animal some minor stimuli such as stroking the eyelid, paw, ear or tail. If there is no response, the intensity of the stimulus can be increased to a pinch of the skin (for many animals between the toes is a good place) – using a small pair of forceps is probably best - and if the animal is not deep enough it will respond by withdrawing the foot (pedal withdrawal reflex). Each species may require a slightly different stimulus area but the paw, eyelid and the cornea are almost universal. The pupil of the eye loses its light reflex and sometimes the eye changes its position in the orbit as the various muscles attached to the globe lose their tension; these signs are useful in cats and dogs but not in small rodents.

An extremely good guide in all species is the rate and depth of breathing. As the animal becomes deeper so the respiratory rate slows and the depth of respiration decreases in amplitude. Furthermore as the intercostal muscles become depressed, so the movements of the diaphragm are more easily seen as a seesaw respiration with the abdomen rising and falling (thoraco-abdominal respiration). The use of an infant respiration monitor can be very useful as it gives an audible bleep at every breath, can be easily heard and changes in rate detected very sensitively. It is particularly useful when an animal cannot be seen under surgical drapes. Heart rate monitors are also used but are often only available for large animals with rates of less than 250 beats per minute (bpm) as they are made for humans and companion animals and not for small rodents, where it is far more difficult to pick up heart rates in excess of 500 bpm. Blood pressure can also be used: as the heart rate slows so it beats with less force, and the blood pressure falls; this can be monitored with an arterial catheter connected to a transducer in larger animals. Pulse oximetry is a non-invasive sensor for tissue oxygenation (colour) and pulse rate. It is a non-invasive sensor and alarms for low/high levels of both can be set.

Table 6-3. Stages of anaesthesia and reflexes to be monitored

Stage and plane of anaesthesia		Reflexes to be monitored
Stage 1	Normal patterns	
Stage 2	Involuntary excitement	Generalised uncoordinated movements; rapid or no breathing, pupil constricts to light
Stage 3	Surgical anaesthesia	
	Plane 1:	Respiration – light, regular but rate slowing; pupil constricted
	Plane 2:	Respiration - moderate rate and regular, thoraco-abdominal
	Most surgical operations can be carried out at this level	moderate amplitude; pupil gradually dilates and loses light reflex; loss of pedal withdrawal reflex.
	Plane 3:	Respiration – deep, regular, thoraco-abdominal minimal amplitude; corneal reflex lost.
Stage 4	Medullary paralysis i.e. over-dosage	Respiration slow gasping; pupil very dilated, eye protruding

Apnoea (failure to breathe) is not uncommon and has to be monitored very carefully and should not be allowed to persist for long i.e. more than 20 seconds or so. If an animal has just been exposed to an anaesthetic gas it may well hold its breath, or if it has just been given an intravenous injection it will involuntarily become apnoeic as the anaesthetic is taken up rapidly by the CNS for a short period before redistribution. In these circumstances there should be a pedal withdrawal reflex and the pupil should be constricted. However, if an animal has been anaesthetised for some time and it stops breathing and the pupil is dilated and there is no pedal reflex, then it must be given artificial ventilation as a matter of urgency because the resulting anoxia will quickly stop the heart. Any slowing of the respiratory rate to less than 50% of resting normal should be corrected.

Lightening of anaesthesia: When animals become light under anaesthesia, i.e. they start to become conscious again, this is often first noticed by movements of a limb or tail. The surgery or experimental procedure should be stopped immediately until the anaesthesia has been adequately deepened. It is therefore very important that animals are able to move their limbs to indicate they are becoming light and so they should not be tied out like a fur pelt left to dry. Only the limbs that need to be positioned for surgical access should be tied down (NB the front legs should always be tied forwards or backwards, never laterally as the brachial plexus may be damaged). Some animals (like rodents, chickens and rabbits) exhibit a phenomenon of tonic immobility that is 'playing dead'. Consequently, it may be thought that an animal is anaesthetised but in reality it has become conscious! The common story is that "the animal just walked off the table from being asleep", but what really happened is that the level of anaesthetic had gradually been getting lighter until the animal, now conscious, decided the moment at which to try to escape. A similar state may also occur at the start of an operation when an animal has not been anaesthetised deeply enough.

Body temperature may drop precipitously in some of the smaller rodents that have very little body fat insulation and need a high metabolic rate to maintain their temperature. In mice e.g. the rectal temperature can drop 1 °C every 5 minutes or so. Consequently it is good practice to monitor body temperature and to provide heat with a warming blanket if necessary, but take care that it is not too warm. The temperature of the blanket should not exceed 40-45 °C as otherwise the skin of the animal may become burned and even the kidneys may become damaged if the animal is on its back (blood appears in the urine). Circulating hot water from a thermostatically controlled water bath of some sort through a blanket is best, although the use of space blankets (light aluminium foil) and bubble wrap are effective at preventing heat loss. Providing some insulation from the metal operating table also reduces heat losses. Conversely some animals may gain excessive heat e.g. from the operating theatre lights, and so they may have to be cooled by exposing part of the body, or using cold water or blowing cold air over them from a fan. If necessary, mechanisms of heat gain should be controlled e.g. by using cold light sources for illumination.

It goes without saying that animals to be anaesthetised should be healthy (unless poor health is part of the experiment) and that a pre-operative check should be carried out. Animals with a lung infection are a serious anaesthetic risk, as are animals with poor liver and kidney function as these organs are frequently involved in metabolising and eliminating an anaesthetic.

2.1.2 Premedication

It is good practice to pre-medicate large species as: it makes induction of anaesthesia and recovery smoother; it will reduce fear and apprehension and so aid stress-free induction of anaesthesia; and it will reduce the amount of anaesthetic required to induce general anaesthesia, so decreasing the undesirable side effects of these agents. If possible premedication should be carried out in their home cage or pen. Tranquillisers and sedatives such as phenothiazines or benzodiazepines or butyrophenones are commonly used. At one time it was thought necessary, in those species that salivate copiously under anaesthesia (e.g. when ether is used, and in sheep), to give anti-cholinergic agents such as atropine, but this has to be used with care as it does not stop the secretion completely, and the residual secretion can become very viscous and block airways. Consequently it may be safer to place the animal head down (and block oropharynx with a swab) and simply collect the saliva in a waste container. Atropine is, however, very useful in species that are difficult to intubate as it can prevent the heart stopping as a result of laryngeal stimulation and a consequential over-stimulation of the vagus nerve. As intubation skills are developed, along with the use of a local anaesthetic spray, atropine can often be avoided.

2.1.3 Neuromuscular blocking agents (NMBs)

These are valuable agents but they should be used for the right reason and are rarely required in routine animal anaesthesia. They paralyse voluntary (striated) muscles in the body at the neuromuscular junction but do not block the passage of sensory impulses to the CNS. Consequently an animal is still able to feel pain but is not able to respond by moving and escaping from it. In fact it is unable to show in an obvious manner that it is feeling pain however, there is a marked autonomic response such as an increase in heart rate and blood pressure and it is good practice to monitor these routinely when using NMBs. NMBs should never be used without a background of a reliable and adequate anaesthetic (see Guidelines on NMBs by the Physiological Society and the UK Home Office). NMBs are useful if respiration needs to be controlled by the anaesthetist, e.g. for thoracotomies, or by scientists for recordings, or if complete muscle relaxation is required. It should be noted that it is also possible to induce apnoea by hyperventilation for a short period (say one minute) and that may meet the need without the use of NMBs.

Artificial ventilation may be needed when surgery takes more than a few hours or if an animal stops breathing when it is too deep then oxygen should be given urgently. It is easier if the animal has been intubated but, if not, compressing the sides of the chest, or applying a face mask or tube to the nose and inflating the lungs manually or with a low pressure oxygen flow can all be helpful.

2.1.4 Starvation prior to anaesthesia

It used to be thought that animals needed to be starved for 24 hours before being given an anaesthetic as they often vomited on induction (and recovery). Much of this was due to the anaesthetics used at that time, such as ether, but nowadays withdrawal of food and water depends more on the species involved. As a general rule food should be withheld for a period that is long enough for the stomach of monogastric animals (dogs, cats, pigs) to be empty and this will take at least 4 to 6 hours. For pigs that are on straw it may take longer, and for ruminants it is impossible to empty the rumen. However, for small rodents it is not good to remove food, and nor is it necessary for two reasons. First, they rely on a steady intake of food to maintain body temperature and so withdrawal of their diet will soon leave them in a catabolic state; secondly, rats, mice guinea-pigs and rabbits seem not to be able to vomit, or it is very rare. Taking into consideration the time it takes for starved animals to eat normally after a procedure or any surgery, withdrawal of food and water may at worst jeopardise that animal's life, and at best delay recovery. No animal need be deprived of water for more than one hour before an anaesthetic providing the stomach is empty as any water will be quickly absorbed.

2.1.5 Choice of anaesthetic

This will depend on the species of animal as well as the strain and its age and sex. It will also depend on the duration of the procedure as well as any potential effects the agent may have on the scientific outputs being measured, for example, some agents used to restrain an animal may affect some physiological parameters. Consequently, a small validation study may be required. Unless there is good evidence do not assume that the traditional anaesthetic agents have been without any scientific effects as few validation studies have been carried out. It is important from a scientific viewpoint that this is done and not simply to rely on custom and practice. Some agents may also adversely affect the animals themselves e.g. carbon dioxide is extremely aversive to animals (and humans) at the levels needed to induce anaesthesia, and some drugs are irritant to tissues such as ketamine and sodium thiopentone. Perhaps, surprisingly, for injectable agents, the same anaesthetic in the same strain, sex, age, source etc of animal may vary in the required dosage rate for no apparent reason and so pilot studies to determine dosage rate should be carried out to ensure a reliable anaesthetic regimen. There are few data on success rates with the various types of anaesthetics, but I suspect that injectables are less efficient than inhalational agents (efficiency based on Number of successful anaesthetic events divided by the Number of attempts), and have a higher mortality.

Some anaesthetics pose safety risks for humans e.g. some gaseous agents and some drugs of addiction. Controls should be in place to scavenge all gaseous anaesthetics as well the issue and auditing of controlled drugs

2.1.6   Gaseous agents of anaesthesia

Gaseous agents, unlike the injectable agents, have the distinct advantage of enabling a rapid recovery from anaesthesia and so an animal will regain its homeostasis quickly e.g. posture, temperature regulation. They are also more flexible to use as anaesthesia can be given for however long the procedure might take. In addition, it is easy to vary the 'dose' and so easy to adjust the depth of anaesthesia. The disadvantages of gaseous agents are that more expensive equipment is required, such as an anaesthetic machine (N.B. this is a one-off expenditure); it is not easy to disinfect the apparatus (but infections are rare); a compressed carrier gas(es) is needed; and waste gases have to be scavenged. Moreover, in some cases, it may obstruct craniofacial procedures, and in other surgeries it may be necessary to reposition an animal during surgery and so a face mask may have to be readjusted. This is not a problem if the animal has been intubated or the face mask can easily be attached to the head, but it can be a problem in small rodents.

Volatile liquids, mainly the fluorinated hydrocarbons are the commonest used gaseous anaesthetics today (see Table 6-4). They have replaced ether and chloroform as the traditional agents for good reasons. First, ether has at least six serious disadvantages: it is irritant to the respiratory mucosa and so it causes bronchial secretions that may block the airways; it is likely to predispose to infection because of this irritant effect; it is aversive to animals and humans; it causes sympathetic nervous system stimulation; it induces liver microsomal enzymes; and it forms explosive mixtures with air and has destroyed at least one laboratory in the UK. Chloroform, on the other hand, is dangerous as it causes liver damage to humans and reputedly affects adversely the breeding of mice. Carbon dioxide also should not be used for anaesthesia as it is very aversive to the animals as well as stimulating the sympathetic nervous system.

Aversion tests undertaken with mice and rats indicate that halothane is the agent of choice with rats whereas for mice it is enflurane (Leach et al. 2002b). Isoflurane is a particularly suitable agent as it has a wide safety margin and is almost exclusively taken up by the lungs and excreted by exhalation i.e. there is very little metabolism by the liver or kidneys. However, it has been found to be more aversive than other agents although not nearly as aversive as carbon dioxide (see Leach et al. 2002b). Isoflurane is relatively cheap, although the cost of any anaesthetic compared with the real costs of an experiment (housing, scientists' and technicians' time, etc) is a trivial matter.

Table 6-4. The common fluorinated hydrocarbon anaesthetic agents

Fluothane/Halothane	Cardio-depressant, cardiac arrhythmogenic, induces liver microsomal enzymes
Enflurane	Similar properties to halothane
Isoflurane	Little biotransformation
Methoxyflurane	Potentially nephrotoxic (and to user), induces liver microsomal enzymes. No longer available
Desflurane	Rapid induction and recovery
Sevoflurane	Rapid induction and recovery

Gaseous agents are first vaporised (in a vaporiser), using a carrier gas such as oxygen. The concentration of anaesthetic is determined by the proportion of vaporised saturated gas mixing with gas that does not go through the vaporiser. The Mean Alveolar Concentration (MAC) is the concentration at which 50% of patients are anaesthetised, and it has been calculated for maintaining anaesthesia for each agent; it follows a normalised standard distribution curve (see Table 6-5). The MAC value will vary according to species, strain, sex and whether a second gas such as nitrous oxide is used. Normally this does not matter, as unlike injectable agents, animals are dosed to effect by increasing or decreasing the level of anaesthetic being delivered. Small animals will normally be induced in an induction box or chamber, and then transferred to a face mask, whereas large animals will have anaesthesia induced by way of an intravenous or other parenteral injection and then intubated and transferred to a face mask or anaesthetic circuit of some sort depending on the length of the scientific procedure (Figure 6-1). Some animals, like pigs, will tolerate a face mask for induction if placed in a sling or other suitable restraint, and anaesthesia slowly induced. Flow rates are adjusted according to the minute volume (respiratory rate x tidal volume, roughly 12 ml/kg), normally measured in litres/minute. As an animal takes in that tidal volume in one third of the respiratory cycle (equal thirds for inhalation, exhalation, and pause) the minute volume has to be multiplied by 3 to meet the volume demand at inspiration, unless there is a reservoir of some sort. If this volume is too low then an animal will inhale some of its expired gases and that can lead to a build up of carbon dioxide. Closed circuits, where carbon dioxide is removed by means of soda lime, provide a more efficient use of gases and anaesthetics and so are used for large animal anaesthesia. In these systems, fresh gases are delivered at a rate sufficient to replace the absorbed carbon dioxide. However, there will need to be a pressure valve to release any excess pressure, and a reservoir bag to meet the tidal volume.

Nitrous oxide may be used in large animals as an additional analgesic agent and as a cheaper carrier gas than oxygen, but normally not at more then 50-67 %. Note that the MAC value for nitrous oxide is so high that it can never be reached. It should never be used on its own to provide anaesthesia, but it can still provide a level of useful analgesia.

Table 6-5. Mean Alveolar Concentration (MAC) values for maintaining anaesthesia in humans and rats for various gaseous anaesthetics (%)

GAS	MAC (human)	MAC (rat)
Halothane	0.75	0.95
Isoflurane	1.15	1.38
Nitrous oxide	105	250

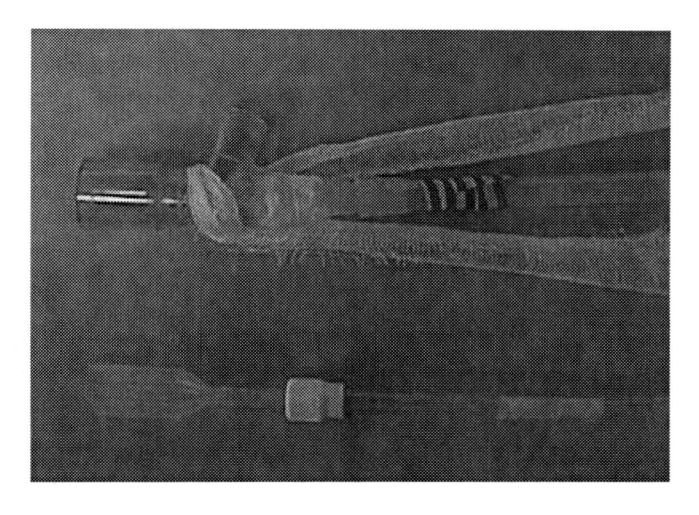

Figure 6-1. Improvised face masks for rats and mice

Anaesthetic circuits for inhalation: It is most important that the dead space between the animal (take this to be at the bifurcation of the bronchus roughly half way along the chest) and the anaesthetic gases is minimal, otherwise an animal will re-breathe its expired gases. The diameter of any connecting tubes should be as large as possible and free from corners so that the resistance to breathing is low.

Endotracheal intubation: Intubation should be carried out when the procedure will last for say more than one hour as it enables artificial respiration to be given easily, and enables the use of more economical anaesthetic delivery circuits such as closed ones (see Figures 6-1 to 6-3 for some common circuits). It can be carried out easily in a range of species from rats to horses. It is important to realise that most of the time mammals breathe through their noses and not their mouths, except when they are panting to lose heat. Picture the horse after a race breathing through flared nostrils – not its mouth. This is an important observation as it helps understand the relevant anatomy for intubation. On looking into the mouth of an animal one cannot see the epiglottis as in a human, only the hard and soft palates. The epiglottis (the entrance to the trachea) is 'hiding' behind or above the soft palate so that the animal breathes through its nose i.e. the epiglottis is directed into the naso-pharynx and not the oro-pharynx. In order to view the epiglottis it has to be 'flipped' into the oropharynx by either stretching the neck or displacing the soft palate with a soft implement, like a gauze swab on the end of forceps, or with a laryngoscope that helps illuminate the oropharynx (in small rodents transcutaneous illumination of the ventral neck is useful). Once that has happened, and the head and neck are in a straight line, intubation is relatively easy. It is often necessary to keep the mouth open in some way and to spray the vocal cords with local anaesthetic, even to give atropine (see above) to prevent excessive vagal stimulation and cardiac arrest.

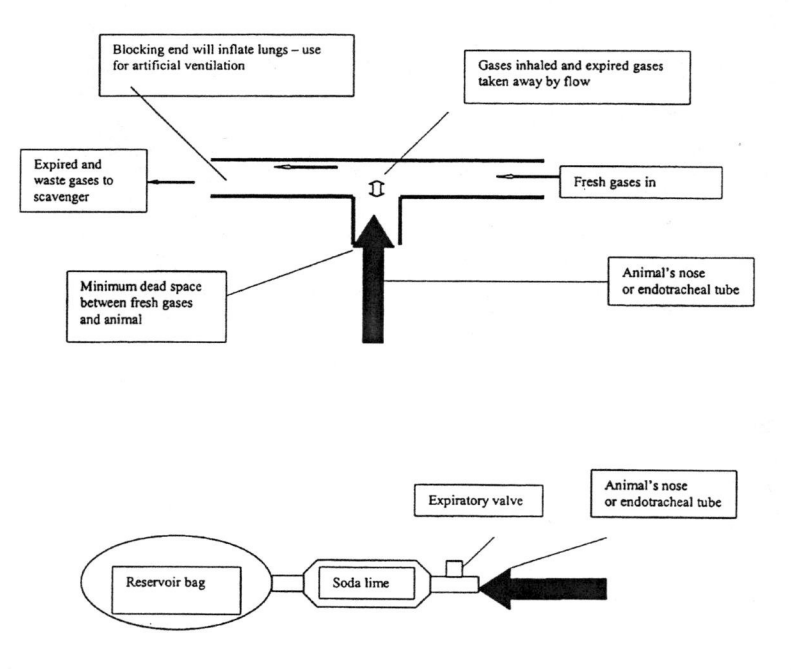

Figure 6-2. T piece (top) and To and Fro (bottom) circuits

EXPERIMENTAL PROCEDURES

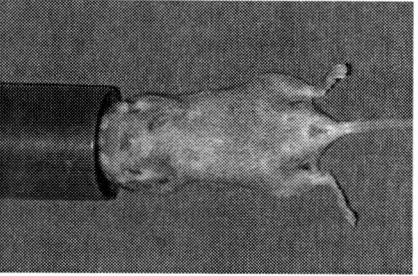

Bains' circuit and scavenger

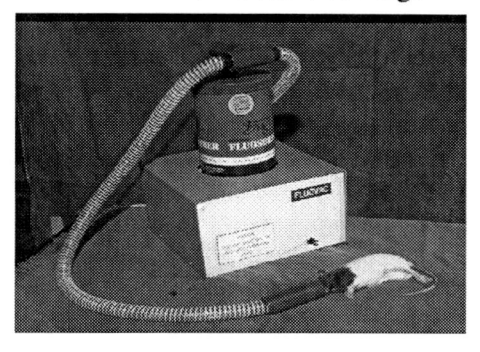

Figure 6-3. Bains' tube circuit: Bains' tube with an inner delivery tube surrounded by an external tube connected to a vacuum pump to the scavenger.

2.1.7   Injectable agents of anaesthesia

There are several types of injectable anaesthetic agents that can be used alone or in combination i.e. as part of a balanced anaesthetic mixture. As most injectable agents depress the respiratory system to some degree, it is important to provide oxygen, especially if the procedure lasts for more than 30 minutes or so. Injectables can be classified in terms of their duration of action as well as by class of compound. I will describe just a few combinations and the reader is referred to standard texts for others as well as for dosage rates, bearing in mind the need for pilot studies. Some agents have very short half-lives as they are rapidly metabolised, and if given by infusion they can provide safe anaesthesia for several hours, evendays, with a relatively rapid recovery at the end of the anaesthetic period.

There are several disadvantages to using injectable agents. For example, the dose has to be calculated in advance and may vary for many reasons and so it is easier to over- or under-dose compared with inhalational agents where animals are dosed to effect. Moreover, once it has been injected it is difficult to reverse an overdose unless there is an antagonist or the animal can be ventilated until the drug has been metabolised. Injectables cannot be dosed to effect unless given by the intravenous route and that requires special skills, especially in small animals. The injection of these anaesthetics may be painful particularly as the intramuscular route is often favoured over the subcutaneous route. In addition, some agents may be irritant such as ketamine, tribromoethanol, and so the welfare of the animal is inevitably compromised when they are used. Furthermore, if the intraperitoneal route is used there is a 10-20 % chance some of it will go into the gut, and then a second injection will often be needed, or another anaesthetic will have to be used, to achieve adequate anaesthesia. This can potentially prolong the period of disorientation for animals and, moreover, it is difficult to calculate how much more to give. The dosage rate of injectables varies according to species, as well as sex, etc. e.g. pentobarbitone can last for 30-60 minutes or more in dogs, cats and rats but only a few minutes in sheep; female rats need a higher dose of alphaxolone-alphadolone than male rats. Sometimes the carrier solution can cause an allergic type of reaction in some species e.g. alphaxolone-alphadolone in dogs due to its Cremphor content. If an anaesthetic has to be repeated, then a tolerance may build up in time leading to an underdosage; a phenomenon not seen with inhalational agents. Because injectables have to be metabolised they are more likely to adversely affect animals with compromised liver or kidney function, or that are immature or diseased, and this may well also affect the science because of their pharmacological effects. With injectables the residual drug effects can persist for long periods and recovery may be prolonged due re-circulation from gut to liver. The variation of strain differences can be considerable, and in mice sleep times have varied 4 fold with identical doses. Finally, many injectable agents are associated with long half-lives and so it may take several hours before animals regain their normal homeostasis and eat and drink normally. Furthermore, if an animal is not been kept warm during the anaesthesia and the body temperature drops significantly, say by 10 °C, then as the metabolic rate will be halved, so the period of recovery from anaesthesia will be doubled.

The advantages of injectables, in general terms, are that the position of an animal can be easily changed if necessary for better surgical access; there is not the initial expense for equipment; and they are probably cheaper to use but compared with the overall real costs of an experiment it is trivial.

Short-acting agents (less than 15 minutes)

Alphaxolone-alphadolone and *propofol* have half-lives of about 7 minutes and, therefore, are relatively safe to use as with an overdosage one can ventilate the animal until the drug has been metabolised. Short acting agents are often used for induction in large animals. Other short acting anaesthetics include *sodium thiopentone* and *methohexitone*: these are best given intravenously but note that thiopentone is irritant outside the vein and, if mis-injected, saline or water should be injected at the site and massaged to minimise tissue necrosis. Propofol is not irritant in this way and also has a great advantage in that it provides a very smooth induction and recovery, with animals seemingly just waking up with good control, even being able to walk without the marked ataxia seen with other agents.

Medium-acting agents (15 minutes to a few hours)

Medetomidine is a modern alpha-2-adrenergic agonist sedative and causes deep sedation with analgesia and has the advantage that it can be reversed by the use of a specific antagonist drug – *atipamezole*. *Xylazine* is another drug in this class. They are often used in combination with *ketamine* that is a dissociative anaesthetic but when given alone causes muscle rigidity. The combination of alpha-2 adrenergic agonist with ketamine gives good anaesthesia for 45 to 90 minutes but causes a marked diuresis, protrusion of the eyeball and the potential for corneal damage due to loss of the eyelid (palpebral) reflex. Moreover it has a prolonged and rather 'restless' recovery time of several hours that can predispose to corneal damage.

Tribromoethanol is still commonly used in transgenic work for historical reasons despite its disadvantages, not least of which is that it causes a peritonitis (which is painful) and on second use it can cause death and paralytic ileus. In rats even its first use is associated with a paralytic ileus and death. One cannot help but conclude that, as there are alternatives (inhalational methods, neuroleptanalgesics, medetomedine and ketamine), it should not be used. Various reasons have been suggested for these adverse side effects, mainly to do with the way the solution of anaesthetic is made up and stored but despite these precautions it is still irritant.

Fentanyl/Fluanisone/Midazolam: Fentanyl (opioid analgesic) and narcotic fluanisone (restraint) given with the tranquilliser midazolam (muscle relaxation) is a commonly used neuroleptanalgesia that works well in many species and is probably the injectable of choice in rodents and rabbits. However, it causes respiratory depression, hypotension and bradycardia and has a prolonged recovery time of several hours. It can be topped up with fentanyl as the other two drugs last for several hours. It can be partially reversed with an analgesic buprenorphine to provide ongoing analgesia after surgery (see section on analgesia).

Sodium pentobarbitone was a commonly used anaesthetic but has the serious disadvantage of lowering the pain threshold, so in order to achieve good analgesia, animals have to be given doses that severely depress the vital centres leading to a high incidence of anaesthetic deaths in some species, such as rabbits. This can be overcome by combining it with an opioid analgesic such as pethidine or fentanyl (or even buprenorphine given long enough in advance) or an inhalational anaesthetic. It still has a place in animal anaesthesia, but not as a single agent. It can be used for induction of anaesthesia in ruminants due to its short half-life.

Long term anaesthesia

Long periods of anaesthesia can be achieved by infusion with short or long acting agents but it is important to supplement with oxygen, and even to artificially ventilate animals if the procedure lasts for more than, say, 2-4 hours. The depth of anaesthesia can be adjusted rapidly by increasing the rate of infusion but recovery may be prolonged unless short acting agents are used. Infusion rates will vary between species and strains and so it is best to set a rate and observe animals closely until a reliable dosage rate is devised. The rate can be adjusted by +/- 20 % increments until suitable depth maintained (usually after half an hour). The rate may have to be lowered after 1-2 hours depending on the type of agent used and the type of surgical stimulation. To prolong anaesthesia, top-up doses can be given in response to signs of returning consciousness, but will result in marked oscillation in the depth of anaesthesia, and some animals may well experience transient awareness and possibly pain, due to insufficient anaesthesia at some point. With any long term anaesthesia, it is important to compensate for insensible water losses by calculating the infusion volume against the animal's physiological requirements (see below).

There are also several non-recovery and long-term anaesthetics and there is often little difference in management of the animal except that it must be maintained for a longer period and any physiological imbalances will become more important with time: e.g. body temperature - hypothermia leading to shock; compromised respiratory function and so use positive pressure ventilation (see above). It is important to insert a venous line to administer drugs and to infuse fluids to provide circulatory support.

Traditionally-used anaesthetic agents include *urethane* which produces stable and long-lasting anaesthesia in several species with maintenance of cardiovascular reflexes, but it is carcinogenic and so should be handled with care and only used if there really is no alternative. *Chloral hydrate* is a weak analgesic so large doses produce other undesirable effects such as respiratory and cardiovascular depression, potentiation of vagal activity, atrial fibrillation, and it enhances the carotid sinus reflex. *Alpha-chloralose* produces long periods of light to medium anaesthesia, but if used at the start of anaesthesia as an induction agent it causes involuntary excitement and so it should only be used after anaesthesia has been induced by a volatile agent.

A combination of *chloralose* and *urethane* can produce long-term anaesthesia with less induction excitement than chloralose alone.

Reversal

It is possible to reverse some of the injectable agents and this will reduce the time to recovery and lessen anaesthetic side-effects such as a reduction in sleep time, a decreased chance of hypothermia, and reversal of respiratory depression. It should be ensured that adequate analgesia is provided after surgery so that animals do not 'wake up' in pain. Atipamizole will reverse the alpha-2 adrenergic agonist but not the other agents in the mixture like ketamine, the benzodiazepines and butyrophenones, and so there may still be marked sedation even excitatory effects.

2.1.8 Intra-operative care

Animals should maintained as far as possible within the usual homeostatic norms during the procedure with respect to temperature, heart function and respiratory rate as well as other physiological parameters. They may have to be given fluids during long operations calculated on the basis of replacing them at 80 ml/kg/24h or 4 ml/kg/h, and the time should be calculated from when an animal has had water withdrawn to when it is able to drink for itself. A skin pinch with a twist will give some idea of the degree of dehydration: it should flip back to its normal flat position quickly (this is a useful post-op sign). In large animal surgery it may be possible to implant a line near the right atrium to measure central venous pressure (CVP) and if it is low, fluids should be given. The colour of the mucous membranes of the eye and mouth, gums or tongue will indicate the state of circulation and oxygenation. If the mucosa is pink then that will indicate good oxygenation. If it is compressed briefly and refills quickly then that indicates that the blood pressure and circulation are likely to be adequate. If there is a sluggish refill time and the heart seems normal then the animal may be hypovolaemic (low CVP) and may need fluids. As mentioned above, it is very important that body temperature is maintained within normal limits. It is also good practice to keep a note of all drugs given the times of administration, routine observations of depth of anaesthesia, respiration and heat rates, temperature and any other noteworthy event, so that the anaesthetic regime can be improved retrospectively.

2.1.9 Recovery rooms, problems on recovery, and post-procedural care

The recovery room should be quiet, warm, give easy visual and physical access of carers to animals. Appropriately trained personnel should monitor the animals closely until they have fully regained consciousness. Suitable post-surgical assessment can be a major factor in assuring the welfare of animals and should be planned in advance. Potential complicating factors include circulatory and respiratory failures, hypothermia, pain and discomfort, infection and wound breakdown. The extents to which these various factors interfere with the welfare of animals will depend largely on the type, length and invasiveness of the surgical protocol. The long-term aim of post-surgical management is to return an animal to its pre-surgical physiological and behavioural state as quickly as possible. Care will commonly include provision of additional warmth and bedding, care of surgical incisions and maintenance of any dressings and, eventually, suture removal. It is important to prepare an on-going plan of observations, to help in the monitoring. Such actions should be under veterinary supervision. The pattern of recovery could be noted and a score sheet of relevant clinical signs drawn up and completed on a regular basis. These record sheets will help in determining the most appropriate anaesthetic and aftercare for the animals and may provide important data for the science.

2.1.10 Emergencies

The common emergencies include respiratory, cardiovascular and body temperature. If an animals stops breathing artificial ventilation has to be applied urgently as if it is not corrected the animal will die from heart failure. If the heart slows or stops the cause has to be quickly found and corrected. In my experience if the heart has stopped it is very difficult to re-start it. Malignant hyperthermia is a potentially lethal condition that occurs in pigs as a result of using halothane (and probably some of the other fluorinated ethers). It is lethal because the cost of treatment exceeds the economic value of the pig, and the experiment should be abandoned as the animal is no longer a standard model with this extra treatment. A more careful selection of the strain of pig is called for.

2.1.11 Local anaesthesia and regional

Local anaesthetic drugs act directly on nerve tissue and block sensory impulses to the CNS, and so an animal does not feel pain. However, local anaesthetics do not provide any sedation and restraint for an animal and may possibly be provided by firm gentle handling or through the use of tranquillisers and sedatives. Local anaesthesia can be used when suturing minor wounds, or when making burr holes in the cranium (which seems particularly sensitive despite anaesthesia), to block nerve supply to a surgical field (regional anaesthesia) and to block spinal nerves (e.g. epidural) to a large area. Examples of local anaesthetics include lignocaine (xylocaine) that will last for 30 –60 minutes and bupivicaine that will last for 6-8 hours. There are also topical anaesthetic creams such as xylocaine gel (good for mucosal surfaces) and EMLA cream (Eutectic Mixture Long Acting) the latter will penetrate the skin and is useful for surface analgesia e.g. for intra-arterial or intravenous injections (rabbits' ears) but it takes some 30 minutes to have its maximum effect.

2.1.12 Anaesthetics for Fish and Amphibia

The most common anaesthetics for animals that live in water are those that can be dissolved and are given by immersion. MS222 is a commonly used agent, but it should be buffered before use to around neutrality. The reader is referred to other text for guidance on the use of other agents (e.g. Ross and Ross 1999) but some of the fluorinated ethers have also been used by dissolving them in the water.

2.2 Analgesia drugs and regimens

There are legal and moral obligations to treat research animals in pain unless that would frustrate the scientific objective. However, some may argue that this is not necessary, giving reasons such as the perception that there are no objective criteria for assessing pain and that pain benefits animals because it immobilises them and prevents them from causing further damage, and concerns about side effects of analgesics. All of these beliefs have been discredited (Morton and Griffiths 1985, Flecknell 1996) and the current consensus is that analgesia should always be provided for animals undergoing surgical procedures unless there is a very strong scientific justification not to do so. It is worth noting that pain treatment can avoid the loss of animals and may well produce better scientific data as pain is an uncontrolled variable and increases post-operative complications. It can lead to an animal failing to eat, making it catabolic, and this may in turn delay in wound healing. The pain itself may increase wound interference e.g. suture line, and if it is thoracic or abdominal pain then respiration may be compromised. There is likely to be a negative impact on an animal's homeostasis. Moreover, the stress on an animal may affect any future human interaction with it and lead to mental distress. Experimentally-induced pain is usually predictable and either avoidable (skilful surgery) or relievable (the use of drugs and good nursing care). One of the major causes of pain is surgery and there is a need for the provision of adequate analgesia for animals until the surgical trauma has healed or is no longer painful.

It goes without saying that the recognition of clinical signs of pain in animals is the first step, and there is considerable debate about how animals show they are in pain. A certain amount of critical anthropomorphism (Morton et al. 1990, OECD 2001) and common sense is required and a good guide is that if a procedure is likely to be painful in humans then assume it is will be so in animals until proved otherwise. Some animals, notably those that are preyed upon, may mask their pain, unlike predatory animals. There are several good texts on this subject (e.g. Flecknell and Waterman-Pearson, 2000) although pain recognition is still the subject of much debate, whereas its treatment is not. It is desirable that animals should not be left in pain, and that post-operative analgesia should be routinely administered to all laboratory animals.

2.2.1 Recognition of pain, discomfort and distress

It is of fundamental importance to be able to recognise when animals are suffering pain and distress so that one can alleviate or reverse such suffering and, if possible, take steps to avoid it in the future: it is the first basic step. Investigators should consider, *a priori*, that procedures that cause pain and distress in humans may cause pain or distress in animals and to look out for the relevant clinical signs (see Morton et al. 1990, Soulsby and Morton 2001, OECD 2001). Members of the care-giving staff are generally familiar with the normal behaviour and temperament of individual animals and therefore are the ones most likely to recognise when an animal is 'not right' (Morton and Griffiths 1985). They may interpret this as some form of emotion (e.g. agitation, stress, pain, fear, anxiety, and so on), but it is important to try to see what signs they are seeing in order to make a scientific assessment of any suffering, and then to provide the right sort of treatment, for example, it is pointless giving analgesics to relieve distress or sedatives to relieve pain. This careful and critical observation of animals leads to descriptors of what the critical abnormal behaviours or other signs are being seen. Some of these signs may be measurable such as change in body weight or rapid breathing, whereas others may be equally reliable and objective such as limping, diarrhoea, closed eyelids and so on (Morton 2000). Just because they cannot be quantified does not make them less useful or less likely to be seen by trained observers. When monitoring animals for signs of pain and distress, it is often useful to assess unprovoked behaviour, and then to look for further signs when the animal is then handled and disturbed. Animals should be checked at regular intervals during the course of an experiment, and particularly at times when any adverse effects are likely to occur. Pain can produce restlessness or inactivity according to the site affected, for example, colic as opposed to musculoskeletal pain respectively. Severe pain, especially if chronic, usually makes animals appear distressed and lethargic, and can be accompanied by vocalisation, and various changes in behaviour. Loss of appetite can be a marker of ongoing pain. An assessment of the level of suffering based on score sheets of clinical signs should be documented retrospectively in order that refinements can be introduced into subsequent studies (see Morton 2000).

Post-surgical pain can largely be prevented by administering 'pre-emptive analgesia' before surgery begins, in conjunction with post-operative supplements. If this protocol is not followed, established pain can only be controlled, and is more difficult to achieve. This is because when pain is uncontrolled, neurones in the dorsal horn of the spinal cord become over sensitised so called 'wind-up' at both a local and central level. This leads to hyperalgesia of the injured site (increased response to noxious stimuli), allodynia (non-painful stimuli become painful) and the affected area is increased. For more information on current techniques for pain assessment in the UK, including downloadable score sheets that can be adapted to specific procedures, see Hawkins (2002). Sharing comprehensive information on how pain and distress were recognised, avoided, prevented or alleviated in an experiment can help others use similar techniques and promulgates good welfare and science. When writing up studies for publication in scientific journals, such information should be included (Morton 1992).

Pain relieving agents should be administered before anaesthesia is terminated so that analgesia can become effective before they regain consciousness and is likely to be necessary for the first 24 to 48 hours, depending on the invasiveness of the surgical procedure. It is important to continuously assess the level of pain until one is confident an animal is no longer in pain, especially when analgesia is discontinued - it is not a one off treatment. There are a number of agents that can be used that include both opioids and Non-Steroidal Anti-Inflammatory Drugs (NSAIDs). Opioids are used in situations in which moderate to severe pain might be expected and some examples include morphine, pethidine, oxymorphone, methadone, butorphanol, buprenorphine, tramadol. Probably the commonest opioid used in animals is *buprenorphine* (0.02 mg/kg i/m) as it has a long half-life of between 4 and 12 hour depending on the species, but it has a slow onset of action and so needs to be given 30 minutes before it is needed. It has been suggested that opioids may cause some respiratory depression but, while this may be true in humans at the normal analgesic dosage rates, in animals it is not seen. It is used to counter the intense barrage from the nociceptors immediately after surgical trauma, particularly in the first 24-48 hours. Shortly after that time (within the first 12 hour or so) the inflammatory response is the major cause of pain and this can be modified by the use of NSAIDs at the time of surgery and may be persisting for several days depending on the actual surgical intervention. Animals can be placed on NSAIDs as well as opiates from day one – or even the day before - in order that they will be effective at a time when the inflammatory response takes place. Some of the commonly used NSAIDs include: carprofen, flunixin meglumine and ketoprofen. Steroids are occasionally used but they have various unwanted side effects; examples include prednisolone and betamethasone. Pain due to infection and inflammatory responses can be controlled by the use of pre-emptive use of antibiotics and even long acting local anaesthetics.

It is worth noting that it is not just surgery where analgesia may be required and that there are many other painful disease conditions that necessitate analgesia e.g. arthritis, ophthalmic conditions.

2.3 Euthanasia

Killing animals is never a pleasant thing to do and scientists often assume it is part of the job of the animal caretakers. It is not. These persons do not come into research animal care to kill animals, any more than scientists do. Several studies have shown substantial psychological impact on those that kill animals (see Arluke 1993, 1996) and this should not be forgotten by those who request others to kill animals for them. Scientists should avoid waiting until animals have built up to large numbers and try to identify animals at an early stage, preferably before weaning, to request they be culled. It is more morally repugnant to kill adult animals that could have been killed at an earlier stage, than when they are pre-weaners. Similarly, to request killing a pregnant animal that could have been killed earlier is reprehensible. While economic pressures reduce the breeding of unwanted animals, there is still much room for greater consideration to be given to this matter.

With the increase in the generation of transgenic animals the numbers that need to be killed is markedly increased because of the inefficiency of the generation of these novel genetically modified animals. After a breeding line has been established it needs to be maintained and if it breeds in a homozygous state then the numbers bred can be matched to demand. If in order to maintain the line, or to decrease animal suffering, heterozygous lines are used then three times as many animals may be bred than needed. If more than one gene is being selected for at one time, then the numbers are even greater. With the advent of more successful use of embryonic stem cells that can be modified *in vitro* and then selected before embryo transfer, the numbers culled when producing a new transgenic or knock-out line can be reduced.

Forecasting the number of animals to be used is obviously important and there tends to be an overestimate of the number required. Coupling this with an unpredictable litter size can compound the problem. So careful breeding records should be maintained and the number of animals requested must be realistic, both of which will lead to fewer surplus animals.

Euthanasia comes from the Greek word *'eu'* having a connotation of good, *'thanatos'* means death. In human terms it tends to mean a good reason to allow someone to die (or even to kill them) but in animal terms it is more about the method used to kill them. For a good gentle death several criteria should be met. An acceptable methods should:

1. cause a rapid loss of consciousness
2. be painless
3. not cause discomfort or distress e.g. through being aversive, requiring excessive restraint (e.g. decapitation), or being unpleasant in some way
4. be aesthetic to the person carrying it out
5. be easy to perform as there is less chance of it going wrong
6. be repeatable without loss of efficiency when large numbers of animals are involved
7. be minimally invasive.

Some methods require more skill than others, and a particular method may be needed for scientific reasons. If a method has to be used that might cause more suffering, then it has to be justified scientifically. In fact if such as method is requested good data should be presented to show that better methods of euthanasia are scientifically poorer, rather than to make any assumptions.

Anyone killing animals should be competent in that technique and generally, physical methods such as neck dislocation demand more skills than say an overdose of an inhalational anaesthetic where animals are simply placed in an induction chamber. Note that different species (and different ages of the same species) may require different methods that are more appropriate to that species e.g. the use of a captive bolt or percussion gun to farmed species compared with rodents where an overdose of anaesthetic may be more appropriate. Whatever methods are used a training strategy and an assessment of the competence should be in place.

In general terms, and given competence of the person carrying it out, an overdose of anaesthetic is acceptable for all species whether given by inhalation, injection or immersion (amphibia and fish). Physical methods leave more room for accidents (e.g. neck dislocation, concussion) and may require some restraint that may cause discomfort or fear (e.g. decapitation). However, it may be possible to first anaesthetise an animal, and some scientists have claimed that this has given more consistent results. More recently the humaneness of using carbon dioxide as an agent for euthanasia and anaesthesia has been questioned and there is now good data showing that it is aversive to most mammals, and in humans it has even been used as a noxious stimulus and described as 'painful' at concentrations above 20% (Danneman et al. 1997; Leach et al. 2002a, 2002b). Carbon dioxide could still be used providing an animal has been rendered unconscious first through e.g. the use of an inhalational anaesthetic. Whatever method is used the death of an animal should be confirmed in some way before the body is disposed of (e.g. onset of rigor mortis, deliberately causing irreversible damage to the brain or the circulation). Placing an animal in the deep freeze for 48 hr (NB in small numbers in a bag) is a convenient way of ensuring an animal does not 'come back to life'.

3. HUMANE ENDPOINTS IN RESEARCH

Many animal experiments are likely to cause some degree of pain but it is important that the level of suffering is no more than that required to meet the scientific objective. No matter how much overall benefit may be gained it is still no excuse for causing avoidable suffering. In some ways it is about selecting the least inhumane endpoint for an experiment. There are several practical ways in which to interpret what is meant by 'humane endpoints' and these will be illustrated below.

1. The scientific objective has been achieved

It should normally be possible to stop an experiment at the point when the data can be interpreted in a scientifically sound manner. For example, when there is a clear statistical difference between control and treated groups. This may occur in experiments designed to look at therapeutic effectiveness of a drug, or the toxicity of a chemical or substance.

2. The animal is no longer 'normal' to give valid data

There may be occasions when an animal's physiological or psychological status is so far from normal that it is very doubtful that the results obtained will truly reflect the scientific variable being investigated. Therefore the animal has departed significantly from normal homeostasis. The scientist has to ask the question whether they are going to include the results in their final analysis if they are outside the range of the other animals in the group? For example, if an animal becomes so dehydrated and catabolic because it is not eating that may well affect the action of the drugs being studied, through delayed clearance rates or metabolism. This may be encountered in repeat dose studies when an animal is so stressed by the dosage procedure, and alternative ways should be sought to achieve the scientific goal. It may also be the case if there is an intercurrent infection in the colony.

3. Suffering, especially death, can be avoided through the use of surrogate endpoints

There are an increasing consideration in the area of safety testing (chemicals and biological products such as vaccines) of whether clinical signs can be used to predict death (the traditional endpoint). While the final stages of death may be painless (when unconscious or moribund) the stages beforehand may well not have been so. Furthermore, in some studies such as infection the suffering may not be about pain, but about extreme discomfort and distress, and animals may well die simply because they cannot reach the food and water. Various papers have been published illustrating the use of biochemical and haematological markers as well as clinical signs (see Table 6-6, Mellor and Morton 1997, Hendricksen & Morton 1999, ILAR 2000, OECD 2001, Morton 2002, Morton & Hau 2002). As an example, in a study on rabies vaccine potency testing where death was the traditional endpoint, it was found that death could be predicted in the test animals when the mice started to show slow circling movements. This saved the mice several days of suffering without jeopardising the scientific objective.

4. Suffering is higher than predicted

When a higher level of suffering is encountered than predicted then if the work has been authorised on the basis of a cost-harm to benefit analysis, then it may have become disproportionate. In other words that level of suffering cannot be justified by the anticipated gains from the experiment.

5. Suffering has breached the absolute limit

In some countries like the UK it is never permitted to cause severe pain or severe distress and so any experiment in which animals are likely to suffer to that degree has to be terminated, or the animal treated in some way, or the animal withdrawn from the study.

Competence of the team: The establishment and delivery of humane endpoints should be a priority for all those involved in animal work and this includes: the animal care staff, scientists and other principal investigators, veterinary technicians, project leaders, those who carry out scientific procedures on animals, and those involved in the experimental design such as statisticians, veterinarians, ethics committee members, law enforcers, and the regulators. All have an ethical responsibility to help determine and implement humane endpoints.

Table 6-6. Clinical signs that can be used as surrogate signs for pending death, pain and distress

*Moribund signs and impending death	Absence of responses Prolonged recumbency Prolonged inability to reach food and water Excessive: weight loss, body condition (emaciation) Severe dehydration or blood loss Signs of irreversible organ failure Laboured breathing Persistent convulsions, diarrhoea, hypothermia Self-mutilation Substantial tumours Treatment related effects indicative of impending death
*Severe pain and distress	Abnormal vocalisation, aggressiveness, posture, reaction to handling, movements Self mutilation, open wounds or skin or corneal ulceration Dyspnoea, fractures Reluctance to move Abnormal appearance Rapid weight loss, emaciation, severe dehydration Significant blood loss
**Clinical signs useful as humane endpoints for toxicology studies	Diarrhoea, debilitating or prolonged Persistent coughing, wheezing, dyspnoea Polyuria, anuria, markedly discoloured urine CNS signs: e.g. tremors, seizures, paralysis Frank bleeding from orifices Distinct icterus, anaemia Persistent self-induced trauma Microbial infections interfering with toxic or carcinogenic responses

* OECD Guidance (2001, Monograph 19)
** Montgomery CA (1990)

4. CONCLUSIONS

This chapter has dealt with various practical aspects of working with laboratory animals and this conclusion tries to draw its seemingly disparate aspects together. The laws in Europe require that all animals should be anaesthetised for the whole of the experiment unless that would mean that the scientific objective could not be achieved, or unless the adverse effects of giving an anaesthetic would cause more animal suffering that not doing so. Most of the time animals are not anaesthetized for the scientific procedure, and in the UK statistics show that this happens in more than 90% of cases. However, it becoming increasingly apparent that procedures such as handling, dosing and sampling, and even the giving of an appropriate anaesthetic suitable for the species and duration of the procedure, are important for both the welfare of the animal as well as for the scientific output. Poor animal welfare can often lead to poor science because of the increased variance of results. Furthermore, the results may be skewed as a direct result of the incidental suffering. In some cases it may not matter scientifically but legally and ethically it does. We should always aim to minimise animal suffering in research and it is because of that imperative we have to be able to recognize when animals are in pain and distress. Only then can we take steps to avoid it, to minimise it, or to relieve it and so act humanely. We should cause only the minimum level of suffering in an experiment and we should cause no more than that to the fewest number of animals consistent with good science. To cause more suffering than is necessary would be to act inhumanely. Sometimes it may be possible to implement early endpoints and so reduce animal suffering, e.g. when the scientific objective has been achieved, but at other times animals have to be killed as they have come to the end of the experiment or occasionally their suffering is too great to be justified. Even killing animals can be done badly and so that too should be done causing as little pain and distress as possible. I hope that this chapter in some small way will provide helpful practical guidance and will act as a stimulus for the reader to always try to promote better animal wellbeing as well as good science.

REFERENCES AND ADDITIONAL LITERATURE

ACLAM (American College of Laboratory Animal Medicine). Anesthesia and Analgesia in Laboratory Animals. Kohn DF, Wixson SK, White WJ, Benson J. eds, Academic Press, NY 10010, USA, ISBN 0-12-417570-8, 1997

Ader R, Plaut S. Effects of prenatal maternal handling and differential housing on offspring emotionality, plasma corticosterone levels, and susceptibility of gastric erosions. Psychosom Med 1968; 30:277-286

Arluke A. Trapped in a guilt cage. New Sci 1993; 134:1815-1818

Arluke A. The Well-being of animal researchers. In The Human Research Animal Relationship. Krulisch L, Meyer S, Simmons RC eds, Scientists Center for Animal Welfare 1996; 7-20

Aldred AJ, Ming CC, Meckling-Gill KA. Determination of a humane endpoint in the L1210 model of murine leukemia. Contemp Top Lab Anim Sci 2002; 41(2):24-27

AVMA. Report of the AVMA panel on Euthanasia. J Am Vet Med Assoc 2000; 218:669-696

AWI. Comfortable Quarters for Laboratory Animals, 9th edition. Reinhardt V, Reinhardt A eds. Animal Welfare Institute, Washington DC 2002

Barnett JL, Hemsworth PH. The impact of handling and environmental factors on the stress response and its consequences in swine. Lab Anim Sci 1986; 36(4):366-369

Canadian Council on Animal Care. CCAC Guideline on choosing an appropriate endpoint in experiments using animals for research, education, testing. Canadian Council on Animal Care, Ottawa, Canada 1998

Claassen V. Neglected Factors in Pharmacology and Neuroscience Research. Biopharmaceutics, Animal characteristics, Maintenance, Testing conditions. In Techniques in the Behavioural and Neural Sciences, Vol 12. Huston JP Series ed, Elsevier, Amsterdam 1994

Close BS, Banister K, Baumans V, Bernoth E-M, Bromage N, Bunyan J, Erhardt W, Flecknell P, Gregory N, Hackbarth H, Morton DB, Warwick C. Recommendations for euthanasia of experimental animals. Commission of the European Communities DGXI Working Party report. Lab Anim 1996; 30:293-316 and Lab Anim 1997; 31:1-32

Danneman PJ, Stein S, Walshaw SO. Humane and practical implications of using carbon dioxide mixed with oxygen for anaesthesia or euthanasia of rats. Lab Anim Sci 1997; 47:376-384

Diehl K-H, Hull R, Morton DB, Pfister R, Rabemampianina Y, Smith D, Vidal J-M, van de Vorstenbosch C. A good practice guide to the administration of substances and removal of blood, including routes and volumes. J Appl Toxicol 2001; 21:15-23

Dean SW. Environmental enrichment of laboratory animals used in regulatory toxicology studies. Lab Anim 1999; 33:309-327

Elvidge H, Challis JRG, Robinson JS, Roper C, Thorburn GD. Influence of handling and sedation on plasma cortisol in rhesus monkeys (Macaca mulatta). J Endocrinol 1976; 70:325-326

Feenstra MGP, Botterblom MHA. Rapid sampling of extracellular dopamine in the rat prefrontal cortex during food consumption, handling and exposure to novelty. Brain Res 1996; 742:17-24

Flecknell PA. Laboratory Animal Anaesthesia, 2nd edition. Harcourt Brace & Company, London 1996

Flecknell PA, Waterman-Pearson A. Pain Management in Animals. WB Saunders, London 2000

Gartner K, Buttner D, Dohler K, Friedel R, Lindena J, Trautschold I. Stress response of rats to handling and experimental procedures. Lab Anim 1980; 14:267-274

Grandin T ed. A Livestock Handling and Transport. CAB International, Wallingford, Oxon, UK 1993

Green CJ. Animal Anaesthesia. Pub. No 8, Laboratory Animals Ltd, London ISBN0-901334-08-1, 1979

Grigor PN, Hughes BO, Appleby MC. Effects of regular handling and exposure to an outside area on subsequent fearfulness and dispersal in domestic hens. Appl Anim Behav Sci 1995; 44:47-55

Hawkins P. Recognizing and assessing pain, suffering and distress in laboratory animal: a survey of current practice in the UK with recommendations. Lab Anim 2002; 36(4):378-395

Healing G, Smith D eds. Handbook of Pre-clinical Continuous Infusion. Taylor and Francis, London 2000

Hemsworth PH, Coleman GJ. Human-Livestock Interactions: the Stockperson and the Productivity and Welfare of Intensively Farmed Animals. CAB International 1998

Hendriksen CFM, Morton DB eds. Humane Endpoints in Animal Experiments for Biomedical Research. Proceedings of the Intnl Conference, 22-25 Nov 1998 Zeist, The Netherlands. Royal Society of Medicine Press, London 1999

ILAR Journal. Humane endpoints for animals used in biomedical research and testing. ILAR J 2000; 41(2)

JWGR (Joint Working Group on Refinement). 1st Report. Removal of blood from laboratory mammals and birds. Lab Anim 1993; 27:1-23

JWGR, eds Morton DB, Jennings M, Buckwell A, Ewbank R, Godfrey C, Holgate B, Inglis I, James R, Page C, Sharman I, Verschoyle R, Westall L, Wilson AB. Refining procedures for the administration of substances. Fourth Report of the BVA-AWF/FRAME/RSPCA /UFAW Joint Working Group on Refinement. Lab Anim 2000; 35:1-41

JWGR, eds Hawkins P et al. Laboratory birds: refinements in husbandry and procedures. Lab Anim 2001; 35(Suppl 1):1-163

JWGR, Morton DB, Hawkins P, Bevan R, Heath K, Kirkwood J, Pearce P, Scott E, Whelan G, Webb A. Refinements in telemetry. Seventh Report of the BVA-AWF/FRAME/RSPCA/UFAW Joint Working Group on Refinement. Lab Anim 2003; 37(4):261-299

Kannan G, Mench JA. Influence of different handling methods and crating periods on plasma corticosterone concentrations in broilers. Br Poult Sci 1996; 37:21-31

Kramer K, van de Weerd H, Mulder A, Van Heijningen C, Baumans V, Remie R, Voss H-P, van Zutphen B. Effect of conditioning on the increase of heart rate and body temperature provoked by handling in the mouse. Altern Lab Anim in press

Lapin IP. Only controls: effect of handling, sham injection, and intraperitoneal injection of saline on behavior of mice in an elevated plus-maze. J Pharmacol Toxicol Methods 1995; 34:73-77

Lascelles D, Waterman A. Analgesia in cats. In Practice 1997; 19:203-213

Leach M, Bowell V, Allan T, Morton DB. Aversion to gaseous euthanasia in rats and mice. Comp Med 2002a; 52(3):249-257

Leach M, Bowell V, Allan T, Morton DB. The Aversion to various concentrations of different inhalational general anaesthetics in rats and mice. Vet Rec 2002b; 150:808-815

Lester SJ, Mellor DJ, Ward RN. Effects of repeated handling on the cortisol responses of young lambs castrated and tailed surgically. N Z Vet J 1991; 39:147-149

Mellor DJ, Morton DB. Humane Endpoints in research and testing. Synopsis of the workshop. In Animal Alternatives, Welfare and Ethics. van Zutphen LFM, Balls M eds. Elsevier Science BV, Amsterdam, The Netherlands. 1997; 297-299

Metz JHM. Effects of early handling in the domestic rabbit (Abstract). Appl Animal Ethol 1984; 11:86-87

Michel C, Cabanac M. Opposite effects of gentle handling on body temperature and body weight in rats. Physiol Behav 1999; 67:617-622

Montgomery CA Jr. Oncological and toxicological research: alleviation and control of pain and distress in laboratory animals. Cancer Bull 1990; 42(4):230-237

Morton DB. A fair press for animals. New Sci 1992 Apr 11; 134(1816):28-30

Morton DB. A systematic approach for establishing humane endpoints. ILAR J 2000; 41(2):80-86

Morton DB. The importance of non-statistical experimental design in refining animal experiments for scientists, IACUCs, and other ethical review panels. In Applied Ethics in Animal Research. Philosophy, Regulation, and Laboratory Applications. Gluck JP, DiPasquale T, Orlans FB eds, Purdue University Press, West Layfayette, Indiana 2002; 149-178

Morton DB, Griffiths PHM. Guidelines on the recognition of pain, distress and discomfort in experimental animals and an hypothesis for assessment. Vet Rec 1985; 116:431-436

Morton DB, Hau J. Welfare assessment and humane endpoints. In CRC Handbook of Laboratory Animal Science Vol. 1. Essential Principles and Practices, 2nd edition. Hau J, Van Hoosier GL Jr eds, CRC Press, Boca Raton 2002; 457-486

Morton DB, Burghardt G, Smith JA. Critical anthropomorphism, animal suffering and the ecological context. Hasting's Center Report Spring Issue on Animals, Science & Ethics 1990; 20(3):13-19

Nicol CJ. Effects of environmental enrichment and gentle handling on behavior and fear responses of transported broilers. Appl Anim Behav Sci 1992; 33:367-380

Norina website: http://oslovet.veths.no

OECD Environmental Health and Safety Publications Series on Testing and Assessment No 19: Guidance document on the recognition, assessment, and use of clinical signs as humane endpoints for experimental animals used in safety evaluation. Environment Directorate, Organisation for Economic Co-operation and Development, Paris November 2000. 2001, http://www.oecd.org/ehs/

Olfert ED. Considerations for defining an acceptable endpoint in toxicological experiments. Lab Anim 1996; 25(3):38-43

Pearce GP, Paterson AM, Pearce AN. The influence of pleasant and unpleasant handling and the provision of toys on the growth and behavior of male pigs. Appl Anim Behav Sci 1989; 23:27-37

Pieretti S, Damore A, Loizzo A. Long-term changes induced by developmental handling on ppain threshold - effects of morphine and naloxone. Behav Neurosci 1991; 105(1):215-218

Podberscek AL, Blackshaw JK, Beattie AW. The effects of repeated handling by familiar and unfamiliar people on rabbits in individual cages and group pens. Appl Anim Behav Sci 1991; 28:365-373

Reese EP. The role of husbandry in promoting the welfare of animals. In Animals in Biomedical Research. Hendriksen CFM, HBWM Köeter HBWM eds, Elsevier, Amsterdam 1991; 155-192

Reid WC, Carmichael KP, Srinivas S, Bryant JL. Pathologic Changes Associated with Use of Tribromoethanol (Avertin) in the Sprague Dawley Rat. Lab Anim Sci 1999; 49:665-667

Reinhardt V. Improved handling of experimental rhesus monkeys. In The Inevitable Bond. Examining Scientist-Animal Interactions. Davis H, Balfour AD eds, Cambridge University Press, Cambridge 1992; 171-177

Reinhardt V. Training nonhuman primates to cooperate during handling procedures: a review. Anim Technol 1997; 48:55-73

Ross LG, Ross B. Anaesthetic and Sedative Techniques for Aquatic Animals, 2nd edition. Blackwell Science, Oxford, UK, ISBN: 0-632-05252-X, 1999

Ryabinin AE, Wang YM, Finn DA. Different levels of Fos immunoreactivity after repeated handling and injection stress in two inbred strains of mice. Pharmacol Biochem Behav 1999; 63(1):143-151

Soothill J, Morton DB, Ahmed A. The HID50 (hypothermia-inducing dose 50): an alternative to the LD50 for measurement of bacterial virulence. Int J Exp Pathol 1992; 73:95-98

Soulsby L, Morton DB. Pain: its Nature and Management in Man and Animals. Royal Society of Medicine Press, London 2001

Svendsen P. Laboratory animal anaesthesia. In Handbook of Laboratory Animal Science, Vol I Selection and handling of animals in biomedical research. Svendsen P, Hau J eds, CRC Press, USA 1994; 311-351

Swanson HH, Poll van-de NE. Effects of an isolated or enriched environment after handling on sexual maturation and behaviour in male and female rats. J Reprod Fertil 1983; 69:165-171

Toth LA. The moribund state as an experimental endpoint. Contemp Top Lab Anim Sci 1997; 36(3):44-48

UFAW (Universities Federation for Animal Welfare) Handbook on The Care & Management of Laboratory Animals, 7th Edition (Volumes 1 and 2). Poole T ed, Blackwell Science Ltd, 1999

UK Home Office: http://archive.official-documents.co.uk/document/hoc/321/321.htm

van den Heuvel MJ, Clark DG, Fielder RJ, Koundakjian PP, Oliver GJA, Pelling D, Tomlinson NJ, Walker AP. The international validation of a fixed dose procedure as an alternative to the classical LD50 test. Food Chem Toxicol 1990; 28:469-482

Van Zutphen LFM, Baumans V, Beynen AC eds. (2001) Principles of Laboratory Animal Science: A Contribution to the Humane Use and Care of Animals and to the Quality of Experimental Results. Revised edition. Elsevier Science Publishers BV, Amsterdam, The Netherlands 2001

Weiss J, Zimmermann F, Zeller W, Bürki K, Meier G. Tribromoethanol (Avertin) as an anaesthetic in mice. Lab Anim 1999; 33:192-193

Wolfensohn S, Lloyd M. Handbook of Laboratory Animal Management and Welfare. Oxford University Press 1994

Wong R, Bradley HW, Jamieson JL. Infantile handling and activity and subsequent wheel running behaviour. Psychon Sci 1967; 7(9):293

Zeller W, Meier G, Burki K, Panoussis B. Adverse effects of tribromoethanol as used in the production of transgenic mice. Lab Anim 1998; 32:407-413

Zeller W, Meier G, Buerki K, Panoussis B. Adverse effects of tribromoethanol as used in the production of transgenic mice: Authors' reply. Lab Anim 1999; 33:193

THE WELFARE OF DIFFERENT SPECIES

Chapter 7

THE WELFARE OF LABORATORY MICE

Vera Baumans
Department of Laboratory Animal Science, University of Utrecht, Utrecht, The Netherlands/
Karolinska Institute, Stockholm, Sweden

1. INTRODUCTION

Animal experiments can only be performed when no alternative is available and when the benefit of the experiment outweighs the animal's suffering. When we use animals, we are legally and morally obliged to safeguard their welfare and minimize discomfort, which will be beneficial for both the animal and the experimental outcome. Discomfort and stress before and during the experimental procedure will lead to non-specific effects due to e.g. endocrinological and immunological changes, resulting in a jeopardized experimental outcome (Van Herck et al.1994).

The mouse (*Mus musculus*) is the most widely used vertebrate species in biomedical research with more than 1000 genetically-defined inbred strains. As a result of the rapid increase in biotechnology new strains expressing novel genetic characteristics are being created. Together with their short reproductive cycle, short life span, small size and low cost of maintenance these features make them suitable models for humans and animals in biomedical research, such as cancer and drug research, vaccine and monoclonal antibody preparation and evaluation of the safety and effectiveness of pharmaceutical products (Baumans 1999).

Genetic mapping in mice began in the early 1900s. Extensive linkage maps and an impressive array of inbred strains are now available to expedite sophisticated genetic research. Laboratory mice have 40 chromosomes that are differentiated by the size and pattern of transverse bands. Inbred strains were first developed in 1909 by C.C. Little and offer a high degree of genetic uniformity. Animals of an inbred strain are homozygous and genetically identical to other mice of the same strain and sex. They are produced by brother-sister matings for at least 20 generations. By contrast, outbred mice are genetically heterogeneous and are often produced by breeding systems that minimize inbreeding (Van Zutphen et al. 2001).

The assumption that all sources of the same inbred strain provide genetically identical mice is not valid, as animals maintained at different institutions for many generations may show genetic drift, although originating from the same source. Therefore, genetic monitoring of colonies is a prerequisite for standardization of laboratory animals resulting in a reduction in the number of animals used.

Inbred mice are generally used for research in fields such as immunology, oncology, microbiology, biochemistry, pharmacology, physiology, anatomy, and radiobiology. They might be highly sensitive to e.g. tumor development or be immunodeficient (the athymic nude or the Severe Combined Immuno-Deficient (SCID) mouse). It is obvious that the welfare of these mouse strains might be easily compromised.

Genetically engineered or modified mice are those with induced mutations, including mice with transgenes, with targeted mutations (knockouts) and with retroviral, proviral or chemically-induced mutations (The Jackson laboratory net pages). Transgenic technology focuses on the introduction or exclusion (knock outs) of functional genetic material in the germ line of an animal, thus changing the genetic characteristics of an organism and its progeny. The most frequently used methods for genetic transformation of the germ line are microinjection of DNA into the pronucleus of fertilized oocytes and the injection of transfected embryonic stem (ES) cells into normal mouse blastocysts, resulting in a subsequent generation of chimaeras. These techniques have led to the rapid development of a variety of animal models, designed for the study of gene regulation and expression, pathogenesis and the treatment of human and animal diseases (e.g. Alzheimer's disease, growth hormone disturbances, poliovirus vaccine testing in humans or mastitis in cows).

The process of transgenesis by microinjection may compromise welfare (Van Reenen and Blokhuis 1993) at the level of the experimental procedures during the process of transgenesis. The donor animals, vasectomised males and foster mothers needed for the production of the transgenic offspring may experience discomfort from procedures, such as early mating (from 3 weeks onwards), anesthesia, surgery and injections.

At the level of integration of the microinjected DNA into the genome, unintentional insertional mutations may occur, resulting in welfare problems. At the level of expression of the introduced gene detrimental side-effects may occur, e.g. the giant mouse with an overproduction of growth hormone, suffering from chronic kidney and liver dysfunction (Poole 1995). It was shown that the presence of a both functional and non-functional microinjected DNA construct increased mortality and body weight of mouse pups in the first 2-3 days after birth, although no significant differences in behavior and morphological development were observed later on (Van der Meer et al. 2001a).

The increase in the use of genetically modified animals has caused an increase in numbers of mice used of more than 23% per year at the University of Washington 1993-2001, not only due to growth in the numbers of these animals in research but also to the large number of mice necessary to create each genetically modified line, such as breeding males, donor females, vasectomised males and pseudopregnant recipient females. Furthermore, non-transgenic and wild-type littermates may be produced that are not suitable for research or further breeding (Dennis Jr. 2002).

From an ethical point of view it can be argued that the integrity of the animal is compromised. Furthermore, concern has been expressed with respect to the patentability of transgenic animals, such as the oncomouse.

It has been suggested that the escape of transgenic animals could impose a risk for animal populations in the wild. However, escapes from modern, well-built laboratories are unlikely and it appears that laboratory animals are not very viable in the wild.

In conclusion, transgenic technology has a great potential for increasing the understanding of the role of genes and may provide suitable animal models for human and animal disease, but the welfare of transgenic animals has to be carefully monitored, at least until the second generation of offspring. A surveillance system e.g. score sheets can be helpful in identifying welfare problems together with humane endpoints in order to euthanize severely affected animals (Mertens and Rulicke 1999, Van der Meer et al. 2001b).

1.1 Animal welfare

Although many definitions exist, animal welfare can be considered as a relative concept, referring to the state of an animal in relation of its ability to cope with its environment (Broom and Johnson 1993). The most important things are whether the animals are physically and psychologically healthy and whether they get what they want. Do they need more space? Is their health improved and do they want more space?

Concerning the animal's health, early warning signs are important. In order to know what the animal wants, preference tests and the animal's behavior in correlation with their choice can provide information (Dawkins 2003, Baumans 1997, Van de Weerd et al. 1997a). Animal well-being is related to a broad behavioral repertoire, and assessment of the animals' state of welfare requires knowledge of the animals' species-specific needs.

2. BIOLOGY AND BEHAVIOURAL NEEDS

The house mouse is a cosmopolitan species, which is commensal with man and is highly adaptable to a variety of environmental conditions. Like most rodents, the mouse is a nocturnal burrowing and climbing animal, which shows a clear circadian rhythm with peaks of activity during the dark period (Figure 7-1).

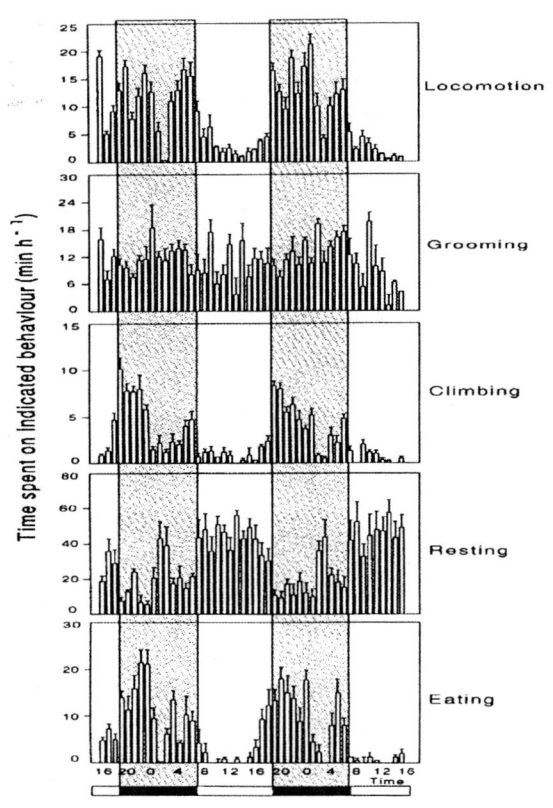

Figure 7-1. Circadian rhythm in some behaviors. Dark period is indicated by black bars (Schlingmann et al. 1998)

Mice are active, highly explorative animals and spend considerable time in the wild on foraging, seeking a wide variety of food. They construct elaborated nests and burrows and form complex social structures. All these behaviors, characterized by a strong motivation, are still present in the laboratory mouse. Preventing them from performing these behaviors can lead to frustration and suffering (Dawkins 1990, Sherwin 2000). As a result, laboratory mice often show abnormal behavior, for example stereotypies (Würbel et al. 1996).

Differences in behavior between inbred strains may reflect behavioral adaptations to different habitats of the feral populations from which the ancestors of the inbred strain originated. For example, BALB/c mice appear to be well adapted to surface living, making superficial nests and showing territorial behavior. C57BL mice are more adapted to living in holes and do not show clear territorial behavior (Van Oortmerssen 1971, Busser et al. 1974). These differences have to be taken into account when studying genetically modified mice with different backgrounds.

2.1 Social behaviour

The social system of mice includes a wide range of social organizations from strong territoriality by reproductively-active males to large colonies with well-developed social hierarchies. The dominant male defends his territory containing resources such as females, food and a nest site. Mice in the wild show rhythmic population numbers, going up to a certain optimum level, followed by a rapid decline and an increase to an optimum again (Baumans 1999).

In almost all laboratories, mice are housed together after weaning in single sex groups. Although this is not natural, male mice in particular will form stable hierarchies in many cases. In other cases fighting may occur and may lead to stress and injuries. A considerable difference in aggression levels between mouse strains has been described (Nevison et al. 1999). Individual housing of this highly social species frequently has been reported to be stressful, resulting in 'isolation syndrome' (Brain 1975). Male BALB/c mice preferred the proximity of another male, either dominant or subordinate (Van Loo et al. 2001a). It was shown that male BALB/c mice are best housed in groups of three animals. Larger group sizes may decrease the chance of a stable hierarchy developing, while pair housing may increase aggression, and lack of social comfort may induce depression-like symptoms in the subordinate. (Van Loo et al. 2001b).

Other factors playing a role in modulation of aggression are the presence of enrichment in the cage, hiding places in particular, and the degree of disturbance in the group. Social housing is also important when mice are submitted to situations which might be stressful, e.g. husbandry or experimental procedures. It was found that after intraperitoneal injections, individual housing slowed down recovery of tachycardia (increased heart rate) in two strains of mice (Meijer et al. submitted).

2.2 Senses

Mice have in principle the same senses as humans, but the importance of each differs.

2.2.1 Hearing

Mice can hear over a broad spectrum of frequencies from 80 Hz up to 100 kHz, but are most sensitive in the range of 15 –20 kHz and around 50kHz. They respond to a range of ultrasonic frequencies; for example the retrieval response of the female is elicited by ultrasonic cries emitted from pups, out of the nest. Common laboratory equipment such as computer screens, running taps and pressure hoses can emit ultrasound (Jennings et al. 1998) and may have considerable effects on the animal's well-being.

2.2.2 Olfaction

The sense of olfaction is highly developed in mice, and is not only used to detect food and predators. Mice also have a wide repertoire of olfactory social signals including urinary odor cues, conveying individuality and status information (Nevison et al. 1999). It should be questioned for this reason whether laboratory mice should be kept in an animal room together with the other sex and / or other species, in particular rats.

Cage cleaning is a necessary routine procedure in laboratory animal facilities. However, removal of the olfactory cues disturbs the social hierarchy of the animals in the cage, often resulting in a peak in aggression among males. Olfactory cues from nesting and bedding material affect aggression in a different way: transfer of nesting material reduced aggression, whereas sawdust containing urine/feces intensified aggression (Van Loo et al. 2000). Mice generally deposit their urine at specific sites in the cage either for hygienic reasons or as signals involved in social communication (Blom 1993, Sherwin 2002).

2.2.3 Vision, touch and taste

Vision: Although mice have good vision, this sense might be less important than others (Sherwin, 2002). They avoid brightly lit areas and, especially in albinos, high light intensities can lead to retinal damage (Van de Weerd and Baumans 1995).

Touch: Mice avoid open spaces and generally keep contact with the wall (thigmotaxis), using their vibrissae (whiskers). Whisker-trimming or barbering, often observed in several mouse strains, where one or more mice trim their cage mates' whiskers may result in a partial loss of the sense of touch (Figure 7-7).

Taste: Wild mice eat a wide range of foods such as seeds, vegetables, fruit, bread (Jennings 1998). In addition to the standard pellets in the laboratory, it might be beneficial to provide mice with such foods, also to meet the need for foraging.

2.3 Life span

Many environmental and genetic factors influence the life span of the mouse. These include diet, number of animals per cage, subclinical infections, husbandry procedures, genetic predisposition to tumors, strain, sex and the presence or absence of deleterious mutant genes. Mice from short-lived strains can be expected to die between 5 and 16 months of age. Mice from long-lived strains often survive to 24, 30 or even 36 months of age (Baumans 1999).

2.4 Reproduction

Sexual maturity in mice occurs very early in life and varies with strain and environmental influences. Ovarian follicle development begins at 3 weeks of age and matures by 30 days. Puberty in males occurs up to two weeks later. Female mice are poly-estrous, spontaneous ovulators and cycle every 4-5 day (Baumans 1999).

Factors such as season, diet, genetic background and environmental factors influence the estrous cycle. The cyclicity of estrus and ovulation are controlled by the diurnal rhythm of the photoperiod. Mating, estrus and ovulation most often occur during the dark phase (Fox et al. 1984). Light cycles of 12-14 hours light and 12-10 hours dark are necessary to maintain regular estrus cycles. Light intensities either too low (5 lux) or too high (250 lux) in animal rooms can affect the estrous cycle. Circadian rhythms are especially sensitive to interruption of the dark phase by short periods of light. (Clough 1984). Elevated environmental temperatures (> 28 °C) reduce fertility, whereas lower temperatures tend to slow growth rates and delay puberty. High noise levels reduce female fertility (Cunliffe-Beamer and Les 1987).

Pheromones and the social environment also affect the estrous cycle. Female mice housed in groups will become di-estrous, an-estrous or pseudopregnant, while the introduction of a male into such a group will synchronize their estrous cycles (the Whitten effect). If the female is housed with a second male within 24 hours after a successful mating, implantation of fertile egg cells will be prevented and the female will return to estrus in 4-5 days (the Bruce effect). These effects are due to pheromones in the urine of the males (Baumans 1999).

Mating can be detected within 24 hours after copulation by the formation of a waxy vaginal plug (a mixture of sperm and secretions from the seminal vesicles and the coagulating glands of the male). The gestation period is 19-21 days, depending on the strain.

The effective reproductive life of a female mouse approaches 2 years but, as litter size decreases with aging, females are usually retired by 1 year of age. Litter size commonly ranges from 1-14 pups, depending on the strain.

Pregnant females build nests for giving birth. Unnecessary disturbance or manipulation of a post-parturient female and her litter should be avoided for the first few days following birth. If handling of neonates cannot be avoided, the risk of cannibalism or rejection of the litter can be reduced by gentle handling of the dam and offspring, placing the dam in a separate cage while the litter is being handled, and wearing plastic gloves in order to prevent the neonates from acquiring human scent.

Cannibalism (which is strain-dependent) can be minimized in most cases by providing a quiet place with reduced light intensity and nesting material and in case mother or pups have to be separated for a while, by rubbing the pups with bedding from the home cage.

Nursing females usually lactate for 3 weeks. The 'milk spot' in the pups (the stomach filled with milk, visible through the transparent skin) indicates sufficient milk uptake in the first 4-5 days after birth (Figure 7-2). If the dam dies or lactation fails when the litter is about 14 days of age, placing a moistened or soft diet and water in the cage may improve survival of pups.

Figure 7-2. Milk spot in infant mice. Photo: P. Rooijmans

Sex, strain, age, reproductive phase and environment can have a dramatic influence on physiological and behavioral data. Body weight and growth curves are also influenced by the above-mentioned factors. Body weights of both sexes increase rapidly during the first 6-8 weeks, but they grow more slowly afterwards until 6 months of age, when a plateau has been reached for a few months, followed by a decline (Baumans 1999).

3. OPTIMAL ENVIRONMENT

The laboratory mouse has partially adapted to captive life, but still shows similarities to its wild counterparts. The animal's environment should cater for physiological and behavioral needs, such as resting, nest building, hiding, exploring, foraging, gnawing and social contacts.

Mice are highly susceptible to predators and are thus likely to show strong fear responses in unfamiliar situations if they cannot shelter, including attempts to flee, biting when handled or sudden immobility to avoid being detected. Careful handling from youth onwards together with conditioning to experimental and husbandry procedures is likely to reduce stress responses considerably (Hurst 1999). For this reason, cages should be provided with shelter / hiding places. Ideally, the animal should feel secure in a complex, challenging environment, which it can control (Poole 1998). Security can be achieved by nestable and manipulable nesting material, hiding places and compatible cage mates (Figure 7-3).

Figure 7-3. Enrichment for mice. Left: after cleaning, right: in use. Photos: M. Meijer

One of the possibilities for improving living conditions of laboratory animals is to provide opportunities for the animals to perform a more species-specific behavioral repertoire by providing environmental enrichment. Environmental enrichment or modifications to the environments of animals in order to improve their biological functioning (Newberry 1995), has been increasingly introduced into laboratory animal research facilities (Olsson and Dahlborn 2002). From a welfare point of view this seems to be a good development as it is generally accepted that with the provision of environmental enrichment the animal's well-being improves.

It has been shown that barren, restrictive and socially deprived housing conditions interfere with the development and function of brain and behavioral functions. Beneficial effects of environmental enrichment have been described in animals with brain damage and disturbed motor function and an increased arborization of dendrites has been seen in the brain (See Mohammed et al. 2002).

A number of aspects of the animal's environment can be identified for enrichment. These include the social environment (conspecifics and human beings), and the physical environment, consisting of sensory stimuli (auditory, visual, olfactory and tactile) and nutritional aspects (supply and type of food). Furthermore, there is the psychological appraisal of the environment with aspects such as controllability and predictability (Van de Weerd and Baumans 1995), which can be increased by structuring the cage with nest boxes, tubes, partitions and nesting material. Van de Weerd (1996) showed that nesting material such as tissue was highly preferred by mice.

Environmental enrichment should meet the animal's needs, be practical, inexpensive and pose no risk to humans, the animals and to the experiment. Enrichment items should be designed and tested, mainly based on knowledge gained in enrichment studies.

There is concern whether or not environmental enrichment conflicts with the standardization of experiments. Standardization increases the reproducibility and comparability of experiments. It aims at reducing unwanted variation caused by animal and environmental factors and at the reduction of the number of animals needed in experiments. Some researchers fear that 'enriched' animals show more variability in their response to experimental procedures because they show more diverse behavior. In complex environments animals are not just responding to one stimulus in isolation but to many variable stimuli at once. This may cause increased variation within subjects (Eskola et al. 1999). However, one might argue that because an animal can perform more of its species-specific behavior in enriched environments, it may be better able to cope with novel and unexpected changes and thus show a more uniform response. Animals from enriched housing conditions are expected to be physiologically and psychologically more stable and they may therefore be considered as more refined animal models, ensuring better scientific results. Results from different studies indicate that the effects of enrichment on the variability of results are dependent on the parameter, type of enrichment and strain (Van de Weerd et al. 2002).

4. HOUSING, FEEDING AND CARE

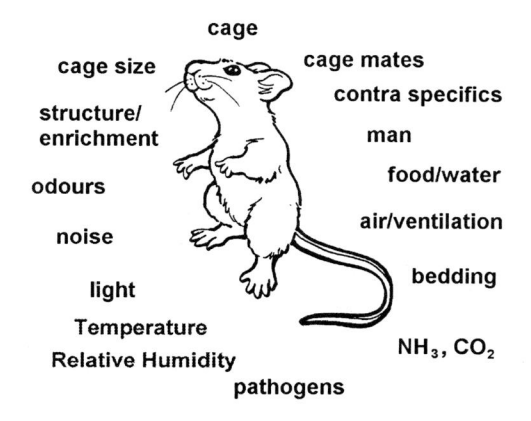

Figure 7-4. The environment of laboratory mouse

Housing systems for laboratory animals have often been designed on the basis of economic and ergonomic aspects (such as equipment, costs, space, workload, ability to observe the animals and to maintain a certain degree of hygiene) with little or no consideration for animal welfare. The environment of an animal consists of a wide range of stimuli, including the social environment of conspecifics (same species), contraspecifics (other species) and humans, and the physical environment such as the cage and its contents (Van de Weerd and Baumans 1995) (Figure 7-4).

Environmental conditions such as housing and husbandry have a major impact on the laboratory animal throughout its life, not only during the experiment itself, but also before and after the experiment. The traditional care and maintenance of laboratory animals does not usually consider the species-specific needs in relation to housing and feeding regimes. The variability in the specific needs is not only different between species but also, due to variability in the genetic background, among strains of a species. However, laboratory mice are usually housed throughout their lives in relatively barren cages and provided with *ad libitum* food, which frequently results in adverse effects on the animal's behavior and physiology and in a shortened life span due to overfeeding and inactivity (Van de Weerd et al. 1994, 1997b, Mattson et al. 2001). Environmental enrichment has been increasingly introduced into laboratory animal research facilities. From a welfare point of view this seems to be a good development as it is generally accepted that with the provision of environmental enrichment the animal's well-being improves (See 'Optimal environment' in this chapter).

Laboratory mice are usually bred and housed in cages made of polycarbonate or polypropylene with solid floors (Figure 7-5). Wire-mesh floors are used if experiments require continuous collection of feces and/or urine or elimination of contact between the animal and bedding material. However, since mice are nesting animals, they prefer solid floors with sawdust for digging, nesting and resting rather than grid floors (Blom et al. 1996). Solid floor cages containing bedding and nesting materials should be used whenever possible (Council of Europe 1997).

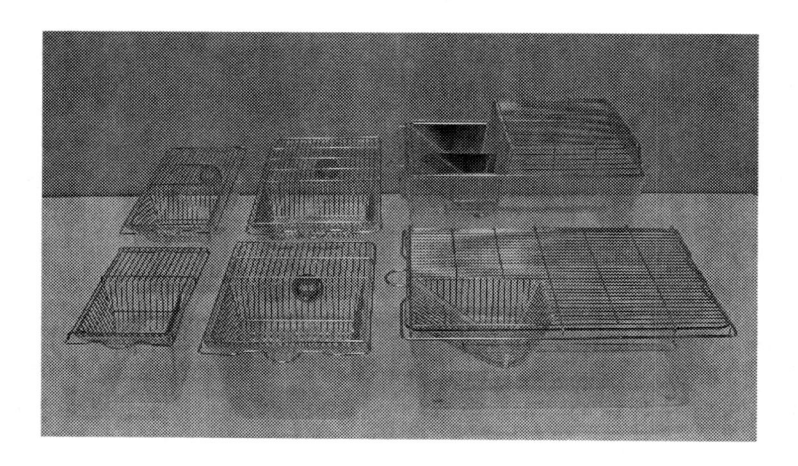

Figure 7-5. Polycarbonate mouse cages. Photo: P. Rooijmans

Bedding is comfortable for resting, absorbs urine and is used by the animals to establish odor patterns. There should be a balance in terms of cleaning between on the one hand increasing the anxiety of the animals by removing familiar odor patterns and on the other the build-up of harmful ammonia. Wood shavings are widely used as bedding material, but fine particles can cause preputial and respiratory disorders. The type of wood can affect physiological parameters in the animal, such as hepatic microsomal enzyme function (Haataja et al. 1989). Hygroscopic material should be avoided as it may cause dehydration in the newborn.

Nesting material, such as paper towels, tissues, and wood wool, can provide shade from lighting, the opportunity to regulate the animal's microclimate, a shelter to hide from conspecifics and the ability to control the environment (van de Weerd and Baumans 1995) (Figure 7-6).

Bedding and nesting material should be in accordance with the mouse's needs - not toxic or harmful to the animal, absorbent, but not dehydrating for neonates, not excessively dusty, economical to use and dispose of, and interfering as little as possible with experiments (Baumans 1999).

Figure 7-6. Nest of BALB/c mice. Photo: P. Rooijmans

4.1 Space recommendations

It is difficult to scientifically specify the minimum cage sizes for maintaining laboratory mice, as much depends on the strain, group size, age of the animals, their familiarity with each other and their reproductive condition. Specifications for laboratory housing of mice are expressed in two documents issued in 1986. One is the European Convention for the protection of vertebrate animals used for experimental and other scientific purposes (Convention ETS 123) from the Council of Europe, with its Appendix A, Guidelines for the Accommodation and Care of Animals (Council of Europe 1986). The other is the very similar European Union Council Directive on the approximation of laws, regulations and administrative provisions of the Member States regarding the protection of animals used for experimental and other scientific purposes (Directive 86/609/EEC) with its Annex II, Guidelines for Accommodation and Care of Animals (European Council Directive 1986).

Article 5.1 of the Convention requires that "Any restriction on the extent to which an animal can satisfy its physiological and ethological needs shall be limited as far as practicable", while Article 5b of the Directive requires them to "be limited to the absolute minimum". These guidelines are mainly based on empirical considerations and are now under revision. According to the NRC Guide for the Care and Use of Laboratory Animals (1996) 'animals should be housed with a goal of maximizing species-specific behaviors and minimizing stress-induced behaviors'. Space recommendations should allow housing of mice in harmonious groups. Increasing the complexity of the cage is more important than increasing floor area as such, as the inclusion of structures will provide opportunities for activity and will increase the usable space. Furthermore, mice partition their space into sleeping, defaecating, urinating and feeding areas, so the available space and structure in the cage should be sufficient. Incentives for activity should be provided in the cage, such as nesting material and climbing possibilities. (See also the Chapter 4 in this book)

4.2 Climate in the animal room

Although environmental measurements such as temperature, relative humidity, NH_3, CO_2, light and noise are taken at room level (macro-environment), it is the micro-environment in the cage that actually affects the animal. Mice 'engineer' their own micro-environment by huddling and building nests and in this way they are able to exert some control over temperature, humidity and light conditions. Furthermore, the micro-environment will vary, depending on the position of the cage on the rack, crowding of the rack, cage type, stocking density in the cage, type and amount of bedding and cleaning frequency. Also the use of filter caps and Individually Ventilated Cages (IVC) will affect the rate of air exchange, build up of ammonia, temperature and humidity (Serrano 1971, Baumans et al. 2002). IVC systems are designed to reduce infection risks, NH_3 and CO_2 concentrations, cleaning frequency, workload and allergens (See Figure 4-4 in Chapter 4). However, the high ventilation rate and less frequent handling of the animals might be a drawback. To protect the animals from draught due to the forced ventilation, nesting material or a hiding place should be provided (Baumans et al. 2002).

When the environmental temperature falls below the thermoneutral zone (where body temperature is regulated by conduction and convection of heat) body temperature is maintained by behavioral adaptation, such as nest building and/or increased metabolic rate. Mice have a relatively large surface area per gram of bodyweight. This results in dramatic physiological changes in their response to fluctuations in the ambient temperature (Fox et al. 1984). There is evidence that the tail has a thermoregulatory function in small rodents (Working Group on Refinement 1993). Providing nesting material gives mice the opportunity to regulate their body temperature (Van de Weerd et al. 1998). Environmental temperature has been demonstrated to influence reproduction, organ weight, food and water intake and hematological parameters. These data indicate that the optimal temperature range for mouse rooms is 20-26 °C (Yamauchi et al. 1983). According to the European Directive the environmental temperature requirements range between 20-24 °C. Slightly higher environmental temperatures (22 to 24 °C) should be considered if the mice have to be housed in wire bottom cages because of greater air exchange between the cage and the room, or if hairless mice are being housed in cages with limited amounts of bedding. Blom (1993) studied preferences of mice for temperature and found that individual BALB/c mice preferred cage temperature of 28 °C, whereas C57BL/6 preferred 24 °C and that preference depended upon the type of cage flooring.

Relative humidity should be kept at 55±10% (Council of Europe 1986). At low humidity levels, mice may suffer from respiratory problems due to dust from bedding and feed, and drying of mucous membranes. At high humidity levels, bedding may fail to evaporate moisture leading to rapid soiling of cages and increased production of ammonia by urease-producing bacteria. Furthermore, infections of the upper airways may occur more easily.

Lighting conditions for the essentially nocturnal mouse are very important. Mice exposed to too high an intensity of light may show retinal atrophy in the long term (Clough 1984). Animals should be subjected to a regular light: dark cycle, generally 12 hours light, 12 hours dark, but there is some debate concerning the beneficial effect on the animal of an artificial 'dawn' and 'dusk' period. A reversed lighting schedule, with lights on at night, can be useful for observing activity of rodents during the day; red light can be used to facilitate daytime husbandry as most rodents are less sensitive to red light than other animals and humans (Spalding et al. 1969). Mice show a preference for a low light intensity and the retina of albino animals may be easily damaged by strong light. Maximum light intensity in the room should not exceed 350 lux and the intensity levels within the cage should be lower, or the animal should be given the opportunity to withdraw to shaded areas, such as a shelter or nest (O'Donoghue 1993).

Many noises audible to the human ear and ultrasound (frequencies higher than the human range) are important for rodents (see also 'hearing'). Mice are more likely to be disturbed by high-pitched sounds and ultrasounds than they are by lower frequencies; care must be taken not to use equipment that emits these ultrasounds, such as electronic devices and computer-screens (Clough 1984, Sales et al. 1999). It has been suggested that a constant background noise, such as radio music has some benefits in facilitating breeding and making animals less excited, although it could also stress some animals. However it may benefit the personnel more, which could have beneficial consequences for the animals in turn (Sherwin 2002). It certainly should not exceed 55 dB (O'Donoghue 1993).

4.3 Social housing

Mice are gregarious animals, therefore it is preferable to keep them in groups rather than in individual housing, but the groups must be stable and harmonious. Allocating animals to new, unfamiliar groups may be a source of intense, stressful conflicts (Brain 1990). Much depends on the animal's familiarity with its cage mates as well as its sex and age at the time of grouping (O'Donoghue 1993) (See also 'Social behavior' in this chapter.) Individually-housed mice are still provided with olfactory, auditory and visual cues from conspecifics in the room, whereas isolated animals are also deprived of this input.

Whisker chewing ('barbering' or 'trimming') occurs more and more frequently in some strains of mice in both sexes (Figure 7-7). Usually one of the mice retains its whiskers and it is assumed that these animals are dominant ones, but this is not conclusive (Van de Weerd et al. 1992). Removal of the barber mouse may lead to another mouse taking over the barbering role. Barbering may not be restricted to whiskers; head and body fur may be involved also. The etiology is not clear. Genetic factors may play a role as well as boredom (Van den Broek et al. 1993).

Figure 7-7. Whisker trimming. Photo: P. Rooijmans

4.4 Food

In ancient Rome dormice were served at important banquets. The most delicious were those fattened on nuts and immobilized in special earthenware jars (Young 1987) - probably the world's first battery mice!

The mouse is an omnivorous animal. The incisors and molars grow continuously and are worn down by mastication. Attention should be paid to malocclusion leading to under-nutrition. Feeding behavior in rodents shows a diurnal pattern with the majority of food consumed during the dark period (Harris et al. 2002). Fasting overnight, which is sometimes part of an experimental protocol, might lead to an increase of activity, resulting in unwanted variation in experimental results.

Food is mostly presented *ad libitum* as pellets in the food hopper on the cage, which prevents soiling of food by the animals. The food rack should be kept sufficiently full, as it is difficult for the animals to gnaw the food when there are only a few pellets left. Restricted feeding has been shown to be beneficial in the long run in terms of reduced morbidity and mortality (Pugh et al.1999, Mattson et al. 2001).

Enrichment related to food, e.g. grain scattered through the bedding, will meet the animal's need for foraging and will prevent boredom (Baumans 2000), although it might interfere with experiments in some cases. It was shown that BALB/c and C57BL/6 mice kept in enriched environments with nesting material weighed more than mice housed under standard conditions, although the latter consumed more food, probably due to the insulating effect of the nest or to reduced boredom (Van de Weerd et al. 1997b)

Water can be provided in a bottle attached to each cage or by an automatic watering system. Water bottles should be changed at least once a week. An automatic watering system supplies water through a valve connected to a piping system and serves a rack of cages from a central reservoir. The water valve can be located either outside or inside the cage. There is always some risk that the valve may become obstructed, resulting in dehydration of the animals or flooding in the cage. Automatic watering systems must be monitored to be certain that the water pressure is adequate and must be regularly cleaned and checked for contamination with bacteria and fungi.

4.5 Transport

Transport of mice, even over short distances is a stressful event for the animal. For example, immune functions are adversely affected by shipment and require at least 2-5 days to return to normal. Reproductive performance can also be decreased. The extent to which the well-being of the animal is compromised depends upon its health status, age (old and very young animals are more susceptible), stocking density in the cage, conditions in the transport vehicle and cage, such as temperature, ventilation, food and water supply, and the acclimatization period (5-7 days) in the new environment.

Mice that are being moved over a short distance within the same facility can be transported in a clean cage or cardboard carton. For longer journeys, the cardboard container should be coated with moisture impervious material. Ventilation openings should be present on at least the two opposite sides and fitted with wire mesh and filters to prevent escape or contamination. Bedding should always be supplied for comfort and to absorb urine. Food can be provided as pellets, water as 'solid' water e.g. agar, apples, potatoes or commercially available jelly.

5. HEALTH AND DISEASE PROBLEMS

5.1 Health

Good health is not only crucial for the animal's well-being, but also for the quality of science, as scientific data from animal experiments can only be reproducible and reliable when obtained from healthy animals. Therefore the health status of the animals should be defined. Microbiological quality control aims at control of the barrier system and the animal itself. Only careful monitoring of animal facilities will provide useful information about the microbiological quality of the animals, which is important as infections may interfere with experimental results (Boot et al. 2001, see also Chapter 3 in this book).

The design of individual monitoring programs in experimental units is dependent on the research objectives and factors such as housing conditions (conventional, isolator, barrier-housed, micro-isolators), the type of experiment (long-term, short-term), frequency of introducing animals and other biological materials, and the importance of a specific pathogen or the likelihood of interference with research.

In addition to welfare considerations, the main aim of health monitoring before and during experiments is to define the microbiological status of the animals in order to take into consideration the presence or absence of certain micro-organisms, lesions and other alterations as experimental variables.

Sometimes the environment has to be protected against contamination by micro-organisms carried by the animal. For that reason, animals can be housed in a 'reversed' version of the classical barrier system, where the animals have to be protected against micro-organisms from the environment.

For short-term experiments, animals can be housed in cages with a protective hood (filter top) or IVC racks, which are only opened within Laminar Air Flow cabinets.

5.2 Diseases

Genetic and environmental factors play an important role in the causation of disease. Genetic factors are involved in the susceptibility to disease, for example differences between mouse strains in tumor incidence. Environmental factors e.g. nutrition, husbandry, infections and other noxious agents can contribute to the causes of disease. Hygiene and improvement of the microbiological status will reduce the incidence of infections considerably. Good management with health monitoring programs, knowledge of the biology of animals and epidemiology of diseases is a prerequisite.

An important concept in the well-being of animals is homeostasis, which implies that the animal is in harmony with its internal and external environment and is able to keep that environment controllable and predictable. When homeostasis cannot be maintained, discomfort or stress may occur, which can become manifest as a disease or abnormal behavior, such as the development of stereotypies.

Stereotypies can be defined as repeated, simple behavioral patterns, which seem to be meaningless and are typical of the individual animal, such as circling movements or constant jumping in the cage. Minimal responses or lack of reaction to stimuli (apathy) can also be considered as abnormal behavior. Together with clinical and physiological parameters, abnormal behavior may indicate that the well-being of the animal is compromised (Baumans 1999).

The signs associated with disease, besides abnormal behavior, include isolation from the group, a hunched-up position, ruffled fur, sunken eyes, reduced growth rate and bodyweight, diarrhea, 'chattering' (a clicking sound associated with severe respiratory infections), labored breathing, cyanotic or pale ears and paws, reduction in the number of offspring born or weaned, reduction in the number of breeding pairs producing offspring or increased mortality (Figure 7-8). However, many infectious diseases of mice cause almost no obvious clinical signs. These latent infections can have a significant impact on experimental results, since they can alter histology or immune responses.

The signs associated with an infection, especially viral infections, vary with strain, age, microflora and immune status of the mouse, previous exposure to the infectious agent and the strain of the infectious agent. To make a correct diagnosis of a disease, specific skills and experience in microbiology and/or pathology are needed.

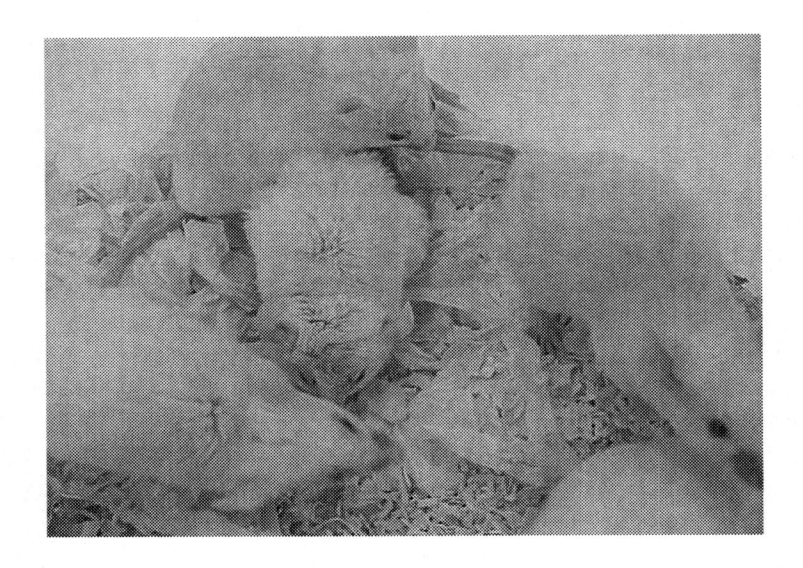

Figure 7-8. A sick mouse in the middle, accompanied by cagemates. Note the ruffled fur and hunched-up position. Photo: P. Rooijmans

A complete diagnostic examination should include: 1) the history of the disease (sex, age, strain, microbiological status, source, use, performed procedures) 2) environment 3) clinical signs such as general condition, respiratory system, circulatory system and locomotion; additional information can be obtained from blood samples, urine, skin scrapings or post mortem and microbiological examination (e.g. histology, detection of antibodies) (Van Dijk 2001).

6. COMMON EXPERIMENTAL TECHNIQUES

6.1 Identification and sexing

Many permanent or temporary methods of identifying individual mice have been developed. For permanent identification tattooing, for example on the tail or toe, may mark mice. Toe clipping is not recommended on humanitarian grounds (Baumans 1999). Ear marking by punching may be obliterated or damaged by fighting. Temporary methods of identification include pen marks on the tail, clipping or plucking unique patterns in the fur, which may be visible for about 14 days, or dyeing the fur with a harmless dye, such as food dyes. Subcutaneous implantation of a microchip is a safe, long lasting, but expensive method of identification. Tail clipping is commonly used in genetically-modified mice in order to take tissue samples from the offspring for identifying their genetic makeup. It can be argued whether this procedure causes lasting harm in terms of pain and disturbance of the thermoregulatory function of the tail.

The sex of the animal can be determined by comparing the anogenital distance, which is larger in males than in females. In females, a hairless strip is visible between anus and genital papilla (Fig. 7-9). Testes can be present in the scrotum, but they can also be retracted through the inguinal canal into the abdomen, especially when the animal is scared or stressed (Baumans 1999).

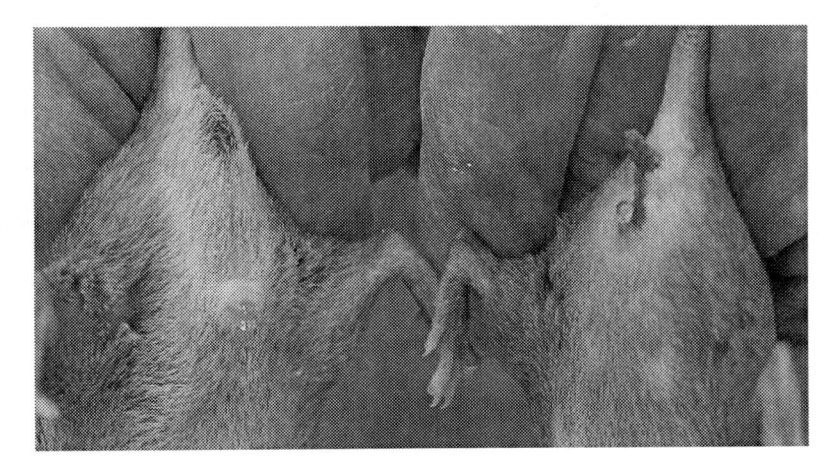

Figure 7-9. Sex difference. Left: male, right: female. Photo: P. Rooijmans.

6.2 Handling and restraint

Mice can be caught inside the cage by grasping the base of the tail with the thumb and index finger and picked up by the tail. The tail must not be held by the tip as this may cause the skin to be sloughed off. If the animal needs to be restrained, it should be placed on a rough surface, such as the cage lid. The loose skin of the neck between the ears can be held between thumb and index finger of the other hand. The mouse is lifted and the tail is secured between the fourth or fifth finger and the palm of the hand. This technique leaves one hand free for injections or other procedures (Figure 7-10). One should be careful not to pull the skin too tightly or the mouse will choke. The color of the mucous membranes of nose and mouth should always be checked for this. Plastic or metal devices can be used to restrain mice for procedures such as tail vein injections, blood pressure measurements or irradiation.

Strains of mice differ in temperament. Some strains are very calm and rarely try to bite or escape (e.g. BALB/c) while others are 'touchy' (e.g. C57BL). All mice go through a 'jumpy' stage at about three weeks of age. Working with mice should always be done quickly, quietly and gently. Training mice to be handled might reduce stress, especially when the weekly handling procedure during cage cleaning cannot be achieved e.g. in IVC systems.

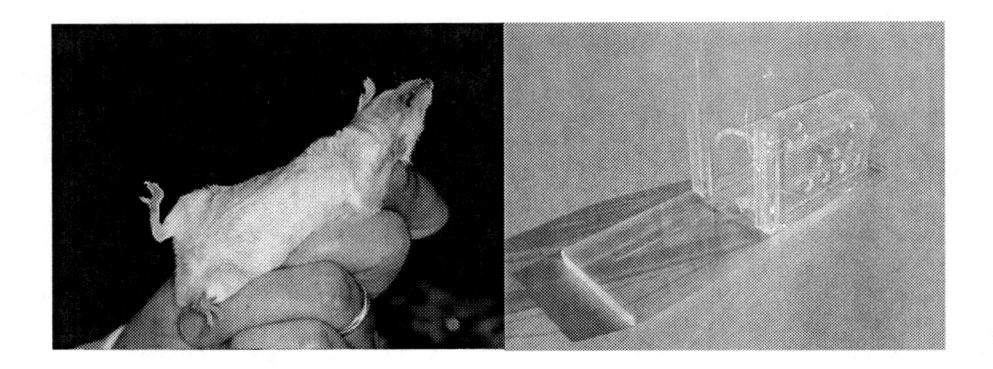

Figure 7-10. Restraining a mouse and restrainer for mice. Photo: P. Rooijmans

6.3 Recording body temperature and heart rate

The rectal temperatures of mice vary between 36.5-38 °C and can be measured by inserting a thermocouple into the rectum. Rectal temperature is influenced by environmental temperature, age, strain and stress (e.g. restraint). Core temperature can also be measured by using telemetry. Although the transmitter, which has to be placed in the abdominal cavity, is still rather big for a mouse, it has been shown that behavior and bodyweight return to normal within 10 days after implantation. (Baumans et al. 2001a). This method is also used to record e.g. heart rate, activity and blood pressure. Once the device has been implanted, measuring seems to be a stress free method in the freely moving animal (Kramer et al. 1993).

6.4 Collection of body fluids

6.4.1 Collection of blood

Blood samples can be obtained from various sites of the body, using a variety of methods. When repeated blood sampling is required, the implantation of indwelling cannulae should be considered. The choice of the method will depend upon several factors, such as the purpose of blood collection (arterial, venous or a mixture of the two), the duration and frequency of sampling and whether or not it concerns a terminal experiment. Samples obtained from various methods will not be identical with respect to cellular or other constituents. Anesthesia may be required for some methods.

Small volumes of blood (for example for a smear) may be obtained by snipping off the very tip of the tail, which will result in a mixture of venous and arterial blood, together with tissue fluid. This method is suitable when small samples are required every few hours. The scab on the tail wound can easily be removed. However, it can be argued that the method might be painful and requires pain relief and that the tail is important for thermoregulation in the mouse (BVA/FRAME/RSPCA/UFAW Joint Working Group, 1993).

The lateral tail vein can be punctured with a 0.6-1.0 mm needle after warming of the tail in warm water, under a heating lamp or in a climatic chamber (Figure 7-11). The mouse can be anaesthetized or placed in a restraining device. The ventral tail artery can be used in the same way. Using razor blades to cut the vein may prolong bleeding and create scar tissue. The lateral saphenous vein can be punctured in the mouse restrained in a piece of cloth (Figure 7-11).

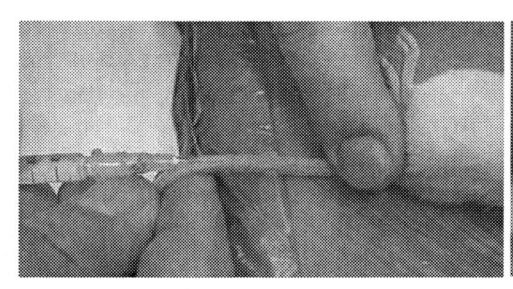

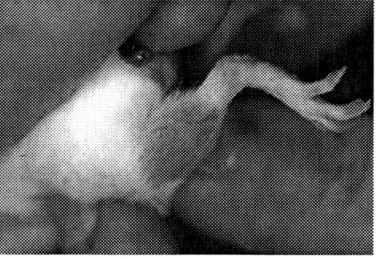

Figure 7-11. Lateral tail vein (left) and lateral saphenous vein (right). Photo: P. Rooijmans

Blood can be aspirated from the jugular vein, mostly by surgical exposure of the vein under anesthesia, which provides a sterile blood sample.

Arterial blood can be obtained from the abdominal aorta, brachial artery or carotid artery under anesthesia.

For puncture of the orbital blood vessels anesthesia is required. The animal is firmly held by the skin at the nape of the neck, which causes distention of the jugular vein. A fine glass tube or a Pasteur's pipette is then placed at the inner canthus of the eye and gently advanced alongside the globe into the vessels. The tube ruptures the vessels and blood can be withdrawn by capillary action. Contamination with tissue fluids and porphyrins from the Harderian gland can occur. It is not possible to take sterile blood samples using this method. Also complications, such as hemorrhage, inflammation and blindness may occur, especially when the same eye is used repeatedly. Moreover, this technique may be aesthetically unpleasant for the operator to perform. For these reasons this method is not considered acceptable in some countries (Baumans et al. 2001b). However, the extent of trauma is proportional to the skill of the technician (Van Herck et al. 1998).

With cardiac puncture, blood is collected directly from the ventricle of the heart of the anaesthetized animal. Puncturing the atrium can be dangerous, due to the risk of leakage into the pericardium, resulting in cardiac arrest and death. When the animal has to survive the procedure, these risks must be taken into account. The total volume of blood in a living mouse is rather constant and is about 8% of its body weight. If the sample volume exceeds 10% of the total blood volume, hypovolaemia and cardiovascular failure ('shock') may occur. As a general rule, a maximum of approximately 8ml/kg body weight can be removed per two weeks from a live mouse (Baumans et al. 2001b; Pekow and Baumans 2002).

Implantation of indwelling cannulae can be used when repeated blood sampling is required. The indwelling catheter can be implanted via the jugular vein into the cranial vena cava and exteriorized at the top of the head, where it is secured with screws and acrylic glue or between the shoulders. The catheter's dead space is usually filled with polyvinyl pyrolidone, in saline with heparin (PVP-solution). Although initially surgery is required, blood sampling itself is considered as not invasive.

To obtain at the end of an experiment, decapitation can be performed using a guillotine or scissors, or by puncturing the aorta under anesthesia. In this way about 30 ml/kg body weight or up to 50% of the total blood volume can be obtained. Concerning decapitation, it has been demonstrated that rats loose consciousness within 3 seconds after decapitation (Haberham et al, 2000). Exsanguination is also possible by removing the eyeball under anesthesia and collecting blood from the eye artery.

6.4.2 Collection of faeces, urine, milk and sperm

Mice often urinate and/or defecate purely as a result of being handled, which may provide the opportunity for collecting small samples of urine and feces. Placing the animal in a plastic bucket will result in urination and urine can be collected with a syringe (Van Loo et al. 2001b). Metabolism cages must be used for the quantitative collection of urine and feces. The animals are housed on a grid above a funnel in which the urine and/or feces are collected and separated. Stress from being housed individually on grid floors should be considered.

Mouse milk can be collected using milking 'machines' with single or multiple teat cups. Prior to milking, the lactating female is separated from her litter for 8 to 12 hours. Then, the mammary glands are washed with warm water and the mouse may be injected subcutaneously with oxytocin to stimulate milk flow.

Sperm for artificial insemination is usually obtained from the epididymides of a recently killed male mouse. However, electro ejaculation of male mice has been described. A balanced salt solution with glucose should be used to dilute mouse semen (Cunliffe-Beamer and Les 1987).

6.5 Administration of substances

Three methods for the administration of substances can be distinguished: via skin, enteral and parenteral (Baumans et al. 2001b).

In case of application to the skin or mucous membranes, the substance is applied in solution or as ointment on shaven skin or mucous membranes, but this method is inaccurate and may cause discomfort to the mouse, when the substance is irritating.

Enteral administration means that the substance is brought into the gastrointestinal tract orally through food or drinking water, which is not very accurate and impossible when the substance is unpalatable, insoluble or chemically unstable. To stimulate the consumption of for example medicines or other substances, these can be administered in fat drippings or yogurt sweets, which are preferred by mice.

Administration via a stomach/esophageal tube is more accurate and can be performed by a curved needle with a blunt end (external diameter 0.8 mm), such as an infant lacrimal sac cannula. The mouse is held firmly by the scruff whilst passing the needle along the palate into the esophagus. The animal should be allowed to 'swallow' the needle to prevent damage to the pharynx and esophagus. As the head is directed upwards, the risk of entering the trachea is minimal. The maximum volume to be administered should be 0.25-0.5 ml (Baumans et al. 2001b).

Parenteral administration refers mainly to injections. The most frequently used injection techniques in the mouse are:
1. *Subcutaneous (s.c.):* usually under the skin of the neck. Resorption of the substance is slow. Maximum volume to be administered is 0.25-0.5 ml.
2. *Intramuscular (i.m.)* injections are usually avoided in the mouse because of its small muscle mass. If necessary, intramuscular injections of 0.05 ml or less may be made into the lateral thigh muscles. The tip of the needle should be directed away from the femur and sciatic nerve.
3. *Intraperitoneal (i.p.)* injections may cause damage to the internal organs. To avoid the urinary bladder the injection should be given slightly off the midline. The needle should neither be inserted horizontally (between the skin and the abdominal wall) nor vertically (risking damage to the kidney). The injection should be given in the lower left or right quadrant of the abdomen. There is only a slight risk of causing damage to the intestines (mainly due to their mobility) (Figure 7-12). To be sure that the procedure has been carried out properly, the plunger must be retracted to determine whether any urine, intestinal contents or blood have been aspirated. The maximum volume to administer is 0.5-1.0 ml (Baumans et al. 2001b, Pekow and Baumans 2002).
4. *Intravenous (i.v.)* injections into the tail vein is usually the fastest and most accurate method, but requires more technical skill. For preparation of the vein see paragraph 'Collection of blood'.

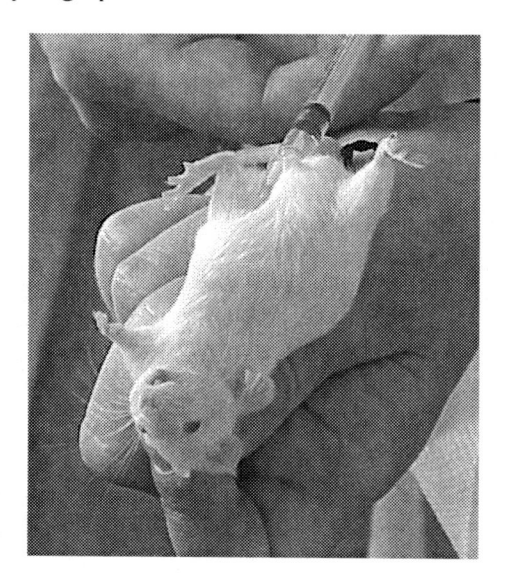

Figure 7-12. Intraperitoneal injection. Photo: P. Rooijmans

Points of special attention when carrying out injections are:
1. Use clean, sharp, short and sterile needles.
2. Use an appropriate needle size. A thin needle causes less pain and prevents the fluid from flowing back. The required thickness of the needle (gauge) will depend upon the viscosity of the fluid. When using very thin needles, there is a risk of breaking. On the other hand larger needle sizes for viscous injections require shorter injection time and because of shorter restraint time appear to be less distressing to the animal (Barclay et al. 1988).
3. Never inject more fluid than the recommended maximum volume.
4. Avoid air bubbles in the injection fluid (which can cause embolism).
5. The injection fluid must be brought to room or body temperature prior to use.
6. Some injection fluids can cause tissue irritation, for example, high or low pH, and should therefore be administered after having been diluted with saline. These substances should, preferably, be given intravenously, as they quickly become diluted in the blood. When using the i.p. route, dilution also occurs, but there is a risk of peritonitis (Baumans et al. 2001b).

6.6　Anaesthesia and analgesia

Principles of anaesthesia/analgesia are covered in Chapter 6 in this book.

Before inducing analgesia, the animal should be in good health and sufficiently acclimatized to an experimental environment. Pre-anesthetic fasting is unnecessary and undesirable in mice, as they show no vomit response. Opthalmic ointment or artificial tears should be used to prevent drying of the cornea during anesthesia.

Pain occurring during surgical procedures can be prevented by the use of appropriate anesthetic techniques. Most anesthetics affect many body systems, and so may interact with experiments. To minimize such interactions, anesthetic regimens should be selected with care, by considering the pharmacology of the drugs involved. On the other hand, pain, *per se*, may also interact with the experimental procedures. Post-operative pain, and pain occurring as a result of non-surgical experimental procedures, can be alleviated by the administration of analgesics. In order to control pain effectively, it is essential to be able to assess the degree of pain that is being experienced by animal.

Post-operative care should be provided, e.g. analgesics and fluids, a quiet, warm place, such as an incubator, intensive care unit or a cage provided with a heating pad or lamp, to maintain body temperature. Observation and attention by skilled personnel are essential. Monitoring of urine output and defecation give an indication of the hydration status and condition of the gastrointestinal tract, respectively. Behavioral scoring systems and food and water intake can be used in the assessment of pain in the animal post-operatively (Roughan et al. 2000).

6.7 Euthanasia

Mice are killed in the laboratory at the end of an experiment, to provide blood and other tissues, to counter suffering when it exceeds an acceptable level, when they are no longer suitable for breeding or when surplus to requirements.

The 1986 Council Directive of the EEC on the protection of animals used for experimental and other scientific purposes (86/609/EEC) requires humane killing of experimental animals, with a minimum of physical and mental suffering. Methods of euthanasia should be painless, achieve rapid unconsciousness and death, require minimum restraint, avoid excitement, be appropriate for the age, species and health of the animal, minimize fear and psychological stress in the animal, be reliable, reproducible, irreversible, simple to administer and safe for the personnel and (if possible) aesthetically acceptable. Some mechanical methods, such as stunning, cervical dislocation and decapitation, although efficiently and fast, may not be aesthetically pleasant. It is important that personnel are trained and experienced in recognizing and confirming death in the animals (EC Working Party reports on euthanasia of experimental animals 1996, 1997).

Acceptable methods of euthanasia in mice are:
1. Stunning by a blow on the head, followed by exsanguination;
2. Cervical dislocation, in which the cervical vertebrae are separated from the skull using a pencil or similar item placed in the neck and followed by a pull on the tail base;
3. Decapitation, by guillotine or scissors;
4. Inhalational anesthetics such as halothane, enflurane and isoflurane in overdose;
5. Carbon dioxide >70%, which may cause excitation and stress, due to irritation of the mucous membranes and the induced hypoxia. The animal is placed in a chamber pre-filled with the gas, which will cause a short lasting distress, but seems to result in a rapid loss of consciousness. However, rats exposed to CO_2 in their home cage in stead of in a gas chamber did not show signs of distress (Hackbarth et al. 2000). There is still discussion as to whether the addition of oxygen contributes to the animal's welfare.
6. Sodium pentobarbitone, i.v. or i.p. The agent should be diluted for i.p. injection to prevent irritation and pain to the peritoneum.

Methods acceptable for unconscious mice include rapid freezing, exsanguination or potassium chloride injection.

If a fetus is removed from the anaesthetized mother, it may be killed by decapitation. Rapid cooling in liquid nitrogen may kill fetuses not anaesthetized prior to removal from the mother. Newborn mice can be killed by stunning or decapitation (EC Working Party 1996, 1997).

7. ASSESSMENT THE WELL-BEING OF THE LABORATORY MOUSE

Well-being is a relative concept, referring to the state of an animal in relation to its ability to cope with its environment (Broom and Johnson 1993). This ability to cope is what we usually try to measure in evaluation of the animal's well-being (see also paragraph 'animal welfare' in this chapter). Assessment of well-being should be ideally performed in a positive way, such as by measuring pleasure, by preference tests and by behavioral observations in the home cage.

In order to measure pleasure, anticipatory behavior expressed by an increase in activity prior to an announced reward may be used (Spruijt et al. 2001). Preference tests aim at the animal's own choice for certain given conditions and combined with behavioral observations during the preference test are useful (Baumans 1997; Blom et al. 1992), providing limitations are taken into account, such as a limited array of options, short-term vs. long-term preferences and previous experience.

Measuring the strength of preference by assessing the strength of motivation for certain options makes preference tests even more valuable (Sherwin and Nicol 1997; Van de Weerd et al. 1998). Observations of the behavior of mice in the home cage can be used for studying e.g. differences in the behavioral repertoire after changing the living conditions of the animals. Animal well-being is in general related to a broad behavioral repertoire, meaning that assessing the animal's well-being requires a thorough knowledge of the animal's species-specific behavior and biology.

Another way of assessing the animal's well-being is the negative approach - measuring the animal's failures to cope, leading to discomfort and/or stress or by measuring pain. Until now, this way has been the most frequently used. For measuring discomfort, behavioral, physiological and post-mortem parameters are suitable. Behavioral parameters such as abnormal behaviors (e.g. stereotypies), posture, sudden fear or aggression, vocalizing, decrease in grooming leading to chromodacryorrhea in small rodents (Figure 7-13) and activity changes can be used. As physiological parameters, for example weight loss, reduced food intake, diarrhea, respiratory and cardiovascular signs, stress-hormone levels, immunological parameters can be used (Baumans et al. 1994). Heart rate, body temperature and blood pressure can be measured stress free in the awake and freely moving animal using radiotelemetry (Kramer 1993). Post mortem parameters are valuable tools to assess the animal's welfare retrospectively, being beneficial for surviving animals. Examples of post mortem parameters are fatty deposits, organ size, infections, stomach ulcers, and dehydration (Baumans et al. 1994).

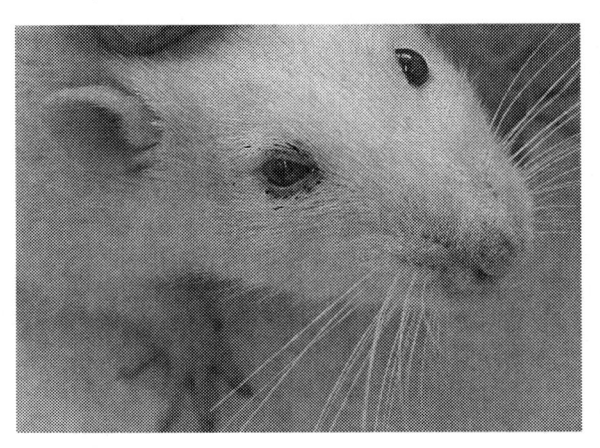

Figure 7-13. Chromodacryorrhea. Photo: P. Rooijmans

REFERENCES

Barclay RJ, Herbert WJ, Poole TB. The disturbance index: a behavioural method of assessing the severity of common laboratory procedures on rodents. Universities Federation for Animal Welfare, Potters Bar, UK 1988

Baumans V. Environmental enrichment: practical applications. In Animal Alternatives, Welfare and Ethics. Van Zutphen LFM, Balls M eds, Elsevier BV, The Netherlands 1997; 187-191

Baumans V. The laboratory mouse. In UFAW Handbook on the Care and Management of Laboratory Animals, Vol 1, 7th edition. Poole T ed, Blackwell Science Ltd, Oxford 1999; 282-312

Baumans V. Environmental enrichment: a right of rodents! In Progress in the Reduction, Refinement and Replacement of Animal Experimentation. Balls M, Van Zeller A-M, Halder M eds, Elsevier, Amsterdam 2000; 1251-1255

Baumans V, Brain PE, Brugère H, Clausing P, Jeneskog T, Perretta G. Report of the FELASA Working Group on Pain and Distress. Lab Anim 1994; 28:97-112

Baumans V, Bouwknecht JA, Boere HAG, Kramer K, Van Lith HA, Van de Weerd HA, Van Herck H. Intra abdominal transmitter implantation in mice: effects on behaviour and body weight. Anim Welf 2001a; 10:291-302

Baumans V, Remie R, Hackbarth HJ, Timmerman A. Experimental procedures. In Principles of Laboratory Animal Science. Van Zutphen LFM, Baumans V, Beynen AC eds, Elsevier, Amsterdam 2001b; 299-318

Baumans V, Schlingmann F, Vonck M, Van Lith HA. Individually ventilated cages: beneficial for mice and men? Contemp Top Lab Anim Sci 2002; 41:13-19

Blom HJM. Evaluation of housing conditions for laboratory mice and rats. The use of preference tests for studying choice behaviour. Utrecht University, Utrecht, the Netherlands 1993

Blom HJM, Van Vorstenbosch CJAVH, Baumans V, Hoogervorst MJC, Beijnen AC, Van Zutphen LFM. Description and validation of a preference test system to evaluate housing conditions for laboratory mice. Appl Anim Behav Sci 1992; 35:67-82

Blom HJM, Van Tintelen G, Van Vorstenbosch CJAHV, Baumans V, Beynen AC. Preferences of mice and rats for type of bedding material. Lab Anim 1996; 30:234-244

Boot R, Koopman JP, Kunstýr I. Microbiological standardization. In Principles of Laboratory Animal Science. Van Zutphen LFM, Baumans V, Beynen AC eds, Elsevier, Amsterdam 2001; 143-165

Brain PF. What does individual housing mean to a mouse. Life Sci 1975; 16:187-200

Brain PF. Stress in agonistic contexts in rodents. In Social Stress in Domestic Animals. Zayan R, Dantzer R eds, Kluwer, Dordrecht 1990; 73-85

Broom DM, Johnson KG. Stress and Animal Welfare. Chapman & Hall, London 1993

Busser J, Zweep A, Van Oortmerssen GA. The Genetics of Behavior. Van Abeelen JHF ed, American Elsevier publishing company, New York 1974

BVA/FRAME/RSPCA/UFAW joint working group on Refinement, Removal of blood from laboratory mammals and birds. Lab Anim 1993;27: 1-22

Clough G. Environmental factors in relation to the comfort and well-being of laboratory rats and mice. In: Proceedings of UFAW/LASA Joint Symposium, Standards in Laboratory Animal Management, Part 1. UFAW, Potters Bar, UK 1984; 7-24

Council of Europe. European Convention for the Protection of Vertebrate Animals used for Experimental and other Scientific Purposes (ETS 123). Council of Europe, Strasbourg 1986

Council of Europe. Resolution on the accommodation and care of laboratory animals. Council of Europe, Strasbourg 1997

Cunliffe-Beamer TL, Les EP. The laboratory mouse. In UFAW Handbook on the Care and Management of Laboratory Animals, Vol. 1, 6th edition. Poole T ed, Blackwell Science Ltd, Oxford 1987; 2785-308

Dawkins MS. From an animal's point of view: Motivation, fitness and animal welfare. Behav Brain Sci 1990; 13:1-9

Dawkins MS. Using behavior to assess animal welfare. Abstract UFAW Symposium Science in the Service of Animal Welfare, April 2003, Edinburgh, UK 2003

Dennis Jr MB. Welfare issues of genetically modified animals, ILAR J 2002; 43(2):100-109

EC Working Party report on euthanasia of experimental animals. Recommendations for euthanasia of experimental animals Part 1. Lab Anim 1996; 30:293-316

EC Working Party report on euthanasia of experimental animals. Recommendations for euthanasia of experimental animals Part 2. Lab Anim 1997: 31:1-32

Eskola S, Lauhikari M, Voipio H-M, Laitinen M, Nevalainen T. Environmental enrichment may alter the number of rats needed to achieve statistical significance. Scand J Lab Anim Sci 1999; 26(3):134-144

European Council Directive. Directive on the approximation of laws, regulations and administrative provisions of the member states regarding the Protection of vertebrate animals used for Experimental and other Scientific Purposes (86/609/EEC) 1986

Fox JG, Cohen BJ, Loew FM. Laboratory Animal Medicine. Academic Press Inc, New York, 1984

Haataja H, Voipio HM, Nevalainen A, Jantunen MJ, Nevalainen T. Deciduous wood chips as bedding material: Estimation of dust yield, water absorption and micro-biological comparison. Scand J Lab Anim Sci 1989; 16(3):105-111

Haberham ZL, Vijn PCM, Van den Brom WE, Hellebrekers LJ. The electro-encephalographic activity observed after decapitation in rats probably does not indicate consciousness. Thesis Utrecht University 2000, 149-160

Hackbarth H, Küppers N, Bohnet W. Euthanasia of rats with carbon-dioxide – Animal welfare aspects. Lab Anim 2000; 34(1):91-96

Harris RBS, Zhou J, Mitchell T, Hebert S, Ryan DH. Rats fed only during the light period are resistant to stress-induced weight loss. Physiol Behav 2002; 76:543-550

Hurst JL. Introduction to rodents. In UFAW Handbook on the Care and Management of Laboratory Animals Vol 1, 7th edition. Poole T ed, Blackwell Science Ltd, Oxford 1999; 262-273

Jennings M, Batchelor GR, Brain PF, Dick A, Elliot H, Francis RJ, Hubrecht RC, Hurst JL, Morton DB, Peters AG, Raymond R, Sales GD, Sherwin CM, West C. Refining rodent husbandry: the mouse. Report of the Rodent Refinement Working Party. Lab Anim 1998; 32(3):233-259

Kramer K, Van Acker SABE, Voss HP, Grimbergen JA, Vijgh WJF, Bast A. Use of telemetry to record ECG and heart rate in freely moving mice. J Pharmacol Toxicol Methods 1993; 30(4):209-215

Mattson MP, Duan W, Lee J, Guo Z. Suppression of brain aging and neurodegenerative disorders by dietary restriction and environmental enrichment: molecular mechanisms. Mech Ageing Dev 2001; 122:757-758

Meijer MK, Kramer K, Remie R, Spruijt BM, Van Zutphen LFM, Baumans V. Stress response to routine experimental procedures in mice: influence of the environment. Submitted

Mertens C, Rulicke T. Score sheets for the monitoring of transgenic mice. Anim Welf 1999; 8:433-438

Mohammed AH, Zhu SW, Darmopil S, Hjerling-Leffler J, Ernfors P, Winblad B, Diamond MC, Eriksson PS, Bogdanovich N. Environmental enrichment and the brain. In: Progress in Brain Research Vol 138. Hofman MA, Boer GJ, Holtmaat AJGD, Van Someren EJW, Verhaagen J, Swaab DF eds, Elsevier Science BV 2002

National Research Council. Guide for the Care and Use of Laboratory Animals, 7th edition. National Academy Press, Washington, 1996; 22

Nevison CM, Hurst JL, Barnard CJ. Strain-specific effects of cage enrichment in male laboratory mice (Mus musculus). Anim Welf 1999; 8:361-379

Newberry, RC. Environmental enrichment: increasing the biological relevance of captive environments. Appl Anim Behav Sci 1995; 44:229-243

O'Donoghue PN ed. International workshop on the accommodation of laboratory animals in accordance with animal welfare requirements. Bundesministerium für Ernährung, Landwirtschaft und Forsten, Bonn, Germany 1993

Olsson A, Dahlborn K. Improving housing conditions for laboratory mice: a review of 'environmental enrichment'. Lab Anim 2002; 36:243-270

Pekow CA, Baumans V. Common nonsurgical techniques and procedures. In Handbook of Laboratory Animal Science, 2nd edition, Vol 1. Hau J, van Hoosier GL eds, CRC Press 2002; 351-390

Poole TB. Welfare considerations with regard to transgenic animals. Anim Welf 1995; 4:81-86

Poole TB. Meeting a mammal's psychological needs: basic principles. In Second Nature, Environmental Enrichment for Captive Animals. Shepherdson DJ, Mellen JD, Hutchins M eds, Smithsonian Institution Press, Washington 1998; 83-94

Pugh TD, Klopp RG, Weindruch R. Controlling caloric consumption: protocols for rodents and rhesus monkeys. Neurobiol Aging 1999; 20:157-165

Roughan JV, Flecknell PA. Effects of surgery and analgesic administration on spontaneous behavior in singly housed rats. Res Vet Sci 2000; 69:283-288

Sales GD, Milligan SR, Khirnykh K. Sources of sound in the laboratory animal environment: a survey of the sounds produced by procedures and equipment. Anim Welf 1999; 8:97-115

Schlingmann F, Van de Weerd HA, Baumans V, Remie R, Van Zutphen LFM. A balance device for the analysis of behavioural patterns of the mouse. Anim Welf 1998; 7:177-188

Serrano LJ. Carbon dioxide and ammonia in mouse cages: effect of cage covers, population and activity. Lab Anim 1971; 21:75-85

Sherwin CM. Frustration in laboratory mice. Sci Cent for Anim Welf Newsl 2000; 22(3):7-12

Sherwin CM. Comfortable quarters for mice in research institutions. In Comfortable Quarters for Laboratory Animals, 9th edition. Reinhardt V, Reinhardt A eds, Animal Welfare Institute, Washington 2002; 6-17; http: //www.awionline.org/pubs/cq02/cqindex.html

Sherwin CM, Nicol CJ. Behavioural demand functions of caged laboratory mice for additional space. Anim Behav 1997; 53:67-74

Spalding JF, Holland IM, Tietjen GL. Influence of the visible color spectrum on activity in mice. Lab Anim Care 1969; 19(2):209-213

Spruijt BM, Van den Bosch R, Pijlman TA. A concept of welfare based on reward evaluating mechanisms in the brain: anticipatory behavior as indicator for the state of reward systems. Appl Anim Behav Sci 2001; 72(2):145-171

The Jackson Laboratory. www.jax.org/jaxmice, 2003

Van de Weerd HA. Environmental enrichment for laboratory mice: preferences and consequences. PhD Thesis, Utrecht University, the Netherlands 1996

Van de Weerd HA, Baumans V. Environmental enrichment in rodents. In Environmental Enrichment Information Resources for Laboratory Animals, AWIC Resource Series No 2 1995; 145-149

Van de Weerd HA, Van den Broek FAR, Beynen AC. Removal of vibrissae in male does not influence social dominance. Behav Processes 1992; 27:205-208

Van de Weerd HA, Baumans V, Koolhaas JM, Van Zutphen LFM. Strain specific behavioural response to environmental enrichment in the mouse. J Exp Anim Sci 1994; 36:117-127

Van de Weerd, HA, Van Loo PLP, Van Zutphen LFM, Koolhaas JM, Baumans V. Preferences for nesting material as environmental enrichment for laboratory mice. Lab Anim 1997a; 31:133-143

Van de Weerd, HA, Van Loo PP, Van Zutphen LFM, Koolhaas JM, Baumans V. Nesting material as environmental enrichment has no adverse effects on behavior and physiology of laboratory mice. Physiol Behav 1997b; 62:1019-1028

Van de Weerd HA, Van Loo PLP, Van Zutphen LFM, Koolhaas JM, Baumans V. Strength of preference for nesting material as environmental enrichment for laboratory mice. Appl Anim Behav Sci 1998; 55:369-382

Van de Weerd HA, Aarsen EL, Mulder A, Kruitwagen CLJJ, Hendriksen CFM, Baumans V. Effects of environmental enrichment for mice: variation in experimental results. J Appl Anim Welf Sci 2002: 5:87-109

Van den Broek FAR, Omtzigt CM, Beynen AC. Whisker trimming behavior in A2G mice is not prevented by offering means of withdrawal from it. Lab Anim 1993; 27:270-272

Van der Meer M, Baumans V, Olivier B, Van Zutphen LFM. Impact of transgenic procedures on behavioral and physiological responses in post weaning mice. Physiol Behav 2001a; 73:133-143

Van der Meer M, Rolls A, Baumans V, Olivier B, Van Zutphen LFM. Use of score sheets for welfare assessment of transgenic mice. Lab Anim 2001b; 35:379-389

Van Dijk JE, Van Herck H, Bosland MC. Diseases in laboratory animals. In Principles in Laboratory Animal Science. Van Zutphen LFM, Baumans V, Beynen AC eds, Elsevier, Amsterdam 2001; 173-195

Van Herck H, Baumans V, de Boer SF. Assessment of discomfort in Laboratory Animals. In Auto-immune Disease Models, a Guidebook. Cohen J, Miller A eds, Academic Press, US 1994; 303-320

Van Herck H, Baumans V, Brandt CJWM, Hesp APM, Sturkenboom JH, Van Lith HA, Van Tintelen G, Beynen AC. Orbital sinus blood sampling in rats as performed by different animal technicians: the influence of technique and expertise. Lab Anim 1998; 32:377-382

Van Loo PLP, Kruitwagen CLJJ, Van Zutphen LFM, Koolhaas JM, Baumans V. Modulation of aggression in male mice: influence of cage cleaning regime and scent marks. Anim Welf 2000; 9:281-295

Van Loo PLP, De Groot AC, Van Zutphen LFM, Baumans V. Do male mice prefer or avoid each other's company? Influence of hierarchy, kinship and familiarity. J Appl Anim Welf Sci 2001a; 4:91-103

Van Loo PLP, Mol JA, Koolhaas JM, Van Zutphen LFM, Baumans V. Modulation of aggression in male mice: influence of group size and cage size. Physiol Behav 2001b; 72:675-683

Van Loo PLP, Van der Meer E, Kruitwagen CLJJ, Koolhaas JM, Van Zutphen LFM, Baumans V. Strain-Specific aggressive behavior of male mice submitted to different husbandry procedures. Aggr Behav 2003; 29:69-80

Van Oortmerssen GA. Biological significance, genetics and evolutionary origin of variability in behavior within and between inbred strains of mice. Behavior 1971: 38:1-92

Van Reenen CG, Blokhuis HJ. Investigating welfare of dairy calves involved in genetic modification: problems and perspectives. Livest Product Sci 1993; 36:81-90

Van Zutphen LFM, Hedrich HJ, Van Lith HA, Prins JB. Genetic standardization. In Principles of Laboratory Animal Science. Van Zutphen LFM, Baumans V, Beynen AC eds, Elsevier, Amsterdam 2001; 129-147

Working Group on Refinement (BVA/FRAME/RSPCA/UFAW joint). Removal of blood from laboratory mammals and birds. Lab Anim 1993; 27:1-22

Wuerbel H, Stauffacher M, Von Holst D. Stereotypies in laboratory mice: quantitative and qualitative description of the ontogeny of 'wire-gnawing' and 'jumping' in ICR and ICR-nu mice. Ethology 1996; 102:371-385

Yamauchi C, Fujita S, Obara T, Ueda T. Effect of room temperature on reproduction body and organ weights, food and water intakes and hematology in mice. Exp Anim 1983; 32:1-12

Young L. Mousetales. Coronet Books, Hodder & Stoughton Ltd, UK 1987

Chapter 8

THE WELFARE OF LABORATORY RATS

Eila Kaliste and Satu Mering
National Laboratory Animal Center, University of Kuopio, Kuopio, Finland

1. INTRODUCTION

As laboratory animals, mice and rats are the most used animal species. According to the latest statistical summary in 1999, 2.6 million rats (27 % of all animals used) and 5.3 million mice (54 %, respectively) were used in the EU member states (Commission of the European Communities 2002). The report lists three main purposes for using rats: 1. Biological studies of a fundamental nature (26 % of all rats), 2. Research, development and quality control of products and devices for human medicine and dentistry and for veterinary medicine (58 %) and 3. Toxicological and other safety evaluations (13 %). Hence, laboratory rats offer an important tool for scientists seeking to clarify the functions of biological systems, as well as to develop new treatments to human diseases and safer living environment.

The use of rats as laboratory animals dates back to the end of the nineteenth century. The first stocks of laboratory rats, e.g. Wistar, Sprague-Dawley, Long-Evans hooded, were developed at the beginning of the twentieth century (Koolhaas 1999). Since then, several outbred stocks, inbred strains, genetically selected lines and, more recently, transgenic lines have been developed all over the world for the purposes of science. After living several generations under laboratory conditions, rats have become well adapted to this environment. In comparison to their wild conspecifics, laboratory rats differ in respect to their behavioural and environmental demands (Inglis and Hudson 1999). Despite these differences, the laboratory rat retains many of its natural needs which should be taken into account in their maintenance and use. One fundamental behavioural need, for example, is the need for safe, concealed areas when resting, since in nature the risk of

2. BIOLOGY AND BEHAVIOURAL NEEDS

2.1 General biology

The Rattus species used in laboratory animal science is the domesticated form of the *Rattus norvegicus*, the Norway rat. It is generally considered a social animal, but may also live a solitary existence (Weihe 1987). *Rattus norvegicus* has shown great adaptability and has been able to survive in a wide variety of habitats in natural environments, though it mainly prefers living in burrows. Breeding in laboratory conditions has led to tame animals, which habituate to repeated stimuli and can be trained to tolerate even unpleasant procedures, such as injections (Weihe 1987).

Figure 8-1. Many strains of laboratory rats are albino.

As a nocturnal animal, the rat is most active in the dark and rests during the light period. In the laboratory, rats have a clear circadian rhythm, which follows the light cycle, corresponding usually to 12/12 hours of light and dark. The most active periods occur at the beginning and end of the dark period. In addition to activity, this rhythm is also known to regulate various physiological variables. For example, the levels of different hormones can vary according to the time of day, as well as heart rate and food intake (Koolhaas 1999). The body temperature during the hours of daylight is 35.6-36 °C increasing during the hours of darkness to 37.8-38 °C (Briese 1998).

The senses of smell and touch are probably the most important senses in rats. These senses enable rats to receive and convey important information about their social and physical environment. For example, such information as the sex, identity, social status and reproductive status of individuals are transmitted by odours (Hurst 1999). Rats can hear sounds in the range of 0.25-80 kHz at 70 dB (Voipio 1997, Barnett 2001) and they emit a large repertoire of ultrasonic vocalizations ranging from 22 to 80 kHz (Koolhaas 1999). Vision is thought to be a less important sense to the nocturnal rat. However, it is known that rats have cones sensitive to visible light and others to ultraviolet, and evidence of dichromatic colour vision in the rat has been found in behavioural discrimination tests (Jacobs et al. 2001).

The average lifespan of laboratory rats varies between 2-4 years, depending on the strain and sex of the animal, its diet and living conditions. For example, the maximum lifespan of Wistar rats under laboratory conditions is about 1200 days for males and 1400 days for females (Koolhaas 1999). The rat reaches sexual maturity on average at the age of 2-3 months. Like other rodents, rats can breed throughout the year. Females have a continuous, short oestrous cycle of 4-5 days, a gestation period of 20-21 days and litter sizes averaging 10 pups in outbred rats. Female rats can reach maximum weights of 400 g and males 800 g, though body weights can vary greatly with the stock or strain.

In nature, the Norway rat is omnivorous, eating a wide variety of plant matter as well as invertebrates and small vertebrates (Nowak 1991). In the laboratory, commercial feed is generally supplied *ad libitum* in pelleted form. The nutritional requirements of rats also varies with sex, physiological condition and age (National Research Council 1995). The major physiological features of the rat's digestive system include a stomach containing both nonglandular and glandular regions, a relatively well developed cecum and the absence of a gallbladder (Bivin et al. 1979). Moreover, rats ingest their own faeces (coprophagy), as a source of such vitamins as vitamin K and biotin.

2.2 Early development: a critical period

Rat pups are affected by their environment even before their birth. Experimental stress manipulations of pregnant females using repeated restrain stress have been shown to exert several influences on development of offspring, including increased anxiety, greater vulnerability to drugs, emotionality and depression-like behaviours (Maccari et al. 2003). These effects result from a dysfunction of the HPA axis, leading to prolonged secretion of corticosteroids after stress and reduced concentrations of glucocorticoid receptors in the central nervous system.

At birth, although rat pups are poorly developed, they mature rapidly during the first postnatal days. The postnatal period (time between birth and weaning) is thought to be the critical factor modulating the phenotype of adult rats. The physiology and behaviour of pups are mainly regulated by the dam and their littermates. Rat pups spend their time huddled in a group with their littermates and the dam, forming social attachment bonds with their group mates (reviewed by Nelson and Panksepp 1998). If removed from this group, rats show separation distress, characterised by an increase in activity, heart and respiration rate, corticosterone release and ultrasonic vocalisations. The effects of separation can be prevented by placing the pup into a warm chamber, by stroking with a paintbrush or tactile contact with a conspecific. This indicates that thermotactile sensory domains may mediate the responses of pups during separation. Olfaction is another important sensory domain affecting pup behaviour. Rat pups seek out and have preferences for odours associated with their mother and the nest. The physiological components of these sensory domains in infant rats include endogenous opioids, oxytocin and norepinephrine.

Experimental manipulations during the postnatal period have been demonstrated to cause marked effects on neurobiological, physiological and behavioural phenotypes in adulthood (reviewed by Pryce and Feldon 2003). The consequences may be detrimental, for example following early isolation of pups (separation of the pup from the dam and litter for one or more hours per day across several postnatal days), or following maternal separation (separation of the mother from her litter for one or more hours per day across several days). In contrast, early handling of pups (picking up the pup and isolating it for several minutes, repeated across several days) has positive effects on adult animals, including greater activity in an open-field test and lower plasma corticosterone responses to stressors. The weekly care routines performed by animal care personnel are one method for conducting early handling. Differences between early handling and animal facility rearing (involving handling 1-2 times per week during cage cleaning and movement of personnel in and out of the animal room) were minimal in contrast to differences between early handled and non-handled rats in pituitary-adrenal stress responses and anxiety in elevated plus maze, open field and the acoustic reflex startle test (Pryce and Feldon 2003).

In the developing rat, maturation of the hypothalamic-pituitary-adrenal (HPA) axis shows a characteristic developmental pattern (Sapolsky and

Meaney 1986). Basal corticosterone levels are high during the final days of gestation and immediately after birth, decrease dramatically during the first two postnatal days, and remain at low levels until approximately day 14, after which the corticosterone levels rise gradually to the level of the adult rat. The period characterised by low corticosterone levels is referred to as the 'stress hyporesponsive period' (SHRP, reviewed by Levine 2001). During this period, the adrenal response to stress is either minimal or non-existent, and levels of the pituitary adrenocorticotropin and hypothalamic corticotropin-releasing hormone remain low. SHRP is essential for normal maturation. Treatments of rat pups with corticosterone during the SHRP is known to result in reduced DNA content and brain size, as well as impaired neuroendocrine function and behaviour in adulthood. The role of the mother is critical during this period of rat development. The mother maintains the hyporesponsiveness of the pup to stress and suppresses the corticosterone responses when the pup is capable of responding. Different aspects of the mother's behaviour regulate different aspects of the HPA system in pups: feeding maintains adrenal insensitivity to ACTH, active licking and grooming suppress neuronal activation and alter the pattern of ACTH-releasing factors; and passive contact inhibits the stress response to novelty.

After weaning, a well-known critical period in rat development occurs at the age of 30-40 days, when social play occurs (reviewed by Vandenschuren et al. 1997). Social play is characterised by behaviours such as punching, chasing, social grooming and sniffing, crawling over or under, charging, boxing, wrestling, pinning and lateral display. Pinning is regarded as the most characteristic posture in social play in rats: one of the animals lays with its dorsal surface on the floor with the other animal stands over it. Social play has been suggested as facilitating the social development of rat pups, since it helps to establish social organisation in the group, develops the ability of animals to express and understand communicative signals and facilitates coping with social conflicts. Male rats play more than females, while females seem to engage more in social grooming. However, no difference has been observed between the sexes in social investigation behaviour. Social play has a high reward value: it can be used as an incentive for maze-learning and conditioned place preference. Exposure to androgens during the neonatal period (up to day 6 of life) is necessary to achieve male-like levels of play.

Isolation during the age characterised by play behaviour (play deprivation) has been found to cause abnormal patterns of social, sexual and aggressive behaviours (Vanderschuren et al. 1997). Although it has no influence on the capacity to perform motor acts, isolation can affect the contextual setting in which aggressive or sexual behaviour are displayed. Rats isolated during weeks four and five and re-housed thereafter in groups, showed marked disturbances in their behavioural and neuroendocrine responses to a social stressor (defeat in the resident-intruder paradigm) in adulthood.

Taken together, these findings clearly show that the early periods are critical to the rat's life, and greatly affect the phenotype in adulthood, especially their sensitivity to stress factors. Therefore, normal development can be best ensured by maintaining a rearing environment that is devoid of disturbances, guaranties rat pups continuous contact with their mother and littermates, and protects pups from experiencing major stress. Frequent handling during cage cleaning provides rat pups with a positive stimulus. After weaning, group housing of rat pups should be continued, at least until the end of the play period.

3. OPTIMAL ENVIRONMENT

The physiological and ethological needs should be used to provide recommendations and guidelines for the housing environments of laboratory rat. However, these needs are not only species- and strain specific, but may also relate to the individual's position and experience within a given social community (Stauffacher 1995). According to Poole (1992), mammals have four major requirements for the environment to meet their behavioural needs: 1) the need for stability and security, 2) appropriate complexity, 3) an element of unpredictability and 4) opportunities to achieve goals. Poole did not include into the requirements the need for social companions, since many mammalian species are solitary and some even prefer privacy. Nevertheless, the Council of Europe (1986) states that 'any restriction on the extent to which an animal can satisfy its physiological and ethological needs shall be limited as far as practicable'. The European Commission's international workshop has set recommendations stipulating that rodent cage environments should satisfy the physiological and ethological needs of resting, grooming, exploring, hiding, searching for food and gnawing (Brain et al. 1993). Leaving these needs unsatisfied over either the short or long term will inevitably lead to poor welfare (Broom and Johnson 1993).

Several guidelines and reference books have been published for the housing environment of laboratory animals (e.g. Council of Europe 1986, Van Zutphen et al. 1993, Home Office 1995, National Research Council 1996, Barnett 2001). Although the recommendations vary somewhat between countries, their common goal is to improve the welfare of animals and promote uniform housing environments.

3.1 Macroenvironment

The five most important features of the laboratory rat's macroenvironment (animal room) are temperature, humidity, ventilation, illumination and noise. For rodent rooms, the recommended range of temperature and relative humidity vary from 20-24 °C and 55±10 % (Council of Europe 1986; EC) to 18-26 °C and 30-70 % (National Research Council 1996; NRC). The ventilation rate should be maintained at 15-20 (EC) or 10-15 (NRC) air changes/hour, the illumination should be kept low (EC) or 325-400 lux (NRC), and the noise level should be minimal (EC) or defined as <50 dBA (Home Office 1989) or <85 dB (NRC). Being highly adaptable, the rat may well survive within all these environments. However, standardised room conditions are needed to ensure reliable research results: reducing variation is one of the main goals in laboratory animal husbandry. The optimal level of the environmental factors in animal rooms depends on housing and husbandry designs. Such factors as the cage type and material, bedding type, ventilation in the cage and the number of animals housed in a cage can greatly influence the conditions inside the cage (microenvironment). Moreover, an animal's needs may vary depending on its physiological state or age. For example, breeding animals may require a warmer environment to ensure the welfare of pups, as well as young rats or rats without hair, such as nudes.

3.2 Microenvironment

3.2.1 Cage size

There has been a tendency in recent years to enlarge and heighten the cages of rats (Table 8-1). A minimum requirement is that the animal should have enough space to turn around and express normal postural adjustments (National Research Council 1996). Obviously, the size needed depends on such factors as the size of animals, whether they are breeding or undergoing experiments. For rats, Lawlor (1990) defined the minimum cage floor size to be that which is large enough to allow the rat to sit or lie without any torsion of the body or the tail. The cage should also be large enough so that enrichment objects can be offered to rats and they can perform different activities with them. Moreover, since rearing is frequently observed during the exploratory behaviour of rats (Lore and Schultz 1989), the cage should be high enough to allow this. A height of 14 cm has been claimed to be too low to allow the upright postures needed in explorative and play behaviours (Brain et al. 1993, Koolhaas 1999), while heights of 23 cm (Brain et al. 1993) or even 30 cm (Lawlor 1990) have been suggested as being more suitable for rats. However, rats also spend most of their time in burrows if given the choice (Boice 1977). Rats also tend to prefer lower cages to higher cages or at least the motivation for rearing did not exceed motivation to be in a shelter (Blom et al. 1995).

Table 8-1. Recommendations for rat caging

	Body weight (g)	Minimum floor area (cm^2/animal)	Minimum cage height (cm)
Council of Europe 2002	<200	200	18
Revision of Appendix A	<300	250	18
Draft	<400	350	18
National Research Council	<100	~110	~18
1996	<200	~150	~18
	<300	~190	~18
	<400	~260	~18
Home Office 1989	<50	100	18
	<150	150	18
	<250	200	18
	<350	250	20
Council of Europe 1986	<100	~120	14
Appendix A	<200	~170	14
	<300	~230	14

Hence, the rat cage should have enough height for rearing, as well as the possibility of withdrawal into lower parts, such as below the food hopper. Indeed, the literature seems to indicate only a limited value of space *per se* for rats and places greater emphasis on the complexity of the space and the advantages of group housing (Bantin and Sanders 1989, Lawlor 1990, Patterson-Kane et al. 2001).

3.2.2 Cage material

The caging material has to be harmless for the animals, inedible, easy to clean and sterilise, resistant to waste products and must be able to withstand attempts to escape (Wolfensohn and Lloyd 1994, Allmann-Iselin 2000).

Polycarbonate or polypropylene cages are usually preferred by investigators over stainless steel cages, because they are lighter, less noisy and thought to give better heat insulation. However, thermoregulatory responses (e.g. metabolic rate, evaporative water loss and core temperature) were not affected by the floor type (acrylic vs aluminium) within the ambient temperature range of 20-28 °C (Gordon and Fogelson 1994). There is also some evidence that rats prefer steel over polycarbonate, if accustomed to stainless steel cages. However, for animals raised in polycarbonate cages, neither of these materials appears to be favoured (Heikkilä et al. 2001).

3.2.3 Cage flooring

Solid-bottom cages are normally recommended instead of grid-floor or wire-mesh cages (Brain et al. 1993, National Research Council 1996). These recommendations are mainly based on the animal's inability to perform species-specific behaviours such as nest building or digging in the grid-floor cages in contrast to solid-bottom cages with bedding material.

Preference tests show that rats prefer to rest in solid-bottom cages with bedding and use grid-floor cages mainly during active periods (Manser et al. 1995 and 1996, Blom et al. 1996, van de Weerd et al. 1996). On the other hand, Nagel and Stauffacher (1994) reported no differences in the resting and exploration behaviours of rats housed in grid-floor cages and those in solid-bottom cages. The effects of cage type on physiological parameters seem to be minimal: no differences have been observed in either the adrenal weight and corticosterone concentrations (Nagel and Stauffacher 1994) or in body weight gain, food consumption or water intake (Manser et al. 1995) of rats housed in these two different floor types. Table 8-2 summarises the data of three experiments studying the effects of cage type both with physiological and behavioural parameters in group-housed Wistar, ANA and AA rats (modified from Eskola and Kaliste-Korhonen 1998, Kaliste-Korhonen et al. 1995, Eskola et al. 2000). The results showed that the cage type had only minor, inconsistent effects on the physiology and behaviour of rats. Our later results confirm these findings (Mering 2000, Mering et al. 2001). In general, rats seem to tolerate quite well housing in grid-floor cages. If housing in grid-floor cages is necessary, the welfare of the animals can be improved by placing a solid resting area (e.g. box or platforms) into the cage (Figure 8-2). Foot lesions are known to occur in older and heavier rats when housed for a long time – over one year - on grid-floor cages (Stark 2001). Placement of the suggested resting area in these cages might prevent this phenomenon.

The husbandry routines differ when rats are housed in solid-bottom or in grid-floor cages: rats housed in solid-bottom cages are transferred once or twice a week to a clean cage, whereas rats housed in grid-floor cages are handled less frequently, since the frequent changing of cages is not necessary due to the changing of trays below the cage. The handling of animals is generally recommended to accustom them to human contact and train them for experimental procedures (Joint Working Group of Refinement 2001). Moreover, with solid-bottom cages, the change to a new, clean cage with clean bedding may function as enrichment when the animals arrange their new home to their liking. On the other hand, the loss of familiar odours due to cage changing has been suggested as a cause for aggression, at least, in mice (reviewed by Van Loo et al. 2000). Stress-like responses were not, however, altered in male rats after adding a small quantity of the soiled bedding to the new cage (Duke et al. 2001).

Table 8-2. Summary of results from three experiments in which the effects of cage type (solid bottom cages = SBC vs grid floor cages = GFC) were evaluated. Numerical values are shown when a significant effect was found. Ne = no effect, - = not tested.

	Wistar[2]		Wistar[1]		AA[3]		ANA[3]	
	GFC	SBC	GFC	SBC	GFC	SBC	GFC	SBC
Physiology								
Food intake (g)	-	-	-	-	167±32	145±29	150±24	126±24
Weight gain (g)	135±5	123±11	ne	ne	ne	ne	ne	ne
Weights (mg) of								
Adrenals	ne	ne	ne	ne	ne	ne	ne	ne
Epididymal adipose tissue	ne	ne	ne	ne	ne	ne	ne	ne
Spleen and thymus	ne	ne	ne	ne	ne	ne	ne	ne
Basal corticosterone (ng/ml)	77±66	279±169	-	-	ne	ne	ne	ne
Cort. change after ACTH	-	-	-	-	259±157	164±60	250±134	429±72
Open-field behaviour								
Activity or defecation (freq)	ne	ne	-	-	-	-	-	-
Standing alert time (%)	ne	ne	-	-	-	-	-	-
Grooming time (%)	0.1±0.4	4±3	-	-	-	-	-	-
Plus maze behaviour								
Rats entered open arms	-	-	-	-	3	7	ne	ne
Time in open arms					ne	ne	ne	ne
Grooming time %	-	-	-	-	3.4±3.2	0.6±1.3	ne	ne
Total entries	-	-	-	-	ne	ne	ne	ne
Play behaviour								
Latency and playing time %	-	-	-	-	ne	ne	ne	ne
Sucrose intake								
Total water+sucrose (ml)	-	-	-	-	ne	ne	ne	ne
% sucrose from total	-	-	-	-	ne	ne	ne	ne

[1]Kaliste-Korhonen et al. 1995, [2]Eskola and Kaliste-Korhonen 1998, [3]Eskola et al. 2000

Figure 8-2. Aspen tube in a grid-floor cage serving as both a solid area to rest on and an enrichment (gnawable) object.

3.2.4 Cage bedding

Bedding in the (solid-bottom) cage is one of the main components of the rat's environment. The purpose of bedding is to absorb moisture and the ammonia from urine and faeces. It should be changed frequently enough to keep animals clean and dry. Several types of beddings are commonly used for laboratory rodents: corn cobs, straw, paper, pine, spruce, aspen, etc. Bedding should be pleasant for animals and give opportunities to fulfil their ethological needs. In individually ventilated cages (IVCs) with cleaning frequency of every 14 days, Alphadri or corn cob substrate may be preferable to wood chip with respect to lower ammonia levels (Ras et al. 2002, RSPCA/UFAW rodent welfare group 2003). In preference experiments, aspen chip is often the most preferred bedding (e.g. Blom et al. 1996, Ras et al. 2002). However, these preferences may have been biased by previous experiences (Ras et al. 2002) and might also be dependent on the strain, age, sex and reproductive condition of the animal (Allmann-Iselin 2000).

Softwood beddings (e.g. pine and red cedar) are known to induce biotransformation enzymes due to their volatile organic compounds (Vesell 1967). The concentrations of these compounds in the bedding can be reduced by autoclaving (Nevalainen and Vartiainen 1996), but even then the concentrations remain much higher than with hardwood. On the other hand, hardwood (e.g. alder and aspen) contains high levels of tannins and alkaloids (Allmann-Iselin 2000). Pelkonen and Hänninen (1997) compared the cytotoxicity and biotransformation inducer activity of beddings using Hepa-1 *in vitro* assay. They found a great variation in both cytotoxicity and inducer activity of the beddings, with pine shaving beddings generally appearing to be highly cytotoxic; whereas, corn cob, rice hulls and straws were practically non-toxic. Furthermore, large differences in cytotoxicity have also been observed according to the country of origin. The importance of the supplier was also emphasised by Allmann-Iselin (2000).

3.2.5 Nesting material

Nesting materials, such as paper towels or wood wool, are highly recommended for rodents, especially for breeding animals. Rowan and Moore (1991) have suggested that pre-weaning mortality can be decreased in some rat strains through the use of proper nesting material. Although no studies have determined which type would be best for rats, studies in mice have found no differences in reproductive performance (number of litters, litter size, neonatal mortality) between aspen wood-wool and paper towel as a nesting material (Eskola and Kaliste-Korhonen 1999a). Provision of nesting material as a method of enrichment is considered to have a more obvious effect on the welfare of rodents than other cage additions, such as toys (Brain et al. 1993). This view conforms with the findings of Manser et al. (1998a,b) showing that rats prefer cages with either a nest box, nesting material or both to empty cages.

3.2.6 Enrichment

The complexity in the cage environment is generally attained by using either environmental tools such as toys or other objects, or cage structures, like shelves, or tunnels. A variety of mazes, tunnels, boxes and other structures are nowadays provided for laboratory rodents (Figure 8-3). As noted by Chamove (1989), the goal of enrichment is to alter behaviour so that it matches as closely as possible the range of the animal's normal behaviour. Purves (1997) also suggests that the goal of enrichment is to make the animals' environment as natural as possible. These suggestions, however, can be criticised in light of the known adaptability of these animals, the fact that natural conditions can also have negative effects (predation and diseases), as well as the counterargument that not all naturally occurring behaviours, nor all aspects of life in a natural setting, are desirable or necessary in laboratory conditions (Clark et al. 1997, Bennett 1999, Dean 1999). In general, however, most scientists do agree that environmental enrichment is beneficial for the well-being of laboratory animals and that it should be applied when ever appropriate or practical.

Figure 8-3. Enrichment objects for rodents

Nesting material has most commonly been provided as a means of enrichment for mice (van de Weerd et al. 1997, Eskola and Kaliste-Korhonen 1999b), while rats have often been provided with items for gnawing (Kaliste-Korhonen et al. 1995, Chmiel and Noonan 1996, Eskola and Kaliste-Korhonen 1998). Nesting material has sometimes been regarded as an unsuitable enrichment for rats, due to the lack of spontaneous nest building behaviour of adult rats (referred by van Loo et al. 2002). Van Loo et al. (2002), however, showed that nesting material is also a suitable type of environmental enrichment for rats, when provided from birth, and rats learn to use it.

Environmental enrichment studies have sought to measure the effects of enrichment on animals and its value by measuring such factors as stress hormones, growth, neurological impacts, behaviour in a home cage or in behaviour tests, and preferences (reviewed by Mering 2000). However, the results in rats have been rather inconsistent and some times contradictory, making it difficult to reach wide general conclusions or even simple recommendations. When a particular enrichment is adapted, a critical evaluation of the suitability and possible consequences should be performed before applying it on a larger extent.

An enriched environment improves the ability to learn and solve problems, i.e. cognitive performance, and it also ameliorates behavioural deficits (Patterson-Kane et al. 1999, Young et al. 1999). The complexity of the housing environment is known to affect the cerebral neurochemistry and anatomy. Enriched housing of rats has been shown to increase a variety of physiological traits, including cortical thickness, the size of neuronal cell bodies, nuclei, and the number of dendrites. These cerebral changes can be produced by different housing throughout the life span of the animal, and their effects can be rapidly discerned (reviewed by Rosenzweig and Bennett 1996). In these studies, however, the significant differences reported have often been found by comparing two extremely different environments: very complex (like 8-10 rats in large cage with toys, activity wheel, etc.) and very impoverished (individual housing in wire-mesh cages in a quiet room) environments. These differences have not been as clear when comparing rats housed in enriched or standard environments (individual or group housing with normal care routines). However, differences have also been reported in studies comparing even these types of environments (Soffié et al. 1999). Hence, it seems most likely that current housing standards, i.e. group housing with bedding, environmental enrichment and frequent handling, is sufficiently adequate to ensure the normal cognitive development of rats.

The environment has a major impact on rat physiology and behaviour, especially in the early periods of life (see 2.2 in this chapter). Varying the rearing environment results in individual variation and, hence, alters responses during experiments. Several studies have shown that environmental enrichment and a complex environment may decrease or increase the variability in research results (Purves 1997, Brain et al. 1995, Gärtner 1998, Eskola and Kaliste-Korhonen 1999c, Eskola et al. 1999, Tsai and Hackbarth 1999, Mering 2000, Mering et al. 2001). Such an increase in the variation of experimental animals leads to an increase in the number of animals needed in an experiment. This conflicts the overall goal of reducing the number of animals required for experiments. However, housing animals in adequately complex environment to ensure their normal development and welfare is thought to be necessary for producing animals with better coping abilities and greater tolerance of stressful manipulations in experiments (Baumans 1997). The well-being of rats with reduced variation might be achieved by rearing the animals in similarly enriched environments, i.e. through the standardisation of the housing environment of laboratory animals, where enrichment tools are consistently included.

4. HOUSING, FEEDING AND CARE

4.1 Group housing

In general, group housing is recommended for laboratory rodents, though care must be taken to ensure that the groups are harmonious and stable

(Brain et al. 1993). Isolating rats may lead to altered emotional or fear responses to novel stimulations (Gentsch et al. 1981, Holson et al. 1991), as well as provoke variations in plasma glucose, triglycerides and total cholesterol levels (Pérez et al. 1997). Moreover, singly-housed rats actively sought social contact with their conspecifis, when given the possiblity (Hurst et al. 1997). Telemetrically monitored singly-housed rats seemed to react more sensitively to changes in their environment and handling procedures than group-housed rats (Sharp et al. 2002, 2003a). If single housing is required, the possibility of seeing, smelling or hearing conspecifics may reduce possible aggressiveness (Hurst et al. 1997).

Group housing may also result in social stress for some of the group members. Laboratory rats do not show strong hierarchical systems when housed under laboratory conditions, i.e. in small groups of same sex in laboratory cages (Grant and Chance 1958, Hurst et al. 1999). In contrast to male mice, fights between cage mates are rare in rat groups. Social classes have, however, been found in rats when they were housed in colonies in open rooms (Hurst et al. 1996) or when males and females were housed together (Blanchard et al. 1995). In the latter situation, the males formed a clear hierarchical order with severe physiological consequences to the subordinate males. Rats as well as mice have been frequently used in studies of social stress (reviewed by Blanchard et al. 2001). The social stress may be induced by dyadic social defeat situations, where the individual is confronted by an aggressive conspecific once or repeatedly. Alternatively, animals may be grouped or maintained in colonies, where social stress is thought to exist e.g. in subordinates due to agonistic interactions with dominants. The markers of social stress within a group are agonistic behaviours, wounding of animals and weight loss in subordinates. Socially-stressed animals also show changes in their behaviour both in their group and in experimental tests: subordinate rats showed reduced activity, enhanced anxiety in an elevated plus maze test, and enhanced risk assessment, reduced aggression and increased defensive behaviour. Moreover, changes in different neurotransmitters have been reported, as well as changes in neuronal structures (Blanchard et al. 2001). When housed in groups, the number of group mates seems to be unimportant, but the rats may well live in groups of different sizes. According to Hurst et al. (1999), the densities of one, three, five or eight individuals had no marked effect on time budgeting and pathophysiological parameters in male or female rats. There were, however, great differences between individual groups in the measured parameters, indicating that the individual composition of groups rather than their size had the greatest impact on the welfare of the rats. Although no clear stress indicators were found in this experiment, escaping behavior (bar chewing, sniffing of cage bars, sniffing out at the cage top and out of the cage) was correlated in both sexes with both aggressive behaviours and pathophysiological responses, suggesting that escaping behaviour could serve as a useful indicator of well-being in rodents.

Female rats do not show aggression toward other individuals, except when they have pups. The tactile and olfactory cues of pups are important to maintain the high levels of aggressive behaviour of dam, called maternal aggression (reviewed by Lonstein and Gammie 2002). Maternal aggression declines as lactation progresses due to changed sensory cues that pups provide as they become older, or reduced exposure of the dam to her pups due to a decrease in the time spent with the pups.

4.2 Husbandry procedures

Husbandry procedures may vary depending on such factors as the type of cages, animal number in cages, feeding and watering systems. In general, the rats are changed once or twice a week into clean cages with a simultaneous change of diet, water and enrichment tools. In the most hygienic systems, caretakers carry a personal protective mask, gloves and clothing to prevent the spread of pathogens.

Normal husbandry procedures, such as cage changes, evoke stress-like responses, typically seen as increased blood pressure, heart rate and behavioural activity (Table 8-3) (Duke et al. 2001, Sharp et al. 2003a). The stressfulness of these procedures for rats is rather questionable, merely the responses indicate a general arousal of the animals to changes in their environment. The variation in the levels of responses, however, may reflect the differing sensitivity of the rats to the changes. There seems to be an interaction between housing (group vs. single) and gender in these responses. Even witnessing such a procedures can produce similar effects in females (Sharp et al. 2003b), though not in males (Duke et al. 2001, Sharp et al. 2002).

After cage changing, the novelty of a new clean cage is suggested as a stress factor contributing to the cardiovascular responses. This could be alleviated by adding a small quantity of the soiled bedding from the previous cage to the new cage, as has been earlier proposed with mice (Van Loo et al. 2000). This has not, however, been found to alter the responses in rats (Duke et al 2001, Sharp et al. 2003a). Based on the cardiovascular responses and behavioural activity, Harkin and co-workers (2002) have concluded that rats take greater notice of changes in their immediate environment (cage change, intruder rat into cage, wet bedding) or direct handling by a human. The changes which occur within the room, like paper towel on the cage lid, white noise, stroboscope light, or room entry caused smaller and shorter responses in rats (Table 8-3).

Table 8-3. Reactions of rats to different manipulations

	Change in heart rate	Duration (minutes)	Home cage activity
Sharp et al. 2002-2003: Holzman Sprague-Dawley rats			
Cage change	↑↑	60 - 90	↑↑
Weigh	↑↑	60 - 90	↑
Injections	↑↑	60 - 90	↑↑
Vaginal smear	↑↑	60 - 90	↑↑
Urine&feces/blood odours	↑	30 - 60	↑
Blank paper towel	↑	30 - 60	↑
Room entry	↑	15-30	
Harkin et al. 2002: Sprague-Dawley rats			
Male intruder	↑↑	150	↑↑
Wet bedding	↑↑	210	↑↑
Subcutaneous injection	↑↑	108	↑↑
Handling	↑	42	No change
Air puff	↑	20	↑
Strobe light	↑	30	↑
Novel odour	↑	4	No change
White noise	No change		No change
Confinement	No change		No change

The human-animal bond has been raised as one of the important factors contributing to animal welfare (Special Issue, ILAR Journal 2002). The human/animal relationship is important both for the well-being of personnel (especially for animal care takers) and laboratory animals. Empathy, sympathy, respect and knowledge of the species will effectively improve the welfare and reduce the suffering of animals (RSPCA/UFAW Rodent Welfare Group 2003).

Housing rodents in individually ventilated cages (IVCs) has increased significantly during the last decade, stimulating interest in their suitability in terms of the animals' well-being. Recently, a rodent welfare group of RSPCA/UFAW discussed this issue (RSPCA/UFAW Rodent Welfare Group 2003). The group also addressed as one of its key welfare concerns the impact of IVCs on the human/animal bond. Particular concerns have been expressed for the microbiological load in cage substrate at different intervals after cage change, as well as the advantages and disadvantages of sealed and non-sealed systems. The threat of power or system breakdowns, leading to a halt in ventilation and increased CO_2 concentrations, can be a serious health risk for the animals, unless adequate alarm systems and emergency backups are available (Krohn and Hansen 2000 and 2002, RSPCA/UFAW Rodent Welfare Group 2003).

4.3 Feeding

The form, content and presentation of the diet should meet the nutritional and behavioural needs of the animal, and uncontaminated drinking water should always be available to all animals. Feed intake can be influenced by the energy content of the diet, environmental factors such as temperature and light cycle, or by other management practices, such as group housing, restrictive feeders and cage design (Keenan et al. 2000). The nutritional needs of rats (National Research Council 1995) are well known by diet producers, and a suitable diet can be chosen to meet situational needs, such as the breeding or maintenance of animals. As rodents, rats gnaw their food, thus the standard pelleted food may be considered to an optimal form for this animal.

The consequence of *ad libitum* feeding is, in many cases, obesity and ultimately decreased survival, leading to an increased incidence of tumours in older animals (Keenan et al. 2000), directly affecting animal welfare. The question is how much should we restrict feeding in rats. According to Keenan et al. (2000), moderate dietary restriction (25% restriction of *ad libitum* feeding) significantly improves survival, controls adult body weight and obesity, reduces age-related renal, endocrine and cardiac diseases, reduces study-to-study variability and increases statistical sensitivity. Conversely, severe restriction (40-50% reduction of the maximum *ad libitum* feeding) is not recommended.

5. EXPERIMENTAL TECHNIQUES

Rats, like all other laboratory animal species, are more or less regularly handled during their lifetime. Experimental procedures usually include the formation of experimental groups, as well as the identification of individuals, administration of test substances, sampling of blood or other specimens and finally, euthanasia. Principles of these procedures for rats are in general the same as for mice, presented in detail in Chapter 7 in this book. The methods used for the procedures, as well as for ensuring of animal welfare, are well documented in the literature with recommendations about best practices (Joint Working Group (JWG) on Refinement 1993 and 2001, Diehl et al. 2001, Pekow and Baumans 2003). Similarly, the methods of anaesthesia, analgesia and euthanasia are well presented in the literature and under continuous development and refinement (e.g. Roughan and Flecknell 2002). Minimising the stress of animals during experimental procedures is the key to preventing their suffering. This is best achieved by well-trained personnel and habituating animals to procedures (JWG 1993, 2001).

Moreover, the reliability of research is based on using animals with normal physiology and behaviour, as well as experiments free from confounding factors, which may affect the results and increase the variation in the data (Poole 1997). The stress reactions of animals clearly belong to these confounding variables, and they should be minimised as far as possible.

The handling of animals in experimental procedures, and even during routine care, causes physiological and behavioural changes in animals (Gärtner et al. 1980, Tuli et al. 1995, Saibaba et al. 1996). Accordingly, to minimise the stress arising from experimental procedures, it is recommended that animals be handled as much as possible during normal care routines in order to accustom them to handling in general (Poole 1997, JWG 2001) (Figure 8-4). For example, van Bergeijk et al. (1990) found that gentled rats had fewer escaping responses than non-gentled rats when approached by a caretaker. They were also easier to restrain and showed fewer signs of distress, as measured by defecation and urination. Moreover, handling of rats twice a week protected them against isolation stress (behavioural and adrenocortical signs of profound fear in an open-field test) when housed in cages with wiremesh floor and without handling during cage change routines (Holson et al. 1991).

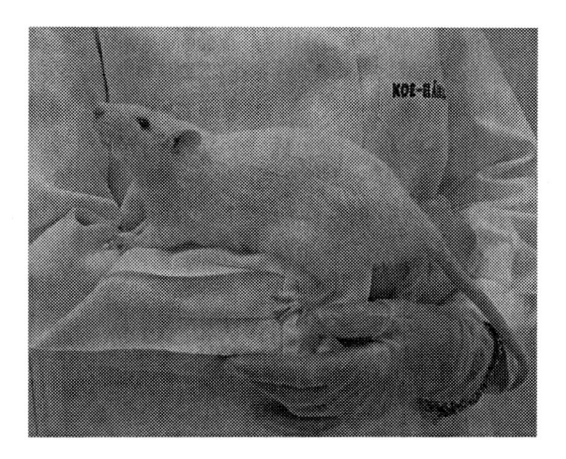

Figure 8-4. Rat should feel secured and comfortable when handled. (Photo: Laboraty Animal Center, University of Turku, Finland)

The immediate reactions of the body to stressful situations include responses of the cardiovascular system characterised by an increase in heart rate and blood pressure (Moberg 2000), as well as an increase of body temperature (Briese 1998). Recently, telemetry methods have been used to measure the cardiovascular reactions of rats to different experimental manipulations (Duke et al. 2001, Harkin et al. 2002, Sharp et al. 2002, 2003a,b), summarised in Table 8-3. Experimental procedures, like subcutaneous or tail vein injections or collection of vaginal smears, did not cause responses greater than those resulting from normal care routines (Sharp et al. 2002, 2003b). This indicates that rats do not consider these experimental manipulations to be very stressful, at least when they are done properly. Moreover, rats housed in groups often, but not always, seemed to have less pronounced responses than rats housed singly, indicating that the single housed rats are more sensitive to stress factors in their environment (Sharp et al. 2002, 2003a,b).

The magnitude of stress in rats as they follow the experimental procedures done to other rats in the same room was also measured: decapitation and necropsy, cage change, restrain and subcutaneous or tail-vein injections caused no significant stress to the male or female animals followed (Sharp et al. 2002, 2003b). Living in groups appears to also diminish these reactions.

It can also be concluded that the effects of experimental procedures can last for about two hours, during which time no experimental measurements should be performed. The greater the procedure, the greater the response and assumed duration might be. In operated mice, the process of recovering from their surgical procedure lasted at least 5-6 days before the normal diurnal rhythm of the physiological parameters had returned (Kramer et al. 1998).

In general, the sensitivity of animals to stress from manipulations is thought to be dependent on their earlier life history. The reactions of animals to the experimental situation has been shown to be primarily affected by factors, such as the physical rearing environment (enrichment or not), social relationships, maternal care, and pre- or postnatal stress (e.g. Baumans 1997, Nelson and Panksepp 1998, Pryce and Feldon 2003).

6. HEALTH AND DISEASES

A competent, well-trained staff is essential to recognise the welfare problems of laboratory animals and to act accordingly when problems have emerged. This has been stated in the Convention text of Council of Europe (1986). The signs of poor welfare are abnormal appearance, piloerection, dehydration, weight loss, red secretion around the eyes and nostrils, posture and reduced locomotion. A well trained care personnel can easily recognise such problems. Animals also have to be checked daily.

6.1 Infections

It is clear that disease-free and pathogen-free animals are important not only for the welfare of the animal but also the success of research, since subclinical infections of animals can drastically increase the variability of results (Beynen et al. 1993), thus threatening the reliability of the research. Ensuring a high standard in modern laboratory animal management requires strict control of hygiene in the animal unit, as well as control of people, animals and equipment coming into the unit. Moreover, regular health monitoring is essential for both breeding and experimental units (see chapter 3 in this book). Following these routines can help guarantee that the laboratory rats are well protected against infections. If an infection is found, usually rederivation procedures are initiated to eliminate the pathogen from the colony (Hansen 2000). The diseases affecting rats are comprehensively listed in the literature with detailed descriptions of the signs, diagnostic methods and treatments (e.g. Percy and Barthold 2001).

6.2 Other health problems

In addition to infectious diseases, laboratory rats may also have welfare problems caused by other factors. Examples of these include (Percy and Barthold 2001):

Malocclusion: overcrowth of the incisor teeth which can lead to difficulties in eating and drinking, and thus a loss in body weight.

Ring tail: annular constrictions of the skin of the tail, leading to dry gangrene of distal tail may be found in preweaning rats when housed in low environmental humidity for longer periods of time.

Dehydration: Rats can rapidly become dehydrated if their watering system is not functioning properly.

In old age, rats may also suffer from such degenerative disorders as nephropathies, myocardial degeneration, polyarteritis nodosa and neoplasms.

6.3 Disease models

Rats are widely used as models for human diseases like cancer, arthritis, and diabetes. For these purposes, several rat stocks or strains have been developed with specific characteristics (e.g. by Hedrich 2000). The welfare of strains produced via spontaneous or induced gene mutations should be ensured as far as possible (van der Meer et al. 1997). Besides genetic modification, disease conditions in laboratory animals can also be induced by experimental procedures or agents. In these kind of manipulations, the welfare or minimal suffering of animals has also to be ensured as far as possible. For this purpose, guidelines have been established for pain recognition and well-being assessment, as well as recommendations for human end points criteria (OECD 2000, Morton and Hau 2003).

7. CONCLUDING REMARKS

The welfare of laboratory rats is the responsibility of humans. As we are dealing with intelligent creatures with the ability to feel pain and suffering and great adaptability to manipulations, we have to ensure their welfare as far as possible throughout their life span.

In conclusion, some basic principles should be fulfilled:

The rearing environment in the early periods of life should be without disturbances and standardised as far as possible. Stress factors during this period may have permanent, negative effects, leading to animals with greater susceptibility to stress. Adequate complexity of the rearing environment ensures the normal development of rats and requires that enrichment be included in the animal's environment.

After weaning rats should be maintained in groups, with enrichment and frequently handled by care personnel in order to provide the rats with a stimulating housing environment and adapt them to experimental procedures.

When rats are taken into experiments, the experimental procedures should be properly performed by competent, confident and well-trained persons, and as much care as possible should be taken to avoid stressing the animals.

In the end, when the euthanasia of laboratory rats is performed, the action should be performed humanely by skilled personnel according to the regulations, causing minimum pain, fear and distress to the animals.

REFERENCES

Allmann-Iselin I. Husbandry. In The Laboratory Rat. The Handbook of Experimental Animals. Krinke GJ ed, Academic Press 2000

Bantin GC, Sanders PD. Animal caging: is bigger necessarily better? Anim Technol 1989; 40:45-54

Barnett SW. Introduction to Animal Technology. 2nd edition, Blackwell Science Ltd 2001

Baumans V. Environmental enrichment. Practical applications. In Animal Alternatives. Welfare and Ethics. Proceedings of the 2nd World Congress on Alternatives and Animal Use in the Life Sciences, Utrecht, The Netherlands. Zutphen LFM, Balls M eds, Elseviere Science 1997; 187-191

Bennett GG. Environmental enrichment for laboratory rodents: Animal welfare and the methods of science. J Appl Anim Welf Sci 1999; 2(4):267-280

Beynen AC, Festing MFW, Van Mortfort MAJ. Design of animal experiments. In Principles of Laboratory Animal Science. Van Zutphen LFM, Baumans V, Beynen AC eds, Elsevier, Amsterdam, The Netherlands 1993

Bivin WS, Crawford MP, Brewer NR. Morphophysiology. In The Laboratory Rat, Vol 1. Baker HJ, Lindsey JR, Wisbroth SH eds, Academic Press, New York 1979; 73-103

Blanchard DC, Spencer R, Weiss SM, Blanchard RJ, McEwen BS, Sakai RR. The visible burrow system as a model of chronic social stress: behavioral and neuroendocrine correlates. Psychoendocrinology 1995; 20:117-134

Blanchard RJ, McKittrick CR, Blanchard DC. Animal models of social stress: effects on behavior and brain neurochemical system. Physiol Behav 2001; 73:261-271

Blom HJM, Van Tintelen G, Baumans V, Van Den Broek J, Beynen AC. Development and application of preference test system to evaluate housing conditions for laboratory rats. Appl Anim Behav Sci 1995; 43:279-290

Blom HJM, Van Tintelen G, Van Vorstenbosch CJAHV, Baumans V, Beynen AC. Preferences of mice and rats for types of bedding material. Lab Anim 1996; 30:234-244

Boice R. Burrows of wild and albino rats: Effects of domestication, outdoor raising, age, experience, and maternal state. J Comp Physiol Psychol 1977; 91:649-661

Brain PF, Büttner D, Costa P, Gregory JA, Heine WOP, Koolhaas, J, Militzer K, Ödberg FO, Scharmann W, Stauffacher M. Rodents. In The Accommodation f of Laboratory Animals in Accordance with Animal Welfare Requirements. Proceedings of an International Workshop held at the Bundesgesundheitsamt, Berlin 17-19 May 1993. O'Donoghue PN, ed, Bundesministerium für Ernährung, Landwirtschaft unf Forsten, Bonn, Germany 1993; 1-14

Brain PF, Palanza P, Parmigiani S. Evaluating pain and distress in laboratory animals. Scand J Lab Anim Sci 1995; 22:18-20

Briese E. Normal body temperature of rats: the setpoint controversy. Neurosci Biobehav Rew 1998; 22(3):427-436

Broom DM, Johnson KG. Stress and Animal Welfare. Chapman & Hall Animal Behaviour Series, Chapman & Hall, London, UK 1993

Chamove AS. Environmental enrichment: A review. Anim Technol 1989; 40:155-178

Chmiel DJ, Noonan M. Preference of laboratory rats for potentially enriching stimulus objects. Lab Anim 1996; 30:97-101

Clark JD, Rager DR, Calpin JP. Animal well-being. III. An overview of assessment. Lab Anim Sci 1997; 47:580-585

Commission of the European Communities. Report from the Commission to the Ccouncil and the European parlament. Third Report from the Commission to the Council and the European parlament on the Statistics on the Number of Animals Used for Experimental and Other Scientific Purposes in the Member States of the European Union, Brussels 2002

Council of Europe. European Convention for the Protection of Vertebrate Animals used for Experimental and other Scientific Purposes (ETS 123). Council of Europe, Strasbourg 1986

Council of Europe. Revision of Appendix A of the Convention. Draft Species specific provisions for rodents and rabbits. Working Party for the Preparation of the Multilateral Consultation of parties to the European Convention for the Protection of Vertebrate Animals used for Experimental and Other Scientific Purposes. Strasbourg, 14 October 2002; http:www.coe.int

Dean SW. Environmental enrichment of laboratory animals used in regulatory toxicology studies. Lab Anim 1999; 33:309-327

Diehl KH, Hull R, Morton D, Pfister R, Rabemampianina Y, Smith D, Vidal JM, Vorstenbosch C. A good practice guide to the administration of substances and removal of blood, including routes and volumes. J Appl Toxicol 2001; 21:15-23

Duke JL, Zammit T, Lawson DM. The effects of routine cage-changing on cardiovascular and behavioral parameters in male Sprague-Dawley rats. Contemp Top Lab Anim Sci 2001; 40(1):17-20

Eskola S, Kaliste-Korhonen E. Effects of cage type and gnawing blocks on weight gain, organ weights and open-field behaviour in Wistar rats. Scand J Lab Anim Sci 1998; 25:180-193

Eskola S, Kaliste-Korhonen E. Nesting material and number of females per cage: effects on mouse productivity in BALB/c, C57BL/6J, DBA/2 and NIH/S mice. Lab Anim 1999a; 33:122-128

Eskola S, Kaliste-Korhonen E. Aspen wood-wool is preferred as a resting place, but does not affect intracage fighting in male BALB/c and C57BL/6J mice. Lab Anim 1999b; 33:108-121

Eskola S, Kaliste-Korhonen E. Enriched environment: Effects on agonistic behaviour, physiology and variation. Abstract in: Third World Congress on Alternatives and Animal Use in the Life Sciences, 29 August – 2 September, Bologna, Italy. Programme and Abstracts. Van Zeller AM, Halder M eds, A special issue of ATLA 1999c; 390

Eskola S, Lauhikari M, Voipio H-M, Laitinen M, Nevalainen T. environmental enrichment may alter the number of rats needed to achieve statistical significance. Scand J Lab Anim Sci 1999; 26(3):134-144

Eskola S, Madetoja M, Taimioja A, Sarviharju M, Kaliste-Korhonen E. Grid floor or solid bottom: which is better? Abstract. In 30th Annual Symposium and Educational Days of the Scandinavian Society for Laboratory Animals Science, May 4-7, Djurönäset, Sverige 2000

Gentsch C, Lichtsteiner M, Feer H. Locomotor activity, defecation score and corticosterone levels during an open-field exposure: A comparison among individually and group-housed rats, and genetically selected rat lines. Physiol Behav 1981; 27:183-186

Gordong CJ, Fogelson L. Metabolic and thermoregulatory responses of the rat maintained in acrylic or wire-screen cages: implications for pharmacological studies. Physiol Behav 1994; 56(1):73-79

Grant EC, Chance MRA. Rank order in caged rats. Anim Behav 1958; 6:183-194

Gärtner K, Buttner D, Döhler K, Friedel R, Lindena J, Trautschold I. Stress response of rats to handling and experimental procedures. Lab Anim 1980; 14:267-274

Gärtner K. Does enrichment of cages enhance the deviation of experimental data and number of experimental animals needed? Studies in mice. In Versuchstierkunde: Mittler zwischen Forschung und Tierschutz. 36. Wissenschaftliche Tagung der Gesellschaft für Versuchstierkunde GV-SOLAS. Universitäts-Krankenhaus Hamburg-Eppendorf, D, 7-10 September 1998; 93-108

Hansen AK. Handbook of Laboratory Animal Bacteriology. CRC Press LLC, Boca Raton 2000

Harkin A, Connor TJ, O'Donnell JM, Kelly JP. Physiological and behavioral responses to stress: what does a rat find stressful? Lab Anim Europe 2002; 2(4):32-40

Hedrich HJ. History, strains and models. In The Laboratory Rat. The Handbook of Experimental Animals. Krinke GJ ed, Academic Press 2000; 3-16

Heikkilä M, Sarkanen R, Voipio H-M, Mering S, Nevalainen T. Cage position preferences of rats. Scand J Lab Anim Sci 2001; 28(2):65-74

Holson RR, Scallet AC, Ali SF, Turner BB. "Isolation stress" revisited: Isolation-rearing effects depend on animal care methods. Physiol Behav 1991; 49:1107-1118

Home Office. Animals (Scientific Procedures) Act 1986. Code of Practice for the Housing and Care of Animals Used in Scientific Procedures. Her Majesty's Stationery Office, London 1989

Home Office. Code of Practice for the Housing and Care of Animals in Designated Breeding and Supplying Establishments. Her Majesty's Stationery Office, London 1995

Hurst JL. Introduction to rodents. In The UFAW Handbook on the Care and Management of Laboratory Animals, 7th edition. Poole T ed, Blackwell Science Ltd 1999; 262-273

Hurst JL, Barnard CJ, Hre R, Wheeldon EB, West CD. Housing and welfare in laboratory rats: time-budgeting and pathophysiology in single-sex groups. Anim Behav 1996; 52:335-360

Hurst JL, Barbard CJ, Nevison CM, West CD. Housing and welfare in laboratory rats: welfare implications of isolation and social contact among caged males. Anim Welf 1997; 6:329-347

Hurst JL, Barnard CJ, Tolladay U, Nevison CM, West CD. Housing and welfare in laboratory rats: effects of cage stocking density and behavioural predictors of welfare. Anim Behav 1999; 58:563-586

ILAR Journal. Special Issue: Implications of Human-Animal Interactions and Bonds in the Laboratory. ILAR J 2002; 43(1)

Inglis IR, Hudson A. Wild rats and mice. In The UFAW Handbook on the Care and Management of Laboratory Animals, 7th edition. Poole T ed, Blackwell Science Ltd, Oxford, UK 1999; 274-281

Jacobs GH, Fenwick JA, Williams GA. Cone-based vision of rats for ultraviolet and visible lights. J Exp Biol 2001; 204:2439-2446

Joint Working Group on Refinement: BVA/FRAME/RSPCA/UFAW. Removal of blood from laboratory mammals and birds. Lab Anim 1993; 27:1-22 and Lab Anim 1994; 28:178-179

Joint Working Group on Refinement: BVAAF/FRAME/RSPCA/UFAW. Refining procedures for the administration of substances. Lab Anim 2001; 35:1-41

Kaliste-Korhonen E, Eskola S, Rekilä T, Nevalainen T. Effects of gnawing material, group size and cage level in rack on Wistar rats. Scand J Lab Anim Sci 1995; 22:291-299

Keenan KP, Ballam GC, Haught DG, Laroque P. Nutrition. In The Laboratory Rat. The Handbook of Experimental Animals. Krinke GJ ed, Academic Press 2000; 57-76

Koolhaas JM. The laboratory rat: In The UFAW handbook on the care and management of laboratory animals, 7th edition. Poole T ed, Blackwell Science Ltd, Oxford, UK 1999; 313-330

Kramer K, Voss HP, Grimbergen JA. Circadian rhythms of heart rate, body temperature, and locomotor activity in freely moving mice measured with radio telemetry. Lab Anim 1998; 27(8):23-26

Krohn TC, Hansen AK. The effects of and tolerances for carbon dioxide in relation to recent developments in laboratory animal housing. Scand J Lab Anim Sci 2000; 27:173-181

Krohn TC, Hansen AK. Carbon dioxide concentrations in unventilated IVC cages. Lab Anim 2002; 36:209-212

Lawlor M. The size of rodent cages. In Guidelines for the Well-being of Rodents in Research. Guttman HN ed, Scientist Centre for Animal Welfare, Bethesda 1990; 19-27

Levine S. Primary social relationships influence the development of the hypothalamic-pituitary-adrenal axis in the rat. Physiol Behav 2001; 73:255-260

Lonstein JS, Gammie SC. Sensory, hormonal, and neural control of maternal aggression in laboratory rodents. Neurosci Biobehav Rev 2002; 26:869-888

Lore R, Schultz LA. The ecology of wild rats: applications in the laboratory. In Ethoexperimental Approaches to the Study of Behaviour. Blanchard RJ, Brain PF, Blanchard DC, Parmigiani S eds, Kluwer, Dordrecht 1989; 607-622

Maccari S, Darnaudery M, Morley-Fletcher S, Zuena AR, Cinque C, Van Reeth O. Prenatal stress and long-term consequences: implications of glucocorticoid hormones. Neurosci Biobehav Rev 2003; 27:119-127

Manser CE, Morris TH, Broom DM. An investigation into the effects of solid or grid flooring on the welfare of laboratory rats. Lab Anim 1995; 29:353-363

Manser CE, Elliot H, Morris TH, Broom DM. The use of a novel operant test to determine the strength of preference for flooring in laboratory rats. Lab Anim 1996; 30:1-6

Manser CE, Broom DM, Overend P, Morris TH. Investigations into the preferences of laboratory rats for nest-boxes and nesting materials. Lab Anim 1998a; 32:23-35

Manser CE, Broom DM, Overend P, Morris TH. Operant studies to determine the strength of preference in laboratory rats for nest-boxes and nesting materials. Lab Anim 1998b; 32:36-41

Mering S. Housing Environment and Enrichment for Laboratory Rats – Refinement and Reduction Outcomes. Kuopio University Publications C. Natural and Environmental Sciences, University of Kuopio, Finland 2000; Series number 114

Mering S, Kaliste-Korhonen E, Nevalainen T. Estimates of appropriate number of rats: interaction with housing environment. Lab Anim 2001; 35:80-90

Moberg GP. Biological response to stress: implications for animal welfare. In The Biology of Animal Stress. Basic Principles and implications for animal welfare. Moberg GP, Mench JA eds, CAB International 2000; 1-21

Morton DB, Hau J. Welfare assessment and humane endpoints. In Handbook of Laboratory Animal Science. Volume 1. Essential Principles and Practices, 2nd Edition. Hau J, van Hoosier GL eds, CRC Press 2003; 457-486

Nagel R, Stauffacher M. Ethologische Grundlagen zur Beurteilung der Tiergerechtheit der Haltung von Ratten in Vollgitterkäfigen. Tierlaboratorium 1994; 17:119-132

National Research Council. Nutrient Requirements of Laboratory Animals, 4th revised edition. National Academy Press, Washington DC 1995

National Research Council. Guide for the Care and Use of Laboratory animals. National Academy Press, Washington DC 1996

Nelson EE, Panksepp J. Brain substrates of infant-mother attachment: contributions of opioids, oxytocin, and norepinephrine. Neurosci Biobehav Rev 1998; 22(3):437-452

Nevalainen T, Vartiainen T. Volatile organic compounds in commonly used beddings before and after autoclaving. Scand J Lab Anim Sci 1996: 23:101-104

Nowak RM. Walker's Mammals of the World, 5th edition. Johns Hopkins University Press, Baltimore, Md 1991

OECD (Organisation for Economic Co-operation and Development). Guidance document on the recognition, assessment, and use of clinical signs as humane endpoints for experimental animals used in safety evaluation.). Environmental Health and Safety Publications Series on Testing and Assessment No 19. ENV/JM/MONO(2000)7, 2000

Patterson-Kane EG, Hunt M, Harper D. Behavioral indexes of poor welfare in laboratory rats. J Appl Anim Welfare Sci 1999; 2:97-110

Patterson-Kane EG, Harper DN, Hunt M. The cage preferences of laboratory rats. Lab Anim 2001; 35:74-79

Pekow CA, Baumans V. Common nonsurgical techniques and procedures. In Handbook of Laboratory Animal Science, Second edition. Hau J, Van Hoosier Jr GL eds, CRC Press 2003; 351-390

Pelkonen KHO, Hänninen OOP. Cytotoxicity and biotransformation inducing activity of rodent beddings. A global survey using the Hepa-1 assay. Toxicology 1997; 122:73-80

Percy DH, Barthold SW. Pathology of Laboratory Rodents & Rabbits, Second edition. Iowa State University Press, Ames, Iowa, USA 2001

Pérez C, Canal JR, Domínguez E, Campillo JE, Guillén M, Torres MD. Individual housing influences certain biochemical parameters in the rat. Lab Anim 1997; 31:357-361

Poole TB. The nature and evolution of behavioural needs in mammals. Anim Welf 1992; 1:203-220

Poole T. Happy animals make good science. Lab Anim 1997; 31:116-124

Pryce C, Feldon J. Long-term neurobehavioural impact of the postnatal environment in rats: manipulations, effects and mediating mechanisms. Neurosci Biobehav Rev 2003; 27:57-71

Purves KE. Rat and mice enrichment. In Animal Alternatives, Welfare and Ethics. Proceedings of the 2nd World Congress on Alternatives and Animal Use in the Life Sciences. Utrecht, The Netherlands 20-24 October 1996. Zutphen LFM, Balls M eds, Elsevier 1997; 199-207

Ras T, van de Ven M, Patterson-Kane EG, Nelson K. Rats' preferences for corn versus wood-based bedding and nesting material. Lab Anim 2002; 36:420-425

RSPCA/UFAW Rodent Welfare Group. Individually ventilated cages and rodent welfare. Report of the 2002 meeting by Hawkins P, Anderson D, Applebee K, Key D, Wallace J, Milite G, Macarthur Clark J, Hubrecht R, Jennings M. Anim Technol Welf 2003; April: 23-34

Rosenzweig MR, Bennett EL. Psychobiology of plasticity: effects of training and experience on brain and behavior. Behav Brain Res 1996; 78:57-65

Roughan JV, Flecknell PA. Buprenorphine: a reappraisal of its antinociceptive effects and therapeutic use in alleviating post-operative pain in animals. Lab Anim 2002; 36:322-343

Rowan AN, Moore DE. Refinement and Rodents: A report concerning the wellbeing of laboratory rodents. Tufts Center for Animals and Public Policy, Report number 9, North Grafton, MA, December 1991

Saibaba P, Sales GD, Stodulski G, Hau J. Behaviour of rats in their home cages: daytime variations and effects of routine husbandry procedures analysed by time sampling techniques. Lab Anim 1996; 30: 13-21

Sapolsky RM, Meaney MJ. Maturation of the adrenocortical stress response: neuroendocrine control mechanisms and the stress hyporesponsive period. Brain Res Rev 1986; 11:65-76

Sharp J, Zammit T, Azar T, Lawson D. Does witnessing experimental procedures produce stress in male rats? Contemp Top Lab Anim Sci 2002; 41(5):8-12

Sharp J, Zammit T, Azar T, Lawson D. Stress-like responses to common procedures in individually and group-housed female rats. Contemp Top Lab Anim Sci 2003a; 42(1):9-18

Sharp J, Zammit T, Azar T, Lawson D. Are "By-stander" female Sprague-Dawley rats affected by experimental procedures? Contemp Top Lab Anim Sci 2003b; 42(1):19-27

Soffié M, Hahn K, Terao E, Eclancher F. Behavioural and glial changes in old rats following environmental enrichment. Behav Brain Res 1999; 101:37-49

Stauffacher M. Environmental enrichment, fact and fiction. Scand J Lab Anim Sci 1995; 22:39-42

Stark DM. Wire-bottom versus solid-bottom rodent caging issues important to scientists and laboratory animal science specialists. Contemp Top Lab Anim Sci 2001; 40(6):11-14

Tsai PP, Hackbarth H. Environmental enrichment in mice: Is it suitable for every experiment? Abstract in International Joint Meeting XII ICLAS General Assembly & Conference, FELASA 7th Symposium, Palma de Mallorca, Balearic Islands, E, May 26-28, 1999

Tuli JS, Smith JA, Morton DB. Corticosterone, adrenal and spleen weight in mice after tail bleeding, and its effect on nearby animals. Lab Anim 1995; 29:90-95

Van Bergeijk JP, van Herck H, De Boer SF, Meijer GW, Hesp APM, van Der Gugten J, Beynen AC. Effects of group size and gentling on behavior, selected organ masses and blood consituents in female Rivm:TOX rats. Z Versuchstierkd 1990; 33:85-90

Vanderschuren LJMJ, Niesink RJM, Van Ree JM. The neurobiology of social play behavior in rats. Neurosci Biobehav Rev 1997; 31:309-326

Van de Weerd HA, Van den Broek FAR, Baumans V. Preference for different types of flooring in two rat strains. Appl Anim Behav Sci 1996; 46:251-261

Van de Weerd HA, Van Loo P, Van Zutphen LFM, Koolhaas JM, Baumans V. Preferences for nesting material as environmental enrichment for laboratory mice. Lab Anim 1997; 31:133-143

Van der Meer, Baumans V, Van Zutphen LFM. Measuring welfare aspects of transgenic animals. In Animal Alternatives, Welfare and Ethics. Proceedings of the 2nd World Congress on Alternatives and Animal Use in the Life Sciences, Utrecht, The Netherlands 20-24 October 1996. Zutphen LFM, Balls M eds, Elsevier 1997; 229-233

Van Loo PLP, Kruitwagen CLJJ, Van Zutphen LFM, Koolhaas JM, Baumans V. Modulation of aggression in male mice: influence of cage cleaning regime and scent marks. Anim Welf 2000; 9:281-295

Van Loo PLP, Chang CH, Baumans V. The importance of learning young – the use of nesting material in laboratory rats. Abstract. In 8th FELASA Symposium, June 17-20, Aachen, G, 2002

Van Zutphen LFM, Baumans V, Beynen AC. Principles of Laboratory Animal Science. Elsevier, Amsterdam 1993

Vesell ES. Induction of drug-metabolizing enzymes in liver microsomes of mice and rats by softwood bedding. Science 1967; 157:1057-1058

Voipio H-M. How do Rats React to Sound? PhD thesis. Scand J Lab Anim Sci 1997; 24 (Suppl 1)

Weihe WH: The laboratory rat. In The UFAW Handbook on the Care and Management of Laboratory Animals. Poole T ed, Longman Scientific & Technical, London, UK 1987; 309-330

Wolfensohn S, Lloyd M. Handbook of Laboratory Animal Management and Welfare. Oxford Univerity Press, Oxford, UK 1994

Young D, Lawlor PA, Leone P, Dragunow M, Furing MJ. Environmental enrichment inhibits spontaneous apoptosis, prevents seizures and is neuroprotective. Nat Med 1999; 5:448-453

Chapter 9

THE WELFARE OF LABORATORY GUINEA PIGS

Norbert Sachser, Christine Künzl and Sylvia Kaiser
Department of Behavioural Biology, University of Münster, Münster, Germany

1. ORIGIN AND BIOLOGY OF THE GUINEA PIG

1.1 The Origin of the Guinea Pig

Guinea pigs (*Cavia aperea* f. *porcellus*, Figure 9-1) are domesticated, conspicuously docile animals, which have spread over the whole world. They are among the most popular pets, raised for fancy and as companions, as well as common laboratory animals in scientific research.

Figure 9-1. The guinea pig (*Cavia aperea* f. *porcellus*). Photo: M. Aulbur

Guinea pigs have a long history as pets and were domesticated about 3000–6000 years ago. They are well adapted to man-made housing conditions but no longer occur in natural habitats. Despite their name, they do not originate from Guinea but from the highlands of South America (Benecke

1994, Gade 1967, Herre and Röhrs 1990, Hückinghaus 1961, Hyams 1972, Wing 1977).

In the middle of the 16th century the Spaniards discovered guinea pigs, which were already domesticated at that time by the Incas in the Andes and were thereafter introduced into Europe. Within the European population the animals rapidly grew into a popular pet (Benecke 1994, Clutton-Brock 1989, Gade 1967, Hyams 1972).

In those early days the animals played an important role in providing the Indians with meat. Even today guinea pigs are important for the consumption of meat in the rural population of South America. On occasions, they have even been used for religious performances or medicinal acts (Benecke 1994, Clutton-Brock 1989, Gade 1967, Herre and Röhrs 1990, Hyams 1972, Weir 1974, Wing 1977). The conditions under which guinea pigs are kept today in South America are not very rigorous; the animals are left to scavenge in and around the huts of the Indians, and it can be assumed that a similar husbandry has been the norm throughout the period of their domestication (Stahnke and Hendrichs 1988, Weir 1974).

The origin of the animal's name is not clear, though various possible explanations have been put forward. In most of the European countries the animals are regarded as small pigs as they squeal and emit squeaky sounds like little pigs and have fat bodies. In addition, they came from across the sea and were, therefore, named "cochon d'Inde" in French or "Meerschweinchen" in German. The first part of the English name "Guinea pig" may derive from the South American countries Guayana or French Guiana because the animals arrived via ships from South America. An alternative explanation could be that the name may have arisen because the ships that transported the guinea pigs to Europe took a stopover in Guinea on the west coast of Africa. Finally the animal could have been named for the coin, guinea, the price of the animal in England in the 17th century (Clutton-Brock 1989, Sutherland and Festing 1987, Wagner 1976). The common name for the guinea pig, cavy, probably derives from its generic name, *Cavia*.

1.2 The wild ancestor *Cavia aperea*

Anatomical and morphological studies leave no doubt that the domestic guinea pig derives from the wild cavy (*Cavia aperea*) that is among the most common and wide spread rodents of South America. This conclusion is mainly drawn from comparisons of the skull and the teeth between both forms. In addition, fertile hybrid progeny of crosses between *Cavia aperea* and *Cavia aperea* f. *porcellus*, which can be bred without problems, confirms this (Figure 9-2, Benecke 1994, Gade 1967, Herre and Röhrs 1990,

Hückinghaus 1961, Künzl and Sachser 1999, Nachtsheim and Stengel 1977, Nehring 1889, Rood 1972, Weir 1972).

The guinea pig thus belongs to the subfamily Caviinae (family Caviidae, order Rodentia) as does its wild ancestor, the wild cavy. All members of the subfamily Caviinae are medium-sized, tail-less rodents that have four digits on the fore feet and three digits on the hind feet. All forms - except the domestic guinea pig - show agouti dorsal pelage and a lighter underside (Rood 1972, Wagner 1976). The subfamily is divided into four genera (*Cavia, Galea, Kerodon* and *Microcavia*), which are widely distributed throughout South America and inhabit a wide range of ecological niches.

Figure 9-2. Cavia aperea (right), *Cavia aperea* f. *porcellus* (centre) and the hybrid of both forms (left) Photo: M. Aulbur

The species *Cavia aperea* has the widest distribution occurring from southern Colombia through Brazil into Argentina. Wild cavies are group living animals (Rood 1972, Sachser et al. 1999). They live in small groups consisting of one adult male and one or several females and their unweaned offspring (Asher and Sachser in press). *C. aperea* is a crepuscular, non-climbing species, which does not dig burrows, but hides and moves through tunnels made in dense vegetation (Guichón and Cassini 1998, Rood 1972, Stahnke and Hendrichs 1988). It inhabits relatively humid habitats with dense ground vegetation (Eisenberg 1989, Mares and Ojeda 1982, Redford and Eisenberg 1992, Stahnke and Hendrichs 1988) and is commonly found in linear habitats, such as fields margins or roadsides (Cassini and Galante 1992, Guichón and Cassini 1998, Rood 1972). The typical habitat of *C. aperea* contains a cover zone with high and dense vegetation, which the animals use as protection from predator attacks (Rood 1972), and an adjacent, more open zone of short vegetation where cavies forage (Cassini 1991,

Cassini and Galante 1992, Guichón and Cassini 1998). Unless their habitat is destroyed, adult *C. aperea* appear to have stable home ranges (Rood 1972). *C. aperea* is a difficult species to study in the field as most social activities occur in tall vegetation and human interference is frequently a problem in the roadside habitats, that it frequents (Rood 1972).

Under semi-natural and laboratory conditions adult male *C. aperea* are highly incompatible in the presence of females; whereas, female *C. aperea* organise themselves into linear dominance hierarchies. The males' body mass is 11% higher than that of non-pregnant females. The male-male competition is assumed to bring about a polygynous mating system (Sachser et al. 1999). Whenever a female comes into oestrus, only one male is present. This male thus mates with several females, whereas every female mates with a single male (Sachser 1998a). Females play an active role in bringing about this species' social and mating system by displaying clear preferences for single males that become favourite social and mating partners (Hohoff 2002).

1.3 Behavioural and physiological consequences of domestication

As illustrated above, the domestic guinea pigs descended from the wild cavy at least 3000 years ago. They were formed from the wild animal (wild cavy) by a gradual transformation process, called domestication, occurring over many generations and centuries (Fox 1978, Haase and Donham 1980, Herre and Röhrs 1990). This process is always accompanied by distinct changes in morphology, physiology, and behaviour (Clutton-Brock 1989, Darwin 1859, 1868, Fox 1978, Hale 1969, Herre and Röhrs 1990, Künzl and Sachser 1999, Price 1984). As a consequence, wild and domestic animals require different conditions and resources to ensure their welfare, though in a biological sense they still belong to the same species (Sachser 2001).

Interestingly, from behavioural observations it appears that the repertoire of the behavioural patterns is similar in domesticated and wild guinea pigs. Thus, domestication has not resulted in the loss or addition of behavioural elements (Künzl and Sachser 1999, Rood 1972, Stahnke 1987).

Distinct differences, however, occur in behavioural frequencies and thresholds (Künzl and Sachser 1999, see Table 9-1): Domestic guinea pigs display lower levels of intraspecific aggressive behaviour and higher levels of sociopositive behaviour (e. g. social grooming) than the wild ancestors. Thus the process of domestication has led to typical traits - reduced aggressiveness, increased tolerance of conspecifics - that are also common to other domesticated species. In addition, overt courtship behaviour is more frequently expressed in guinea pigs, and a lower threshold for vocalisation is found. Last, guinea pigs are distinctly less attentive to their physical environment and show much less exploration behaviour than their ancestors (Künzl and Sachser 1999, Künzl et al. 2003).

Table 9-1. Behavioural consequences of domestication: Comparison between wild (Cavia aperea) and domestic guinea pigs (*Cavia aperea f. porcellus*). For reference and original data see Künzl and Sachser 1999 and Künzl et al. 2003.

Behaviour	Wild	Domestic
Aggressive	+	-
Sociopositive	-	+
Courtship	-	+
Vocal	-	+
Attentive	+	-
Explorative	+	-

+ high frequency   - low frequency

Owing to these differences, social interactions between members of the wild and the domesticated forms proceed in a completely different manner. When domestic guinea pigs are kept in breeding groups of one adult male and several adult females, the mature sons and daughters will integrate rather peacefully into the social system of the groups and all animals will cohabit in a non-aggressive and non-stressful way (Sachser 1998a). The resulting large mixed-sex groups can even be regarded as the most appropriate way to keep domestic guinea pigs (see below). When adult wild cavies are kept in breeding groups of one male and several females, a completely different picture emerges: the daughters integrate into the linear dominance hierarchy of the females. In contrast, the father and his sons become rather incompatible when the sons attain sexual maturity. Then, in most cases, the sons must be taken out of the groups because otherwise the father will injure or even kill them (Sachser 1998a). Thus, it is nearly impossible to keep adult males together in the presence of females

Domestic guinea pigs have higher body weights than wild cavies. Since guinea pigs have been bred for food for centuries, this increase in body weight is not surprising.

We studied the endocrine stress reaction of wild cavies and domestic guinea pigs to an environmental challenge: the animals were caught and a blood sample was taken. Cortisol concentrations determined from this sample were designated as initial values. Immediately after the first sample, the individuals were placed into an unfamiliar enclosure. At 60 and 120 min after collection of the first sample a second and third sample were taken to determine cortisol concentrations. Wild cavies reacted with a greater increase in their serum cortisol concentrations than did domestic guinea pigs (see Figure 9-3).

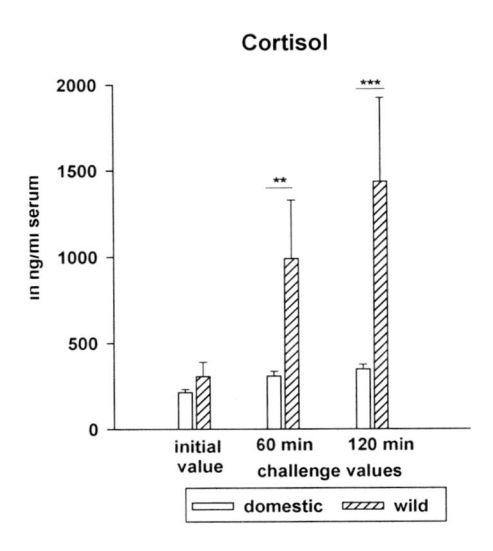

Cortisol

Figure 9-3. Serum cortisol concentrations in male wild and domestic guinea pigs. Values are given as means and SEM. Initial values: concentration before the males were placed into an unfamiliar clean cage; 60- and 120 min challenge values: 60 and 120 min after taking the initial value. * $p<0.05$, ** $p<0.01$, *** $p<0.001$. For reference and original data see Künzl et al. 2003.

Furthermore, catecholamine concentrations were determined from blood samples taken immediately after catching the animals in their home cages. Again, the hormone concentrations were distinctly higher in the wild than in the domesticated form.

Thus at an endocrine level, domestic guinea pigs respond to a stressor with a distinctly reduced activation of the pituitary-adrenocortical and the sympathetic-adrenomedullary systems compared with their wild counterparts. In addition, significant lower cortisol levels in response to adrenocorticotropic hormone (ACTH) application point to a generalised reduction in stress responsibility in the domestic guinea pigs (Künzl 2000).

In general, this trait can be regarded as the physiological correlate of the reduced alertness, nervousness, and sensitivity of the domesticated animals compared to their wild counterparts. While the reduced stress response obviously helps domestic animals to adapt to artificial housing conditions, it is counter selected in wild animals in their natural habitats by natural selection (Künzl and Sachser 1999). The decreased reactivity of the organism's stress axes can be regarded as a physiological mechanism that helps domesticated animals to adjust to man-made housing conditions. As a consequence, under man-made housing conditions, good welfare can be achieved much easier in domesticated animals than in their wild ancestors.

1.4 Biological characteristics of the domestic guinea pig

New-born guinea pigs are precocial. They look like small-sized adults. They are fully furred, their eyes are open and the teeth are fully developed. On the day of birth the young guinea pigs start to eat solid food and drink water, although lactation lasts for 2-3 weeks. Young males reach their sexual maturity within 2-3 months of age (body mass about 500 g), while young females may reach sexual maturity at less than one month of age (with a body mass of about 300 g).

Guinea pigs are fully grown at the age of 8 to 12 months. Under laboratory conditions they can live for 3 to 4 years and reach a body mass ranging from 800 to more than 1000 g. Adult guinea pigs measure up to 30 cm in length (Sutherland and Festing 1987).

Female guinea pigs show a postpartum oestrus, thus becoming receptive immediately after giving birth. If they do not become pregnant at that time, females show periodical oestrus cycles of about 16 days. They show some behavioural changes like increased locomotor activity and increased frequencies of marking behaviour during their oestrus cycle (Birke 1981). The young are born after a gestation period of around 67 days. Usually 1-4 young are born in one litter (Sachser 1994a, Sutherland and Festing 1987).

Guinea pigs do not show a day-night cycle of activity. Instead, they are characterised by an ultradian rhythm, that is, alternating phases of activity and rest that lasting for about 2-3 hours. Thus the activity of these animals is not dependent on the light-dark regime (Sachser et al. 1992).

The vision of guinea pigs is characterised by two types of cones in the retina, having peak sensitivities of about 429 nm and 529 nm. This suggests a retinal basis for a colour vision capacity. Behavioural tests show that guinea pigs have dichromatic colour vision with a spectral neutral point centered at about 480 nm (Jacobs and Deegan 1994). However, they may have poor depth perception.

In contrast, the auditory range of guinea pigs appears to be better than that of humans, especially at high frequencies. Although guinea pigs have a maximum sensitivity ranging between 500 and 8,000 Hertz, they are known to be responsive to an even broader range of frequencies between 125 to 32,000 Hertz (Harper 1976). However, the upper limit of their auditory sensitivity most likely approaches 40,000 to 50,000 Hertz.

Olfactory perception in guinea pigs plays an essential role in the social behaviour of these animals (Beauchamp et al. 1979, 1980, 1982, Martin and

Beauchamp 1982). Male guinea pigs mark individual females with their anal glands. Guinea pigs also mark the environment with their anal glands. Urine has a high communicative value for this species. Young guinea pigs, for example, discriminate between maternal urine and urine of an unknown lactating female (Jäckel and Trillmich 2003).

2. ASSESSMENT OF WELFARE IN GUINEA PIGS

Similar to other mammals, the degrees of stress and welfare of guinea pigs can be determined from 1) the general appearance of the animal, 2) body weight changes, 3) its behaviour and 4) physiological parameters (e.g. Broom and Johnson 1993).

1) General appearance of the animals: The fur should be sleek and not scrubby, the claws should be short, the eyes clear and lucent, not clotted and dull, and the nose as well as the anal region should be clean.

2) Body weight change: This parameter is a nonspecific, but extremely reliable indicator of animal welfare. The body weight of infant guinea pigs should increase consistently, whereas the body weight of adult animals should remain relatively constant. Modifications in their environment can cause guinea pigs to lose body weight. If an adult guinea pig loses more than 10% of its body weight within three days of changing environment, the former housing conditions should be restored immediately.

3) Behaviour: In general, good welfare is indicated by normal frequencies of courtship, comfort, feeding, drinking and locomotory activity. Juveniles should regularly display play behaviour (frisky hops). In contrast, high degrees of stress are signalled by high frequencies of aggression, apathy and absence or reduced frequencies of feeding, drinking and comfort behaviour. Such behavioural patterns are often paralleled by extreme neuroendocrine stress responses. Moreover, guinea pigs typically use specific vocalisations to signal stress. For example, juveniles emit the so-called distress call, a high-pitched whistle, which indicates excitement or anxiety. The distress call can be repeated several times in a call-period. Distress calls are most frequently encountered upon separation of the juveniles from their mothers. Another stress indicator is the frequent occurrence of the vocalisation referred to as chirp. This call is emitted in situations of discomfort (Rood 1972). Chirps are a rapidly repeated series of high-pitched birdlike notes at the same frequency. The animal assumes an attend posture and the body twitches with each note. Chirps may be given in a continuous series lasting 10 or 15 min.

4) Physiological parameters: Degrees of stress can be diagnosed reliably from endocrine parameters. Mammals are characterised by two stress axes:

the pituitary-adrenocortical and the sympathetic-adrenomedullary systems (Henry and Stephens 1977, Sachser 1994a, von Holst 1998). The pituitary-adrenocortical and the sympathetic-adrenomedullary systems play a major role in helping individuals adjust to their physical and social environment. The activation of each of these systems provides the organism with energy and shifts it into a state of heightened reactivity that is a prerequisite for responding to environmental changes in an appropriate way. Although the short-term or moderate activation of both systems represents an adaptive mechanism to cope with conflict situations, the long-term hyperactivation of both the pituitary-adrenocortical and the sympathetic-adrenomedullary system is related to the etiology of irreversible injury and even death (Henry and Stephens 1977, von Holst 1998). Serum glucocorticoid concentrations represent a good indication of the activity of the pituitary-adrenocortical system. The most common method to assess degrees of stress in guinea pigs is to determine concentrations of serum cortisol from blood samples (in guinea pigs cortisol is the main glucocorticoid: Fujieda et al. 1982, Jones 1974). Good indicators for the activity of the sympathetic-adrenomedullary system include serum concentrations of catecholamines (epinephrine, norepinephrine), the heart rate and adrenal tyrosine hydroxylase activity (Sachser 1994a, von Holst 1998). However, it is necessary to sacrifice the animals in order to determine adrenal tyrosine hydroxylase activity. Aside from endocrinological stress parameters, other factors like immune parameters can contribute to the diagnosis of stress and welfare in guinea pigs.

3. SOCIAL LIFE AND WELFARE: BASIC RESEARCH

Much of research has focused on the "social life" and sociophysiology of guinea pigs (e.g. Beauchamp 1973, Berryman 1978, Berryman and Fullerton 1976, Bilsing et al. 1987, Jacobs 1976, King 1956, König 1985, Kunkel and Kunkel 1964, Levinson et al. 1979, Rood 1972, Sachser 1986, 1990, 1994ab, Sachser & Hendrichs 1982, Sachser et al. 1998). Important conclusions have been drawn from this research concerning which social housing conditions can be recommended and which conditions should be avoided. In this section, we present the findings from basic research. In the following section, we derive recommendations for the social housing conditions of guinea pigs.

3.1 Social organisation of domestic guinea pigs

Various forms of social interactions can be observed in guinea pigs. On the one hand, animals compete with each other in agonistic encounters, thus allowing dominance relationships to be established and individuals to stratify into relative social positions. On the other hand, social interactions can also proceed in a sociopositive and/or sexual way, which may result in the establishment of social bondings. The overall structure of dominance relationships and social bondings constitute the animals' social organisation.

Low and high population density
We studied the effects of increasing population density on the patterns of social interactions and reproductive success. A small number of guinea pigs (four males and two females) were placed in a 16 m² enclosure. The animals were allowed to reproduce freely, and after 20 months there were about 50 individuals in the colony. However, even at such high population density, the reproductive success of females (that is, the number of surviving offspring/time) did not decline, and the number of fights between individuals did not distinctly increase (Sachser 1986, 1994b). The question thus arises: which mechanisms allow guinea pigs – in contrast to many other mammals (Christian 1975) – to cope so effectively with high population numbers? The answer to this question is shown in Figure 9-4: Guinea pigs change their patterns of social interactions and their social relationships, that is their social organisation, when population numbers increase (Sachser 1986).

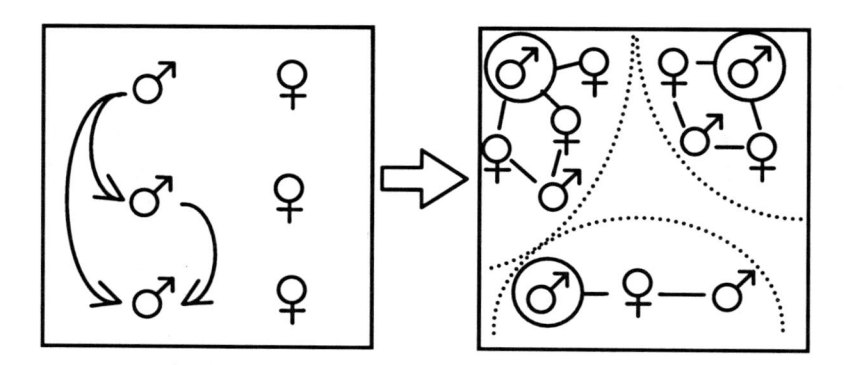

Figure 9-4. Social organisation in guinea pigs at high and low densities. Low density in the left side: arrows among males indicate direction of aggressive behaviours. High density in the right side: lines between males and females indicate individual social bondings. Alphas (circled males) dominate non-alphas (noncircled males). Dotted lines: the borders of "territories" (Sachser 1986).

At low population numbers (for example, three males and three females) the social organisation is characterised mainly by a linear dominance hierarchy among the adult males. Subordinate males retreat whenever a higher ranking conspecific approaches; this largely precludes threat displays and fights. Individuals of identical rank are never found. The highest-ranking male shows much more courtship behaviour towards each of the females than any other male, and he is probably the father of the offspring. Social bondings do not exist between males and females. Among the females there is also a linear rank order (Thyen and Hendrichs 1990). However, their agonistic interactions are less pronounced than those among males. Fighting and threat displays do not occur between the sexes.

When individual numbers increase, guinea pigs change their social organisation. Groups of 10-15 or more split into subunits, each consisting of one to four males and one to seven females. The highest-ranking male of each subunit, the alpha, establishes long-lasting social bondings toward all females of his subunit. The alphas guard and defend their females during oestrus, and they sire more than 85% of offspring, as shown by DNA fingerprinting (Sachser 1998a). The lower ranking males also have bondings with the females of their subunits, that is, they interact predominantly with these animals. Alphas of different subunits respect each other's bondings, that is, they do not court other alphas' females even if these are receptive. In general, individuals belonging to a given subunit live in an area that does not overlap with the area of other subunits. It is in these areas that most of the social interactions are displayed, and where the individuals have their resting and sleeping places. The alphas defend the borders of these areas during their females' oestrus (Sachser 1986, 1994b).

Social organisation at high population numbers is thus characterised as follows: a) splitting of the whole group into subunits provides all individuals with social and spatial orientation; b) escalated fighting is rare because alphas respect the male-female bondings of other alphas; c) the individuals' different social positions are stable over long periods (months), and the basic patterns of social organisation are independent of individual animals. Thus the change in social organisation from a strictly dominance-structured system at low population numbers to a system in which long-lasting bondings are predominant at high numbers can be regarded as a mechanism for facilitating adjustment to increasing population density.

The change of the social organisation
What are the causes for this change? At low density the highest-ranking male monopolises all females. However, the costs, for example, in time and energy spent in agonistic encounters and in maintaining exclusive access to all females within the whole area, increases with the number of competitors

and females. It is economical to defend all females only as long as the net benefits (in terms of reproductive success) exceed net costs. When the relationship between benefits and costs becomes unprofitable, an alternative behaviour yielding a higher reproductive success is preferable, that is, controlling a certain number of females within a certain area and respecting the same behaviour pattern displayed by other males. Thus the cause for change in social organisation is considered to be the highest-ranking male's change in reproductive strategy to maximize his fitness (Sachser 1986).

At low population numbers the highest-ranking male probably sires most offspring. This reproductive advantage results from his dominance. At high population numbers the alphas reproduce with "their" females. This reproductive advantage results from two mechanisms: a) alphas respect the "ownership" of other alphas, and b) they dominate non-alphas in agonistic interactions. Thus at low and high population numbers a polygynous mating system is typical for domestic guinea pigs.

Physiological consequences of social stratification

When male guinea pigs living at different population numbers are compared, they show increased activity of the sympathetic-adrenomedullary system at high densities. The activity of the pituitary-adrenocortical system, in contrast, is not affected by individual numbers. Thus, a male living in a large colony shows cortisol concentrations that are not higher than those for a male living in a small group or together with only one female. These endocrinological data support the behavioural findings: an increase in density does not necessarily contribute to an increase in social stress for the individuals as long as a stable social environment is maintained by social mechanisms (Sachser 1990, 1994a).

At high and low population numbers males take different social positions, which remain stable for months. Alphas, for example, clearly dominate nonalphas of the same subunit. These dominance relationships are independent of place and time. Alphas bite more often and are less often bitten than nonalphas, and they display far more courtship and sexual behaviour than do the lower-ranking males (Sachser 1990). Surprisingly, despite these clear differences in behaviour and status, alphas and nonalphas do not differ significantly in their activity of the pituitary-adrenocortical and the sympathetic-adrenomedullary systems (Sachser 1987, 1994b, Sachser et al. 1998), indicating that having low social status does not necessarily entail a higher degree of social stress than does high social status. Established social relationships resulting in predictable behaviour are seen as the main reason for this, since all individuals live in a stable social organisation.

3.2 Social support

In most mammals mothers are important bonding partners for their offspring. In some species, bondings can also occur between adult individuals, as happens in the guinea pig (see above). The most likely function of bonding is to maintain constant availability or accessibility to a bonding partner (Bowlby 1972). On a physiological level the presence of a bonding partner can buffer the individual's neuroendocrine responses in stressful situations, thus reducing any increase in pituitary-adrenocortical and sympathetic-adrenomedullary activities (Kawachi and Berkman 2001, Sachser et al. 1998). This phenomenon, referred to as social support, has been shown to be one of the most important factors buffering stress responses.

In guinea pigs, bonding exists between mothers and offspring. The presence of the mother has a stress reducing effect on the infant (Hennessy 2003, Hennessy and Ritchey 1987). In an experiment conducted by Sachser et al. (1998), 14-day-old male guinea pigs were removed from a colony and then placed into an unfamiliar cage either singly or together with the mother. Infants separated from their mothers showed a distinct increase in serum cortisol concentrations. However, when they were placed into an unfamiliar cage together with their mother no stress response was found. Not only does the mother exert a stress-reducing effect but so do siblings of the same litter, as well as familiar and even unfamiliar lactating females (Figure 9-5; Sachser et al. 1998).

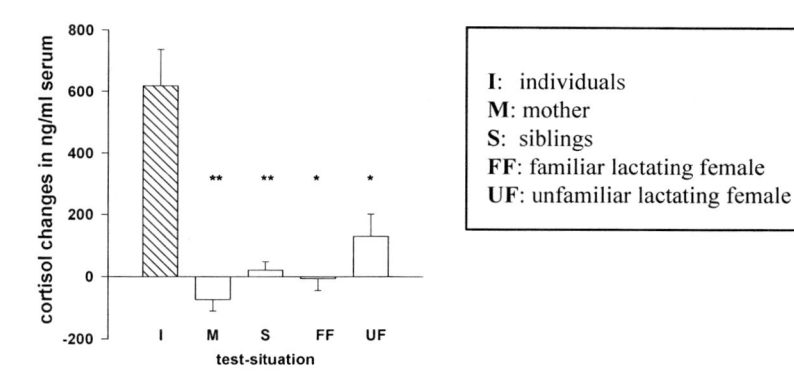

Figure 9-5. Cortisol responses of two-week-old male guinea pigs (N=5-10) which were placed into an unfamiliar enclosure a) individually, together with b) the mother, c) two siblings of the same litter, d) a lactating female from the same colony, e) a lactating female from an unfamiliar colony. Blood samples were taken immediately before and 2 h after the beginning of the experiments. Changes in cortisol titers are given as means ± SEM. Initial values amounted to 319 ± 16 ng/ml serum. Differences to situation I: * $p < 0.05$, ** $p < 0.01$. For reference and original data see Sachser et al. 1998.

Both the social constellation and the spatial environment seem to influence the neuroendocrine and behavioural changes experienced by an organism in a threatening situation. Guinea pig pups, for example, express more distress calls in an unfamiliar than in a familiar environment irrespective of whether their mothers are present or absent (Pettijohn 1979). Furthermore, no increase in the cortisol concentrations of infant guinea pigs have been observed after separation of infants from their mother when staying in their familiar enclosure together with familiar group members (Wewers et al. 2003).

Male-female bonding is very strong in colonies of guinea pigs (see above; see also Jacobs 1976, Sachser 1986). An individual colony-living male has three categories of females: a) his bonded females with whom he has the most amicable interactions, b) females living in the same colony, with whom he is familiar but has no social ties, and c) unfamiliar females living in a different colony, which he has never encountered. Interestingly, the male's endocrine stress response when placed in an unfamiliar cage, that is, his increase in activity of the pituitary-adrenocortical system, is sharply reduced when a bonded female is present. In contrast, the presence of an unfamiliar female or of one with whom he is merely acquainted has little effect. Thus the effect of various types of relationships differs considerably, with substantial social support being provided only by the bonded partner (Sachser et al. 1998).

Furthermore, for female guinea pigs living in large mixed-sex colonies, it has been found that the presence of the bonding partner leads to a sharp reduction in the acute stress response. Thus, in guinea pigs, not only can the female bonding partner provide social support to her male partner but the male bonding partner can also give social support to his female. Nevertheless, the reaction of female guinea pigs differs from that of the males in the presence of a familiar conspecific who is not the bonding partner: their stress response is lower than when tested individually. Thus, for females, but not for males, social support can also be provided by a familiar social partner (Kaiser et al. 2003a).

3.3 Effects of social experiences on physiology and behaviour

The above results show that guinea pigs establish complex and long-lasting stable social structures in which the individuals assume different social positions. Established social relationships result in predictable behaviour, which provide all members of the social system with a high sense of security. Bonding partners provide social support in challenging situations. As a result, an increase in population numbers and stratification into different social positions do not seem to adversely affect the animals' welfare and health. The question arises as to which factors enable guinea pigs to organise themselves in such a non-stressful and non-aggressive way even at high densities? The answer is: a) a great tolerance toward conspecifics, which has been acquired during the process of domestication (see above); b) the ability to establish and to respect dominance relationships; and c) the ability to establish and to respect social bondings. However, whether these abilities are realised depends on the social conditions under which the individuals were reared.

When two adult males that grew up in different large colonies are placed into an unfamiliar enclosure in the presence of an unfamiliar female they quickly establish stable dominance relationships without displaying overt aggression. No significant changes are found in pituitary-adrenocortical or sympathetic-adrenomedullary system activities, either in the dominant or in the subdominant male (Sachser and Lick 1991). However, such a "peaceful" stratification into different social positions requires that the opponents have been engaged in agonistic interactions with older dominant males around puberty, as is the case in individuals reared in colonies. In such encounters, they experience the role of a subdominant individual, thus acquiring the social skills needed to adapt to conspecifics in a non-aggressive and non-stressful way (Sachser 1993, Sachser and Lick 1991; evidence that the time around puberty is decisive for social development is summarised in Sachser et al. 1994).

In contrast, a male, who grows up singly, or with a female, is prevented from experiencing agonistic interactions around puberty (since in this species no fighting and threat displays are found between the sexes). Thus, these social skills cannot be learned.

When two males, both reared in such a way, confront one another in the presence of an unfamiliar female in an unfamiliar enclosure, high levels of aggressive behaviour are displayed, and escalated fighting is frequent. During the first days of the confrontation no stable dominance relationships are established. To avoid irreversible injuries and even death in the losers, we had to stop about half of the experiments. Distinct and persistent increases in pituitary-adrenocortical system activities were found mainly in the subdominant males (Sachser and Lick 1991, Sachser et al. 1994). There is also evidence suggesting that fighting ability is not what determines the outcome of such contests, but rather the winners are those males that succeed in establishing a bond with the female.

The crucial role of social experiences has also been shown in a study taking a different approach (Sachser and Renninger 1993). Males reared individually or in colonies were introduced singly into unfamiliar colonies of conspecifics for a period of 20 days. Colony-reared males easily adjusted to the new social situation. On the first day they explored the new environment but did not court any female, thereby preventing attacks from the male residents, that had already established bondings with the females. In the course of the following days they gradually integrated into the social network of the established colonies and could even gain a higher social position than they had had in their native colonies. In the new colonies, no changes could be discerned in either their body weights or in the activities of the pituitary-adrenocortical and sympathetic-adrenomedullary systems on the 1st, 3rd, 6th, 19th, and 20th days (see Figure 9-6). In contrast, when individually-reared males were placed into a colony of conspecifics, all intruders lost appreciable body weight and some individuals even died, though they had not been injured or attacked by the residents. These males were frequently involved in threat displays and fighting (Sachser and Renninger 1993, Sachser and Lick 1991).

In contrast, female guinea pigs are able to adapt to unfamiliar conspecifics independently of their social rearing conditions (Kaiser et al. 2003d). Thus, in females the social rearing conditions play not such an important role in their integration into an unfamiliar colony of conspecifics like in males.

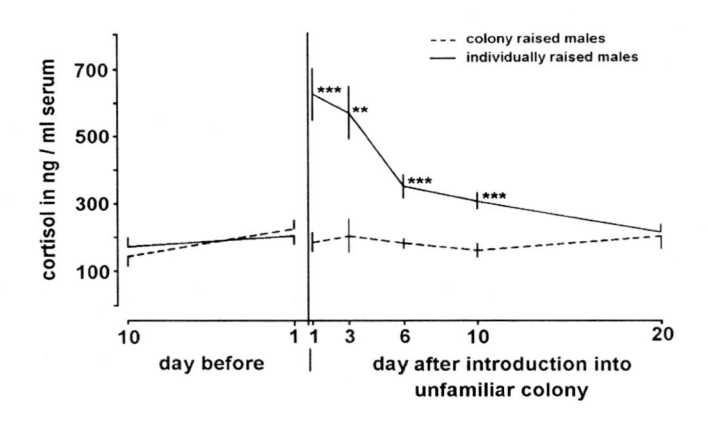

Figure 9-6. Serum glucocorticoid concentrations in males reared individually (dotted lines; N=6) and in colonies (solid lines; N=6) before and after transfer into an unfamiliar colony for 20 days. **$p<0.01$; ***$p<0.001$. For reference and original data, see Sachser and Renninger 1993.

3.4 Effects of the prenatal social environment on physiology and behaviour

Recently, we discovered that even prenatal social factors can have a profound effect on the physiology and behaviour of adult animals (Sachser and Kaiser 1996, Kaiser and Sachser 1998, Kaiser et al. 2003b,c). In this experiment, offspring were compared whose mothers had either lived in a stable social environment during pregnancy and lactation or in an unstable social environment during this period. The stable social environment was created by keeping the group composition (one male, five females) constant; in the unstable social environment situation every third day two females from different groups were exchanged. The subjects studied were male and female offspring of mothers who had either lived in a stable social environment during pregnancy and lactation (SE daughters or SE sons) or in an unstable social environment during the same period of life (UE daughters or UE sons). After weaning, groups of SE and UE daughters and SE and UE sons were established consisting of two females or two males each. At adulthood, the spontaneous behaviour of the offspring was recorded in the home cages.

Surprisingly, the UE daughters were characterised by a distinct behavioural masculinisation: they displayed behavioural patterns – intensive naso-anal sniffing, rumba - which are essential parts of the male courtship behaviour that in mixed-sex groups of guinea pigs are never shown by females (Sachser and Kaiser 1996, Kaiser et al. 2003b, Figure 9-7).

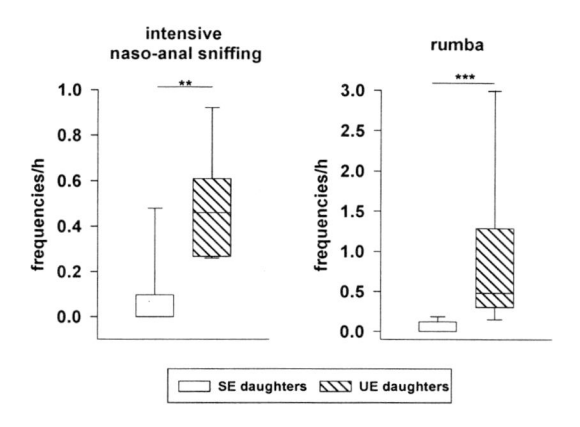

Figure 9-7. Frequencies/h of male-typical courtship behaviour (intensive naso-anal sniffing, rumba) of SE and UE daughters. SE daughters: female offspring of mothers who had lived in a stable social environment during pregnancy and lactation; UE daughters: female offspring of mothers who had lived in an unstable social environment during pregnancy and lactation. Represented as medians, 10, 25, 75 and 90% quartiles (each box N=10). **p<0.01, ***p<0.001. For reference and original data, see Kaiser et al. 2003b.

This behavioural masculinization corresponded to significantly higher serum testosterone concentrations in UE than SE daughters. The activity of the pituitary-adrenocortical system did not differ between the two categories of females. However, significantly higher adrenal tyrosine hydroxylase activities, as an indication of the activity of the sympathetic-adrenomedullary system, and higher adrenal weights in UE than in SE daughters indicated higher degrees of stress in the UE daughters (Kaiser and Sachser 1998). Interestingly, the period of lactation seemed to be of no impact on this phenomenon (Sachser and Kaiser 1996).

These findings corresponded to different distributions of sex hormone receptors in the brain: UE daughters showed an upregulation of androgen-receptor and estrogen-receptor-α in specific brain areas in the limbic system compared to SE daughters (Kaiser et al. 2003b).

What were the effects in male offspring? UE sons showed a behavioural infantilisation that was accompanied by significantly decreased adrenal tyrosine hydroxylase activities (Kaiser and Sachser 2001). Furthermore, UE sons showed a downregulation of androgen receptors and estrogen receptor-α in specific parts of the limbic system such as the medial preoptic area of the hypothalamus and the hippocampus, compared to the SE sons (Kaiser et al. 2003c). Thus, clear evidence is provided that early social stress induces changes in endocrine, autonomic, and limbic brain-function, which is mirrored by changes in male and female adult social behaviour.

4. RECOMMENDATIONS CONCERNING THE SOCIAL HOUSING OF GUINEA PIGS

In this section, we give recommendations concerning the social housing of guinea pigs, which are based on findings from basic research on the guinea pigs' social life and welfare. Table 9-2 summarises the housing conditions recommended for guinea pigs and those that should be avoided.

a) Solitary housing

Domestic guinea pigs were derived from the wild cavy (*Cavia aperea*) and still bear the heritage of their wild counterparts. Thus, we should not expect that the domestic animals would be able to adjust easily to all types of artificial housing. Rather, they require at least some of the essentials that have been selected for in their wild ancestors. Dogs, for example, were derived from group-living wolves, a species with strong social bonding among group members. As in their wild ancestor, to achieve good welfare, dogs still require the availability of a social bonding partner, which can be a member of the same species or a human. Cats were derived from a solitary, territorial species and even in the domesticated form, stable spatial structures seem to be more important for good welfare than the permanent presence of interaction partners, irrespective of whether these are cats or humans (Sachser 2001). Although very little information is available concerning the behaviour of wild cavies in their natural habitat, it is known that they live in social groups (see 3.1 in this chapter). Since the ancestor of the guinea pig is a social living wild species, it seems likely that solitary housing would not be appropriate for its domesticated counterpart.

Correspondingly, the following experiment provides evidence that individually housed guinea pigs less effectively cope with stress situations than those living in social groups. Blood samples were taken from the ear vessels of solitary reared males and males reared together with a female. After the blood sampling the males were put into an unfamiliar empty enclosure or were returned to their home cages. After 20 minutes a further blood sample was taken. Solitary-reared males showed an increase in stress hormone concentrations (cortisol) irrespective of whether they were placed into their home cage or into an unfamiliar enclosure. In contrast, pair-reared males showed a lower increase when placed into their home cage. Thus, the combination of a familiar female and a familiar enclosure can reduce the stress reaction, though the familiar enclosure itself is insufficient to achieve the same effect (Sachser 1994a).

Table 9-2. Beneficial and detrimental housing conditions for guinea pigs

Housing condition	recommended	to be avoided
Solitary		X
Guinea pig and (pygmy) rabbit		X
Pair: 1 male, 1 female	X	
Harems (1 male, several females)	X	
Female groups: 2 females	X	
> 2 females	X	
Male groups: 2 males	X	
> 2 males		X
Small mixed-sex groups	X	
Large mixed-sex groups	XX	

b) Housing with a (pygmy) rabbit

Although it has been popular to house guinea pigs together with a pygmy rabbit, results from preference tests suggest that guinea pigs should be housed together with conspecifics rather than kept together with a (pygmy) rabbit: In a first test, 30-day-old guinea pigs could choose between two enclosures: an empty enclosure and an enclosure where a female conspecific (age around 130 days) was living. In a second test, the young guinea pigs could choose between an empty enclosure and an enclosure containing a pygmy rabbit. The tests were conducted for 12 days. The test environment consisted of two similar enclosures (each 80x50x50 cm), which were connected by a tunnel (length: 10 cm). Each enclosure contained a hut as well as food and water *ad libitum*. The adult female guinea pigs as well as the pygmy rabbits were too large to pass through the tunnel. Two light barriers installed in the tunnel were used to measure the frequencies of the young guinea pigs' movements between the cages as well as the duration of their stays in both cages. We found clear differences. While guinea pigs spent only 40% of their time together with the rabbit, about 80% of the time was spent together with the older female guinea pig (Sachser 1998b). Furthermore, rabbits and guinea pigs showed different daily rhythms and had different communication signals. These results suggest that the housing of guinea pigs together with pygmy rabbits should be avoided.

c) Housing in pairs (1 male, 1 female)

This housing condition is appropriate for guinea pigs, since no fighting and threat displays are found between the sexes in this species. Moreover pair reared males can cope more effectively with adversity than can individually housed conspecifics (see above).

d) Housing in harems (1 male, several females)

This housing condition is also appropriate for guinea pigs. Although agonistic interactions can occur between females, such aggression is rare and of low intensity: escalated fights and bites almost never occur.

e) Female groups

Housing in all-female groups can be recommended. The number of individuals is limited only by the enclosure size (at least 0.25m²/animal), since females can be housed without problems in large groups. Although in all-female groups levels of aggression are slightly higher than in groups with one male and several females (Thyen and Hendrichs 1990), aggression is rare and of low intensity. As is also the case in groups of one male and several females, escalated fights and bites almost never occur. Moreover, females living in large all-female groups are not subjected to high degrees of stress.

f) Male groups

Agonistic behaviour is rarely or never found in groups consisting of two males. In contrast, from the age of about 3-4 months escalated agonistic interactions frequently occur in groups of more than two males. Additionally, animals that live in groups of two or four individuals show lower concentrations of the stress hormone cortisol than those living in groups of six or twelve (Beer and Sachser 1992). Thus, good welfare seems to be promoted in groups of two males, while indicators of social stress regularly occur in larger groups. Therefore, we recommend - if mixed sex housing is not possible - that male guinea pigs be kept in groups of two.

What is the reason for the escalated fights in all-male groups consisting of more than two males? We investigated groups consisting of two, four, six and twelve males. In all these groups the males showed courtship and sexual behaviour towards each other in the same way as males usually court females, that is, even pseudo-copulations occur including mounts and ejaculations. To our surprise, not only is homosexual behaviour exhibited but specific males also displayed female-typical behaviour. These individuals, for example, not only tolerate the courtship and mounting by others, but also actively display defense-urine-spraying, a behavioural pattern which is only typical for females, and usually serves to keep off courting males. These males are designated as "pseudofemales". The other males of the groups behave towards these pseudofemales as they would toward real females: they compete with each other for the pseudofemale. The pseudofemales do not take part in aggressive conflicts and rarely receive aggressive behaviour. Moreover, physiological stress indicators show that pseudofemales are subjected to lowest stress in the groups (Beer and Sachser 1992).

g) Mixed-sex groups of a few males and females

Housing in small mixed-sex groups can also be recommended. However, the formation of such a group with animals of unknown origin should be avoided since it frequently leads to intensive threat and fighting behaviour, as well as extreme stress responses and injuries (see above).

h) Mixed-sex groups of many males and females

The housing condition that we favour most for guinea pigs is the large mixed sex colony. In this housing type, both dominant and subordinate animals can live a good life and are not subjected to high degrees of stress. All animals are able to learn the social skills necessary for cohabitation with conspecifics (see above). However, such a large mixed-sex group should have a varied age structure. The best is to allow a small mixed-sex group to increase to the desired size. The creation of a large group consisting of animals coming from pairs or solitary housing condition as well as animals of unknown origin should be avoided since it frequently leads to intensive threat and fighting behaviour, extreme stress responses and health problems.

5. GENERAL RECOMMENDATION

5.1 Feeding

The food has to be stored under cool, dry conditions and protected against contamination. It should be available *ad libitum*. Guinea pigs should be fed daily at a fixed time. Access to food has to be guaranteed for all animals. Furthermore it is important to ensure the availability of vitamins, especially of vitamin C, since guinea pigs are susceptible to vitamin C deficiency. If the vitamin C content of the food is insufficient, vitamin C or Na-ascorbat can be added to the drinking water (0.5g/l; note, however that dissolved vitamin C decomposes within 24 hours). It is also important to feed hay several times per week. Besides fulfilling dietary requirements, hay can also serve as material for play. Furthermore, guinea pigs must be able to regularly engage in gnawing behaviour to prevent overgrowth of their incisors. For this purpose hard food pellets, carrots or wooden branches are suitable.

5.2 Bedding

The bedding material should be dry and absorbent. It should be free of toxic residue (e.g., timber preservative), vermin and infectious components. Dust should be reduced to a minimum, and the bedding material should be changed once a week.

5.3 Environmental factors

Draught-free ventilation must be installed. The relative humidity should be maintained at around 50-60% and the optimal temperature at 18-22 °C (according to the European convention, Appendix A 20-24 °C (Council of Europe 1986). However, guinea pigs used as pets can also be housed outside throughout the whole year. In such a case, access to dry huts must be ensured for all animals. Additionally, heaters must be installed during frosts.

Guinea pigs should be kept neither in constant light nor in constant darkness. A light-dark rhythm corresponding to 12 h light and 12 h dark as well as a natural light-dark rhythm are suitable (e.g. if the animals live in outdoor pens).

Guinea pigs should be protected against loud noise, especially clanking sounds since these can cause panic reactions and injuries.

5.4 General Housing

Guinea pigs can be kept in either pens or cages. Good experiences have been achieved with ground housing. In this case, bedding (such as wood shavings) - is necessary. However, it should be noted that flagstones can be irreversibly contaminated with urolithic acid.

It is also important to provide sufficient space. Overcrowded conditions in cages can lead to endocrine stress reactions and higher frequencies of aggression. The minimum cage size should not be less than 2500 cm² per animal (according to the European convention Appendix A 600 cm² for adult guinea pigs, Council of Europe 1986). A height of about 15 cm (as in Makrolon cages type III: 39x23x15 cm) does not provide sufficient space, as it can impair the play behaviour (frisky hops) of juveniles and prevents adult individuals from fully rising. We suggest a minimum height of 30 cm.

Specific forms of gratings (e.g., in metabolic cages) can also cause problems and cages with grid floors are inappropriate, since they are often the cause of diseases (e.g., pad abscess). Especially young animals are vulnerable, because their extremities can be injured in wide meshed grids. Structural divisions, such as huts, should also be offered, to provide all animals with a place of refuge.

5.5 Care and Handling

Guinea pigs should regularly undergo a health check, involving examination of the appearance (skin condition, claws, eyes, nose, and anal region) and regular measurement of the body weight. Furthermore, behavioural abnormalities should be recorded. Quarantine should last for 5-15 days. They should be picked up gently with both hands: one hand should firmly hold the shoulder, and the other hand should support the hindquarters.

Because guinea pigs are highly sensitive to changes in their environment, it is important to habituate them slowly to new conditions. The combination of several factors - each of them separately having no negative effects - can often result in strong stress responses and severe health problems. For example, the combination of a new cage and a new drinking bottle should be avoided. New drinking bottles should be offered together with the old one until animals begin to drink from the new one. Moreover, unfamiliar food should be mixed with a familiar one, and the proportion of the old food should be incrementally reduced. When transferring animals to a new cage, no further modifications should be made simultaneously, including providing them with new food or removing a social partner. An animal's ability to adjust to a new situation can be most reliably determined from its drinking and feeding behaviour. If the animals are feeding and more notably drinking, then they can adapt to the new situation. In contrast, refusal of water and food, scrubby fur and apathy point to extremely poor welfare.

5.6 Experimental techniques

Blood samples can be taken from the guinea pigs' ear vessels in a non-stressful way (Sachser and Pröve 1984). The blood sampling procedure is performed in the following way. The animal is caught, and to increase blood flow, a small quantity of salve to increase blood circulation is put on the upper side of one ear for about 30 seconds and then carefully removed. A cold-point-lamp is put behind the ear to increase the visibility of the blood vessels. With a disposable sterilised needle (0.6 x 25 mm) the marginal ear vessels are punctured and the blood is collected in heparinised capillaries. Blood samples have to be taken within a maximum time of 5 min (glucocorticoid concentrations do not increase until 5 min in response to a stressor). In guinea pigs blood sampling by puncturing the ear vessels is a non-stressful event that does not lead to an increase in cortisol concentrations (Sachser 1994a). Thus, there is no need to anaesthetise the animals.

Recently, non-invasive techniques have been developed to assess degrees of stress. For example, glucococorticoids can be determined from saliva. In guinea pigs saliva can be sampled very easily by putting a piece of cotton wool into the mouth of the animal for several minutes.

Administration of substances and other techniques are presented e.g. by North (1999) and Pekow and Baumans (2003).

6. CONCLUSIONS

Guinea pigs (*Cavia aperea* f. *porcellus*) were domesticated 3000-6000 years ago in the highlands of South America. They derived from the wild cavy (*Cavia aperea*) – a social living wild species - and still bear the heritage of their wild counterparts. Based on findings from basic research on the guinea pigs' social life and welfare the following recommendations can be made: (1) Solitary housing and housing of large all-male groups (> 2 males) should be avoided. (2) Guinea pigs should be housed together with conspecifics rather than kept together with a (pygmy) rabbit. (3) Keeping of pairs (1 male, 1 female), harems (1 male, several females), all-female groups, groups consisting of two males as well as mixed-sex groups of a few males and females can be recommended. (4) The housing condition that we favour most is the large mixed-sex colony of many males and females. (5) Because guinea pigs are highly sensitive to changes in their environment, it is important to habituate them slowly to new conditions. If the animals are feeding and more notably drinking, then they can adapt to the new situation. In contrast, refusal of water and food, scrubby fur and apathy point to extremely poor welfare.

ACKNOWLEDGEMENT

We thank Matthias Asher and Ken Pennington for critical comments on the manuscript.

REFERENCES

Asher M, Sachser N. Social system and spatial organization of wild guinea pigs /cavia aperea) in a natural low density population. J Mammal in press

Beauchamp GK. Attraction of male guinea pigs to conspecific urine. Physiol Behav 1973; 10:589-594

Beauchamp GK, Criss BR, Wellington JL. Chemical communication in Cavia: responses of wild (*C. aperea*), domestic (*C. procellus*) and F1 males to urine. Anim Behav 1979; 27:1066-1072

Beauchamp GK, Wellington JL, Wysocki CJ, Brand JG, Kubie JL, Smith AB. Chemical communication in the guinea pig: urinary components of low volatility and their access to the vomeronasal organ. In Chemical Signals: Vertebrates and Aaquatic Invertebrates. Müller-Schwarze D, Silverstein R eds, Plenum Press, New York 1980; 327-339

Beauchamp GK, Martin IG, Wysocki CJ, Wellington JL. Chemoinvestigatory and sexual behaviour of male guinea pigs following vomeronasal organ removal. Physiol Behav 1982; 29:329-336

Beer R, Sachser N. Sozialstruktur und Wohlergehen in Männchengruppen des Hausmeerschweinchens. In Aktuelle Arbeiten zur artgemäßen Tierhaltung 1991, KTBL-Schrift 351, Darmstadt 1992; 158-167

Benecke N. Der Mensch und seine Haustiere. Konrad Theiss Verlag GmbH & Co, Stuttgart 1994

Berryman JC. Social behaviour in a colony of domestic guinea pigs. Z Tierpsychol 1978; 46:200-214

Berryman JC, Fullerton C. A developmental study of interactions between young and adult guinea pigs. Behaviour 1976; 59:22-39

Bilsing A, Schneider R, Astrath S. Sozialverhalten von Hausmeerschweinchen *Cavia aperea* f. *porcellus* in Abhängigkeit von sozialen Erfahrungen. Zoologische Jahrbücher (Physiologie) 1987; 91:113-120

Birke LIA. Some behavioral changes associated with the guinea-pig oestrus cycle. Z Tierpsychol 1981; 55:79-89

Bowlby J. Attachment. Penguin, Middlesex, England 1972

Broom DM, Johnson KG. Stress and Animal Welfare. Chapman and Hall, London 1993

Cassini MH. Foraging under predation risk in the wild guinea pig *Cavia aperea*. Oikos 1991; 62:20-24

Cassini MH, Galante L. Foraging under predation risk in the wild guinea pig: the effect of vegetation height on habitat utilization. Ann Zool Fenn 1992; 29:285-290

Christian JJ. Hormonal control of population growth. In Hormonal Correlates of Behavior vol. 1. Eleftheriou, BE, Sprott RLS eds, Plenum Press, New York 1975; 205-274

Clutton-Brock J. A Natural History of Domesticated Mammals. Cambridge University Press, Cambridge 1989

Council of Europe. European Convention for the Protection of Vertebrate Animals used for Experimental and other Scientific Purposes (ETS 123). Council of Europe, Strasbourg 1986

Darwin C. On the Origin of Species by Means of Natural Selection. John Murray, London 1859

Darwin C. The Variation of Animals and Plants under Domestication. John Murray, London 1868

Eisenberg JF. Mammals of the Neotropics. Volume 1: the Northern Neotropics. University of Chicago Press, Chicago 1989

Fox MW. Effects of domestication in animals: A review. In The Dog: Its Domestication and Behaviour. Fox MW ed, Garland STPM Press, New York, London 1978; 3-19

Fujieda K, Goff AK, Pugeat M, Strott CA. Regulation of the pituitary-adrenal axis and corticosteroid-binding globulin-cortisol interaction in the guinea pig. Endocrinology 1982; 111:1944-1949

Gade DW. The guinea pig in andean folk culture. Geogr Rev 1967; 57:213-224

Guichón ML, Cassini MH. Role of diet selection in the use of habitat by pampas cavies *Cavia aperea pamparum* (Mammalia, Rodentia). Mammalia 1998; 62:23-35

Haase E, Donham RS. Hormones and domestication. In Avian Endocrinology. Epple A, Stetson MH eds, Academic Press, New York 1980; 549-565

Hale EB. Domestication and the evolution of behaviour. In Behaviour of Domestic Animals, 2nd edition. Hafez ESE ed, Williams and Wilkings, Baltimore 1969; 22-42

Harper LV. Behavior. In The Biology of the Guinea Pig. Wagner IE, Manning PJ eds, Academic Press, New York, San Francisco, London 1976; 31-52

Hennessy MB. Enduring maternal influences in a precocial rodent. Dev Psychobiol 2003; 42:225-236

Hennessy MB, Ritchey RL. Hormonal and behavioral attachment responses in infant guinea pigs. Dev Psychobiol 1987; 20:613-625

Henry JP, Stephens PM. Stress, Health and the Social Environment. A Sociobiological Approach to Medicine. Springer, New York, Heidelberg, Berlin 1977

Herre W, Röhrs M. Haustiere - zoologisch gesehen. Gustav Fischer Verlag, Stuttgart, New York 1990

Hohoff C. Female choice in three species of wild guinea pigs. PhD thesis, University of Münster, Germany 2002

Hückinghaus F. Zur Nomenklatur und Abstammung des Hausmeerschweinchens. Z Säugetierkd 1961; 26:108-111

Hyams E. Animals in the service of man: 10 000 years of domestication. J.M. Dent and Sons Ltd, London 1972

Jacobs GH, Deegan JF. Spectral sensitivity, photopigments, and color vision in the guinea pig (Cavia porcellus). Behav Neurosci 1994; 108: 993-1004

Jacobs WW. Male-female associations in the domestic guinea pig. Anim Learn Behav 1976; 4:77-83

Jäckel M, Trillmich F. Olfactory individual recognition of mothers by young guinea-pigs (*Cavia porcellus*). Ethology 2003; 109:197-208

Jones CT. Corticosteroid concentrations in the plasma of fetal and maternal guinea pigs during gestation. Endocrinology 1974; 95:1129-1133

Kaiser S, Sachser N. The social environment during pregnancy and lactation affects the female offsprings' endocrine status and behaviour in guinea pigs. Physiol Behav 1998; 63:361-366

Kaiser S, Sachser N. Social stress during pregnancy and lactation affects in guinea pigs the male offsprings' endocrine status and infantilizes their behaviour. Psychoneuroendocrinology 2001; 26:503-519

Kaiser S, Kirtzeck M, Hornschuh G, Sachser N. Sex specific difference in social support - a study in female guinea pigs. Physiol Behav 2003a; 79:297-303

Kaiser S, Kruijver FPM, Swaab DF, Sachser N. Early social stress in female guinea pigs induces a masculinization of adult behavior and corresponding changes in brain and neuroendocrine function. Behav Brain Res 2003b; 144:199-210

Kaiser S, Kruijver FPM, Straub RH, Sachser N, Swaab DF. Early social stress in male guinea pigs changes social behaviour, and autonomic and neuroendocrine functions. J Neuroendocrinol 2003c; 15:761-769

Kaiser S, Nübold T, Rohlmann I, Sachser N. Female, but not male guinea pigs adapt easily to a new social environment irrespective of their rearing conditions. Physiol Behav 2003d; 80:147-153

Kawachi I, Berkman LF. Social ties and mental health. J Urban Health 2001; 78:458-467

King JA. Social relations of the domestic guinea pig living under semi-natural conditions. Ecology 1956; 37:221-228

Kunkel P, Kunkel, I. Beiträge zur Ethologie des Hausmeerschweinchens. Z Tierpsychol 1964; 21:602-641

König B. Maternal activity budget during lactation in two species of *Caviidae* (*Cavia porcellus* and *Galea musteloides*). Z Tierpsychol 1985; 68:215-230

Künzl C. Verhaltensbiologische Untersuchungen zur Domestikation des Meerschweinchens. PhD Thesis, University of Münster, Germany 2000

Künzl C, Sachser N. The behavioural endocrinology of domestication: a comparison between the domestic guinea pig (*Cavia aperea* f. *porcellus*) and it's wild ancestor the wild cavy (*Cavia aperea*). Horm Behav 1999; 35:28-37

Künzl C, Kaiser S, Meier E, Sachser N. Is a wild mammal kept and reared in captivity still a wild animal? Horm Behav 2003; 43:187-196

Levinson DM, Buchanan DR, Willis FN. Development of social behavior in the guinea pig in the absence of adult males. Psychol Rec 1979; 29:361-370

Mares MA, Ojeda RA. Patterns of diversity and adaptation in South American hystricognath rodents. In Mammalian Biology in South America. Mares MA, Genoways HH eds, Pymatuning Laboratory of Ecology, Special Publications No. 6, Pittsburgh, Pennsylvania 1982; 393-432

Martin IG, Beauchamp GK. Olfactory recognition of individuals by male cavies (*Cavia aperea*). J Chem Ecol 1982; 8:1241-1249

Nachtsheim H, Stengel H. Vom Wildtier zum Haustier. Verlag Paul Parey, Berlin, Hamburg 1977

Nehring A. Über die Herkunft des Hausmeerschweinchens. Sitzungsberichte der Gesellschaft naturforschender Freunde zu Berlin 1889; 1:1-4

North D. The Guinea-pig. In The UFAW Handbook on the Care and Management of Laboratory Animals, seventh edition, volume I Terrestrial Vertebrates. Poole T ed, Blackwell Science Ltd 1999; 367-388

Pettijohn TF. Attachment and separation distress in the infant guinea pig. Dev Psychobiol 1979; 12:73-81

Pekow CA, Baumans V. Common nonsurgical techniques and procedures. In Handbook in Laboratory Animal Science, second edition, vol I Essential Principles and Practices. Hau J, Van Hoosier GL eds, CRC Press 2003; 351-390

Price EO. Behavioral aspects of animal domestication. Q rev biol 1984; 59: 1-32

Redford KH, Eisenberg JF. Mammals of the Neotropics, vol 2, the Southern Cone. University of Chicago Press, Chicago 1992

Rood JP. Ecological and behavioural comparisons of three genera of Argentine cavies. Animal Behaviour Monographs 1972; 5:1-83

Sachser N. Different forms of social organization at high and low population densities in guinea pigs. Behaviour 1986; 97:253-272

Sachser N. Short-term responses of plasma-norepinephrine, epinephrine, glucocorticoid and testosterone titers to social and non-social stressors in male guinea pigs of different social status. Physiol Behav 1987; 39:11-20

Sachser N. Social organization, social status, behavioural strategies and endocrine responses in male guinea pigs. In: Hormones, Brain and Behavior in Vertebrates. Balthazart J ed, Comparative Physiology, Karger, Basel 1990; 9:176-187

Sachser N. The ability to arrange with conspecifics depends on social experiences around puberty. Physiol Behav 1993; 53:539-544

Sachser N. Sozialphysiologische Untersuchungen an Hausmeerschweinchen. Gruppenstrukturen, soziale Situation und Endokrinium, Wohlergehen. Parey, Berlin 1994a

Sachser N. Social stratification and health in non-human mammals - a case study in guinea pigs. In Social Stratification and Socioeconomic Inequality. Ellis L ed, Praeger, Westport 1994b; 2:113-121.

Sachser N. Of domestic and wild guinea pigs: studies in sociophysiology, domestication, and social evolution. Naturwissenschaften 1998a; 85:307-317

Sachser N. Was bringen Präferenztests? In: Aktuelle Arbeiten zur artgemäßen Tierhaltung 1997. KTBL Schrift 380, Darmstadt 1998b; 9-20

Sachser N. What is important to achieve good welfare in animals? In Coping with Challenge: Wwelfare in Animals including Humans. Broom DM ed, Dahlem Workshop Report 87, Dahlem University Press, Berlin 2001; 31-48

Sachser N, Hendrichs H. A longitudinal study on the social structure and its dynamics in a group of guinea pigs (*Cavia aperea* f. *porcellus*). Säugetierkundliche Mitteilungen 1982; 30:227-240

Sachser N, Pröve E. Short-term effects of residence on the testosterone responses to fighting in alpha male guinea pigs. Aggress Behav 1984; 10:285-292

Sachser N, Lick C. Social experience, behavior, and stress in guinea pigs. Physiol Behav 1991; 50:83-90

Sachser N, Lick C, Beer, R, Weinandy R. Tagesgang von Serum-Hormonkonzentrationen und ethologischen Parametern bei Hausmeerschweinchen. Verhandlungen der Deutschen Zoologischen Gesellschaft 1992; 85:120.

Sachser N, Renninger S-V. Coping with new social situations: the role of social rearing in guinea pigs. Ethol Ecol Evol 1993; 5:65-74

Sachser N, Lick C, Stanzel K. The environment, hormones, and aggressive behaviour: a 5-year-study in guinea pigs. Psychoneuroendocrinology 1994; 19:697-707

Sachser N, Kaiser S. Prenatal social stress masculinizes the females' behaviour in guinea pigs. Physiol Behav 1996; 60:589-594

Sachser N, Dürschlag M, Hirzel D. Social relationships and the management of stress. Psychoneuroendocrinology 1998; 23:891-904

Sachser N, Schwarz-Weig E, Keil A, Epplen JT. Behavioural strategies, testis size, and reproductive success in two caviomorph rodents with different mating systems. Behaviour 1999; 136:1203-1217

Stahnke A. Verhaltensunterschiede zwischen Wild- und Hausmeerschweinchen. Z Säugetierkd 1987; 52:294-307

Stahnke A, Hendrichs H. Meerschweinchenverwandte Nagetiere. In Grzimeks Enzyklopädie Säugetiere. Grzimek B ed, Kindler Verlag, München 1988; 314-357

Sutherland SD, Festing MFW. The guinea-pig. In The UFAW Handbook on the Care and Management of Laboratory Animals, Sixth Edition. Poole TB ed, Churchill Livingstone, New York 1987; 393-410

Thyen Y, Hendrichs H. Differences in behaviour and social organization of female guinea pigs as a function of the presence of a male. Ethology 1990; 85:25-34

Von Holst D. The concept of stress and its relevance for animal behavior. In Advances of the Study of Behavior. Lehman DS, Hinde R, Shaw E eds, Academic Press, London, New York 1998; 1-131

Wagner JE. Introduction and taxonomy. In The Biology of the Guinea Pig. Wagner IE, Manning, PJ eds, Academic Press, New York, San Francisco, London 1976; 1-4

Weir BJ. Some notes on the history of the domestic guinea pig. Guinea pig newsletter 1972; 5:2-5

Weir BJ. Notes on the origin of the domestic guinea-pig. Symposium of the Zoological Society of London 1974; 34:437-446

Wewers D, Kaiser S, Sachser N. Maternal separation in guinea pigs: a study in behavioural biology. Ethology 2003; 109:443-453

Wing ES. Animal domestication in the Andes. In Origins of Agriculture. Reed CA ed, Paris:Mouton Publishers, The Hague 1977; 837-859

Chapter 10

THE WELFARE OF LABORATORY RABBITS

Lena Lidfors[1], Therese Edström[2] and Lennart Lindberg[3]
[1]*Department of Animal Environment and Health, Swedish University of Agricultural Sciences, Skara, Sweden;* [2]*Astra Zeneca R & D, Mölndal;* [3]*National Veterinary Institute, Uppsala, Sweden*

1. INTRODUCTION

Rabbits were the fifth most commonly used mammalian laboratory animal after mice, rats, guinea pigs and pigs in Sweden during 2002 (CFN 2003). According to the latest statistics for the EU member states, 227 366 rabbits were used during 1999 (Commission of the European Communities 2003). Both domesticated rabbits and European wild rabbits may be used for experimental research, but there are several problems in keeping and breeding the European wild rabbit (Bell 1999). Today the most common breeds used are the New Zealand White (NZW), the Dutch and the Half Lop (Batchelor 1999). Most laboratories buy these breeds as health defined (previously called Specific Pathogen Free) from accredited breeders (Townsend 1969, Eveleigh et al. 1984).

Rabbits are used for many different purposes with a large number being used for antibody production, but also for orthopaedics and biomaterials (Batchelor 1999). The rabbit is especially suitable for studies on reproduction (Batchelor 1999). Rabbits are also used for cardiac surgery, and studies of hypertension, infectious diseases, virology, embryology, toxicology, experimental teratology (Hartman 1974), arteriosclerosis (Clarkson et al. 1974) and serological genetics (Cohen and Tissot 1974).

Laboratory rabbits are by tradition kept individually in small cages with restricted food availability. This has led to several physiological problems related to the fact that they move too little, as well as behavioural disorders. Over the past 10-15 years many laboratories have improved the housing for rabbits, both in cages and also by introducing floor pens and group housing. However, there are still some aspects to be addressed in rabbit housing especially in relation to the fact that they can be very aggressive when kept in groups.

The aim of this chapter is to present the most recent knowledge about the laboratory rabbit's biology, behavioural needs, optimal environment, housing, feeding, care, handling, health and experimental techniques, in order to ensure their optimum welfare.

2. BIOLOGY AND BEHAVIOURAL NEEDS

2.1 Domestication

The laboratory rabbit originated from the European wild rabbit (*Oryctolagus cuniculus*, Figure 10-1) (Harcourt-Brown 2002). Rabbits were already kept in fenced hunting areas by the Romans 2000 years ago, and domestication appears to have started around the year 500 AD by the French. European wild rabbits were introduced to Great Britain by the Normans soon after the Conquest of 1066 (Meredith 2000). The domestication most probably occurred in the Monasteries, where several different breeds were developed.

Figure 10-1. European wild rabbits in a fenced grass area in England used for behavioural ecology research (photograph by Lena Lidfors 1995)

During the first half of the 20th century many new rabbit breeds were developed, and today there are 76 recognized breeds of fancy fur and Rex rabbits (British Rabbit Council 1991). The New Zealand White (NZW) was accepted as a breed in 1925 according to American Standards (Batchelor 1999). This rabbit has been bred for fast growth for meat production and their white pelts which can be dyed any colour, but it has also become a very commonly used laboratory breed. The negative impact of this breeding emphasis is that laboratory New Zealand White rabbits relatively easily become obese, and food is often restricted in singly housed animals to prevent this. Other rabbit breeds more suitable for laboratory research have been developed more recently, taking into account the purpose of a certain research area, for example the Half Lop, which has long ears suitable for repeated blood samples during anti-body production.

2.2 General biology

The natural diet of rabbits consists of grass and herbs, but also fruit, roots, leaves and bark (Cheeke 1987). The rabbit needs coarse fibre, not just grass (Brooks 1997, Meredith 2000). The rabbit is caecophagous (often called coprophagous), i.e. it needs to ingest the smaller, soft and green coated faeces pellets produced by the caecum 4-8 hours after feed intake in order to survive (Brooks 1997). The rabbit takes the soft faecal pellets directly from its anus, whereas the harder pellets are placed on specific latrines close to the territory borders (Donnelly 1997). The rabbit has a light and fragile skeleton that only makes up 7-8 % of the body weight (Donnelly 1997).

Rabbits can adjust to different ambient temperatures, but have problems with too high temperatures and therefore seek shade from sun light in their burrows or under bushes. They have a 190° field of vision, but can not see the area beneath their mouth. They use their whiskers and the sensitivity of the lips as well as scent and taste during foraging (Meredith 2000). The whiskers are also used during orientation in the burrows and dens. Scent and taste is more important in identifying members of the own breeding group than vision. They have good hearing, and their big ears occupy about 12 % of the total body area (Meredith 2000).

Reproduction in the European wild rabbit is seasonal and appears to be determined by the interaction of a number of environmental factors, for example daylength, climate, nutrition, population density and social status (Bell and Webb 1991, Bell 1999). Sexual maturity occurs at the age of 5-7 months in males and 9-12 months in females, depending on climate (Myers et al. 1994). Domestic rabbits are sexually mature at 4-6 months of age (Falkmer and Waller 1984), but the small breeds are sexually mature earlier than the medium and large breeds (Bennet 2001). Before mating the rabbits circle around each other, parade side by side, jump over each other and sniff around the genital region (Lehmann 1991). Hafez (1960) found seven degrees of sex drive which ranged from aggressive with immediate mounting and ejaculation to offensive reaction with general smelling of the skin, biting and no ejaculation. The buck mating takes only seconds, after which he dismounts to the side or backwards (Bennett 2001). When allowing mating the doe raises the hind quarters. The ovulation is induced by the mating and the sperm is waiting until the egg appears after 8-10 hours, when conception takes place (Bennett 2001). The doe can mate immediately following parturition, and if the young are removed after delivery she will be sexually receptive for at least 36 days (Hagen 1974).

The doe is pregnant for between 28-30 days (Bell 1999) and 30-32 days (Batchelor 1999). Some days before the birth the doe digs an underground nest either within the main warren or in a separate breeding "stop" dug specifically for the purpose (Bell 1999). These are short, shallow burrows dug at a distance of 10-50 yards from the warren. High ranking females have been found to often give birth to their young in a special breeding chamber dug as an extension to the warren, whereas some of the subordinate females are chased away from a warren and forced to drop their litters in isolated breeding "stops" (Mykytowycz 1968). The survival rate is much higher for the high ranking than the low ranking females. The doe collects and carries grass to her burrow and shortly before giving birth she plucks her own fur from the belly, sides and dewlap, and places it on top of the grass (Gonzáles-Mariscal et al. 1994, Bell 1999). The better the construction of the nest the higher the survival rate of the young (Canali et al. 1991).

The rabbit foetuses are born with no or only little hair cover, and are deaf and blind (Batchelor 1999). They each weigh around 50 g at birth and gain about 30 g per day (Falkmer and Waller 1984). The size of the litter is 3-6 young for European wild rabbits, and 4-12 young for domesticated rabbits (Batchelor 1999). At birth the young rabbits are very sensitive to cold, but as their fur grows out within a couple of days they become less sensitive (Batchelor 1999). The eyes open at 10-11 days (Kersten et al. 1989, Batchelor 1999), or 9-12 days (Falkmer and Waller 1984). Hearing develops at the same time, but no reference on this has been found. When the young are born the doe leaves the burrow and covers the entrance with soil, urine marks it, and then leaves (Mykytowycz 1968). She returns to the burrow once a day, and then digs herself into the burrow and nurses her young (Bell 1999). In a study on Dutch Belted rabbits the nursing took place in the early morning and lasted for 2.7-4.5 minutes (Zarrow et al. 1965). When given the opportunity to retrieve their young, does did not perform this behaviour (Ross et al. 1959). The doe closes the burrow and or breeding "stop" for about 21 days, after which the young emerges onto the surface (Bell 1999). The young are very mobile at four weeks of age, and leave their breeding stop soon after emergence (Lloyd and McCowan 1968). At that time they start seeking forage, but they continue to suckle for some more weeks. The does reach maximum milk production two weeks after giving birth, this declines during the fourth week and they may lactate for an additional 2-4 weeks. At 8 weeks of age the young are consuming approximately 90% of their intake in the form of plant proteins (Hagen 1974). In commercial rabbit breeding the young are usually weaned and separated from the doe at 6-7 weeks of age (Hagen 1974, Bennett 2001).

NZW does in semi-natural enclosures behave in a similar way towards their young as the wild rabbit, but when the young are about 18 days old their mother no longer closes the entrance to the breeding stop properly and the young are nursed outside (Lehmann 1989). The mother-young relationship was quite loose, and the mothers were not preferred social partners for the young except for suckling attempts. Nursing was unlikely after four weeks when the doe littered again within a few days, but suckling attempts occurred up to 60 days (Lehmann 1989).

2.3 Natural behaviour

The behaviour of the European wild rabbit (Bell 1999, Figure 10-1) and the behaviour of NZW rabbits kept in a free-range enclosure have been studied (Lehmann 1991, Stauffacher 1991). Studies of natural, free-ranging and enclosed populations in Australia, New Zealand and United Kingdom have shown that European wild rabbits live in small, stable, territorial breeding groups (Parer 1977, Gibb et al. 1978, Cowan 1987, Bell and Webb 1991). The breeding group is generally described to defend its territory, a core area with a warren within a larger home range, by patrolling the borders and scent-marking (Bell 1999). However, in some studies it has been found that breeding groups only occupy single warrens (Myers and Schneider 1964, Bell 1977), whereas in other studies of free-living populations of rabbits most breeding groups used multiple warrens (Dunsmore 1974, Parer 1977, Wood 1980, Daly 1981, Cowan 1987). During the peak feeding periods at dawn and dusk members of neighbouring breeding groups may move out from their territories to forage in communal grazing areas (Bell 1980). The number of entrances per warren was found to be 11.5 ± 8.0 SD, with a range of 1-37 (Cowan 1987). The different breeding groups together make colonies of up to 70 rabbits. The social unit is the breeding group, which consists of 1-4 males and 1-9 females (Meredith 2000).

Within breeding groups separate, stable, linear dominance hierarchies are formed within each sex (Bell 1983). The strict rank order is maintained by rabbits keeping a fixed distance from one another and exhibiting submissive behaviour. Early studies of the European wild rabbit in Australia revealed that a substantial portion of the rabbits' daily activity was filled with direct or indirect aggression (Mykytowycz and Rowley 1958, Mykytowycz and Fullager 1973). It was concluded that this behaviour underlies the social and territorial organisation, which are the main factors affecting the numbers of free-living populations of this species (Mykytowycz 1960). In some studies does have been found to be more aggressive than bucks (Southern 1948, Myers and Poole 1961), or found to fight as strongly as males (Lockley 1961). Females mainly fight over breeding burrows (Cowan and Garson 1985), and are more aggressive towards juvenile females (91% of interactions) than juvenile males (Cowan 1987). Males are tolerant to females, young and sub-adults in their breeding group, and have even been observed to interrupt aggressive interactions between females. When the number of rabbits living together increased, fighting increased dramatically (Myers 1966). In nature young males normally move to a new social group before the start of their first breeding season, while young females stay on to breed in their breeding group (Parer 1982, Webb et al. 1995).

Lehmann (1991) has described 20 different social behaviours of New Zealand White rabbits kept in a semi-natural enclosure, and how the social behaviours change with age. Indifferent contacts were body-to-body, nose-to-body, nose-to-nose contacts and anogenital nuzzling. Amicable behaviour comprised cuddling up and allogrooming. Subdominant behaviour was crouching, retreating and fleeing, whereas aggressive behaviour was circling, nudging, attacking and chasing. Actual fights with aggressive leaping and ripping were observed only 7 times. Mykytowycz and Hesterman (1975) made paired encounters between male and female European wild rabbits and New Zealand White rabbits in a home pen, and found that aggression was equally prevalent in both sexes and inter-sexual fighting occurred just as frequently as fighting between members of the same sex. The domesticated NZW rabbits were fighting just as much as the wild rabbits, but paired domestic females and domestic males paired with wild rabbits of both sexes fought less frequently and less viciously than other paired rabbits.

Rabbits have three specialised scent glands, i.e. in the anal region, in the groin and under the chin (Mykytowycz 1968). The territory of rabbits is scent marked by placing faeces in dunghills, and about 30 such dunghills have been located in a typical warren. The rabbit also marks its territory by pressing the under-chin against structures in its environment so that droplets from its sub-mandibular glands are forced through pores of the skin. Scent marking is more intensive by males than females and more in dominant than sub-dominant individuals, and it is correlated with larger anal and sub-mandibular glands in dominant males (Mykytowycz 1968). The most dominant animals of a warren were found to possess the heaviest anal glands. Males scent mark females and young rabbits of their breeding group by spraying urine on them. Females scent mark their young, attack other young within the same breeding group, and may chase and even kill young from other breeding groups. Females may attack even their own young if they have been smeared with foreign urine (Mykytowycz 1968).

The rabbit has a home-range for performing foraging behaviour. The size of the home-range varies depending on food availability, age of the rabbit, the rabbit's status within the breeding group and number of rabbits in the group (Donnelly 1997). The home-range of wild European rabbits has been estimated to be 5 ha (Myers et al. 1994), 0.4-2.0 ha (Cowan and Bell 1986), or 8 ha (Vastrade 1987). Cowan (1987) found that that the mean asymptotic male range was 0.71 ha and female range 0.44 ha.

The rabbit is mainly a nocturnal animal, emerging from the burrows in late afternoon (Fraser 1992). The old bucks emerge first, about four hours before sunset (Mykytowycz and Rowley 1958), and by sunset 90% of the rabbits have emerged (Fraser 1992). The rabbits are visible on the ground during 11-14 hours of the diurnal cycle (Mykytowycz and Rowley 1958). When the rabbits are above ground they spend about 44% of their time eating, 33% inactive, 13% moving and 10% on other activities (Gibb 1993). During semi-natural conditions young NZW rabbits were active for an average of 30% of the daytime. Feeding on pellets and grazing took a third each of the active time, and the remaining third was spent exploring, gnawing, intensive locomotion and, for the older rabbits, engaging in sexual behaviour (Lehmann 1989). The rabbit's choice of habitat depends on the opportunities to find shelter and protection, and where the soil is loose it digs burrows, and where the soil is more compact it seeks protection in dense vegetation (Kolb 1994). In case of danger a rabbit can stamp with its hind feet, thus causing the other rabbits to flee underground (Black and Vanderwolf 1969). If it is too late to flee the rabbit can freeze, i.e. stop in its movement and be completely motionless. A rabbit can, if caught by a predator, emit a high distress scream (Cowan and Bell 1986), which may cause the predator to release its prey. Apart from this and some low sounds during mating and mother-young care rabbits are silent animals.

The movement patterns of rabbits consist of hopping, crawling and intensive locomotion, i.e. running, start-and-stop, jumping, double and capriole (Kraft 1979, Lehmann 1989). Hopping is usually performed when rabbits move over longer distances, whereas crawling is performed when feeding on grass or exploring on the spot and during social encounters (Lehmann 1989). Rabbits perform comfort behaviours such as licking and scratching themselves, shaking the body, rubbing against objects and stretching their body.

Several studies have compared the behaviour of wild European rabbits with those of domesticated strains of rabbits (reviewed by Bell 1984). Both wild rabbits and domestic strains have reproductive seasonality, suckle their young only once every 24 h, show two main feeding periods at dawn and dusk, form breeding groups with separate linear dominance hierarchies among male and female members, and reproduce successfully with the female digging breeding burrows, building nests for their young and covering the entrance to the burrows with soil (Bell 1984). The difference found is that domestic rabbits rest more above ground during the day (Stodart and Myers 1964) and that males chin-mark more often (Kraft 1979) also in unfamiliar territory (Mykytowycz 1968). New Zealand White females had fewer days in anoestrus than wild rabbits, and therefore produced more litters with a larger mean size (Stodart and Myers 1964). When European wild rabbits were brought into the laboratory they failed to breed, and females born into the laboratory as a result of egg transfer from wild to domestic mothers retained their nervous disposition and failed to mature sexually (Adams 1987).

2.4 Abnormal behaviour

When rabbits are kept alone in traditional small cages several abnormal behaviours may arise. The most obvious abnormal behaviour is stereotypy, e.g. wire-gnawing, excessive wall-pawing (Lehmann and Wieser 1985, Wieser 1986, Bigler and Lehmann 1991, Loeffler et al. 1991, Stauffacher 1992). When many rabbits gnaw on the wire of their cages there might be a high sound level in the animal room (Figure 10-2). Digging may be constrained by the solid floor of the cage (Podberscek et al. 1991). The stereotypic behaviour substitutes for natural behaviours which cannot be performed in standard laboratory housing (Stauffacher 1992) and may indicate frustration, anxiety or boredom, and develop in stages involving a progressive narrowing of the behavioural repertoire (Gunn 1994). Individually caged rabbits show stereotypies, such as somersaulting, no full hops, less activity than group-penned rabbits, and less marking and investigatory behaviour than in group pens (Podberscek et al. 1991). Social isolation can induce the physiological symptoms of stress, which are relieved by the presence of conspecifics (Held et al. 1995).

There are indicators of boredom in rabbits such as hunched posture (Gunn and Morton 1995), inertia (Metz 1984), and a staring coat and dull eyes (Wallace et al. 1990). Prolonged inactivity associated with unresponsiveness, may, like stereotypies, be associated with brain chemistry changes which make the problem seem less bad (Broom 1988). Gunn and Morton (1995) mention problems with under-grooming which may lead to a staring coat, and over-grooming which may lead to hair-balls which in turn may cause intestinal stasis (Jackson 1991) and lead to death by gastric trichobezoars (Wagner et al. 1974). Under-eating, causing weight loss and over-eating causing obesity are other behavioural problems (Gunn and Morton 1995).

Figure 10-2. A male New Zealand White rabbit gnawing on the wire of its cage (photograph by Lena Lidfors 1994).

The very limited freedom of movement in caged rabbits leads to changes in locomotion which prevent hopping, thus causing changes in the locomotion apparatus (Lehmann 1989, Stauffacher 1992). It is mainly in the femur proximalis and the vertebral column that changes of the bone structure occur, and the changes consist of thinner and less strong hollow bones (Lehmann 1989, Drescher and Loeffler 1991a,b). Lehmann (1989) showed that growing rabbits perform almost no hopping and intensive locomotion mainly during playing when kept in cages compared to an outdoor enclosure. Crawling, which was the most common movement pattern, was slightly less common in the cages. These rabbits often performed interrupted jumps, where the hindlegs only were lifted slightly and then put down again, thus not being used as in normal hopping. Abnormal postures may also be shown due to spatial restriction for lying stretched out during resting or performing stretching behaviour (Gunn 1994).

Rabbits may show restlessness, e.g. afunctional bouts of activity with disconnected elements of feeding, comfort, resting, alertness and withdrawal behaviour alternating with locomotion, which leads to space-time organization disorder as well as panic (Lehmann and Wieser 1985, Bigler and Lehmann 1991, Stauffacher 1992).

An abnormal behaviour related to reproduction is disturbed sexual behaviour which leads to low conception rates (30-70 %) (Stauffacher 1992).

One reason for this may be that for mating the doe is placed in the bucks' cage a few days to several weeks after she last gave birth. The mating performed then has been described more as a rape than normal mating behaviour (Stauffacher 1992). However, the doe may be aggressive against the buck if he is placed in her cage (Bennett 2001). Around giving birth the doe may show disturbed nesting behaviour and nesting stereotypies which may lead to rearing losses (Wieser 1986, Wullschleger 1987, Loeffler et al. 1991). The doe may also show disturbed nursing and cannibalism caused by restlessness of the mother and her pups which may also lead to rearing losses (Bigler 1986, Brummer 1986, Stauffacher 1992).

2.5 Behavioural needs

Based on the previously presented research on the abnormal behaviours performed by housed rabbits, the behaviour of Wild European rabbits and free-ranging New Zealand White rabbits many researchers involved in this area have had an insight to the behavioural needs of rabbits. Based on this research and on several discussions in international workshops suggestions have been made on the following behavioural needs and behaviours involved in performing them:

1) Locomotion and exercise: crawling, hopping, running, jumping, quick changes of direction.
2) Control, "security and safety": withdrawal, digging, burrowing, sit upright, rearing, sniffing, "stamping".
3) Foraging: eating, drinking, caecophagi, gnawing, manipulation of resources, searching, exploration, exploitation of resources.
4) Behavioural rhythm: extended periods of rest and grooming, activity.
5) Companionship and social contact: different social behaviours.
6) Demonstrate presence; scent marking by anal and submandibular glands.
7) Female choice in mating; circling, parading, jump over and sniff genital region before mating.
8) Nest and burrow with constant temperature: digging, carrying nesting material, plucking fur.
9) Separate from litter: nursing once/twice a day, close nest entrance.
10) Warmth and protection: huddle in hair-nest for new-born rabbits.
11) Learn how to cope: play and exploration in young rabbits.

As some of the suggested needs lead to both physical and behavioural problems if they are not performed, e.g. foraging, movements, other are more questionable, for example digging. There is a need for more research to establish exactly which behavioural needs are so important that one must provide them in the rabbit housing, i.e. essential needs.

3. OPTIMAL ENVIRONMENT

3.1 Light and noise

Laboratory rabbits are recommended to be kept in animal rooms with a regular light:dark cycle and isolated from external lighting fluctuations (Batchelor 1999). Some laboratory facilities have introduced artificial dawn and dusk periods, usually working for 30 minutes before full light comes on in the morning and 30 minutes before the light is shut down in the evening. However, there is still some debate as to the relative merit of creating an artificial dawn and dusk period (Batchelor 1999).

As rabbits are more nocturnal than diurnal a partially reversed lighting schedule can be established to observe their activity (Batchelor 1999), but rabbits may also switch to a more diurnal pattern of activity due to noise or scheduled feeding in the laboratory (Jilge 1991).

The optimal light intensity in rabbit rooms is recommended to be 200 lux at one metre above the floor (Iwarsson et al. 1994). Too high levels of illumination can result in retinal degeneration in some albino mammals, and this may include NZW rabbits (Batchelor 1999). Solid side caging will reduce the amount of light, and where in the rack a cage is situated may also have an impact on light intensity. If rabbits are housed in floor pens, shelves and boxes may provide hiding places also from high light intensity.

The rabbits hearing range thresh-hold level is 75-50 000 hertz with the best hearing at 2 000 to about 9 000 hertz (Iwarsson et al. 1994). The rabbit is sensitive to high sound frequencies (Milligan et al. 1993), but also sudden noise may scare rabbits and lead to injures. Background music masks sudden loud sounds, and is claimed by some to result in lower excitability (Batchelor 1999).

3.2 Temperature and humidity

The mean room temperature for rabbits is recommended to be 18 °C with a range of 15-22 °C (GV-SOLAS 1988). The Lower Critical Temperature is -7 °C and the Higher Critical Temperature is 28.3 °C (Spector 1956). Rabbits only have sweat glands on their lips, and lower capacity to ventilate through their mouth than dogs (Donnelly 1997). Wild rabbits avoid high temperature and stay away from direct sunshine, and during the day they stay in the cooler burrows. The ears of the rabbit are highly vascular and function as ventilators during high ambient temperature. In the rabbit room a humidity level of 55 ± 10 % is recommended (Batchelor 1999).

3.3 Ventilation and hygiene

In the animal room a ventilation rate of 15-20 air changes per hour is recommended (Iwarsson et al. 1994). Air changes lower than this may be satisfactory if the cleaning routines is of high standard and the stocking density is low (Batchelor 1999).

Ammonia level should never exceed 10 ppm. When measuring ammonia level in a modified farm animal building with group housed rabbits the highest level was 1-2 ppm (Batchelor 1999). High ammonium levels can inactivate the cilia in the air ways of rabbits. High levels of CO_2 may become a problem for the rabbits if the ventilation is not working properly.

Due to normal changes of the fur 2-3 times per year rabbits loose hair in the laboratory unit. The hair may fly around in the animal room, and eventually ends on the floor. Especially during handling of the rabbits hair may be released.

Rabbits kept in cages should be moved to a clean cage once a week, but the waste pan under the perforated floor could be changed more often. If rabbits are kept on wire net floor and a band can be moved under the cages this should be cleaned once per day. Many laboratories today have dishwashers where the whole rack is washed in warm water with disinfectants and dried at high temperatures thus working as an autoclave. The feed is usually pseudo-pasteurized (70 °C for 10 minutes) at the producers. In order to reduce the risk of bringing pathogens into the laboratory, hay and any environmental enrichment should be autoclaved or irradiated with cobalt, but the latter is not so common.

Rabbits kept in floor pens should be taken out of the pen after which the pens should be thoroughly cleaned and disinfected. Batchelor (1999) suggest that this should be done at least once per month. However, the laboratories we know of carry out this cleaning at least once per week and more often when needed in order to maintain a high level of hygiene. If cleaning is not done often the rabbits can become infected with coccidiosis. Cleaning changes the olfactory environment for the territorial rabbits and may actually be stressful for them (Batchelor 1999).

3.4 Transport

The rabbits should be in good condition before transport. Usually rabbits are transported in containers made of strong cardboard, fibreboard, fibreglass or wood with wire mesh windows for ventilation, and equipped with a filter in summer time (Swallow 1999, Batchelor 1999). The floor of the container should be waterproof and covered with absorbent material, for example saw dust. For more detailed information about the general requirements and species requirements when transporting animals see Swallow (1999).

Regulations for transport of animals may differ for the EU (according to the Convention of Europe) and for different countries over the world. In England and Sweden, when transporting rabbits of less than 2.5 kg the smallest area per animal should be 1000 cm^2 in non-filtered crates (winter) and 2 000 cm^2 in filtered crates (summer) (Swallow 1999, SJVFS 2000:133). In Sweden minimum container height should be 24 cm, and in England it should be 21 cm. In the Guide to the Care and Use of Experimental Animals from the Canadian Council of Animal Care (Olfert et al. 1993) it is stated that the rabbits should be transported in disposable containers with sufficient space to allow the animals to stand, lie down and turn around. Transporting rabbits in compatible pairs, for example siblings, has been practised by a breeder in Sweden with good results.

The transportation vehicles should be equipped with thermometers and ventilation that can cool down the air during warm weather and provide heating during cold weather. A source of water, for example a gel, and feed, for example a carrot, an apple or straw, sufficient for the duration of the transportation has to be provided (Batchelor 1999, Olfert et al. 1993). Swallow (1999) suggests that rabbits should not require feeding for journeys of less than 24 h duration, but if unforeseen delays occur rabbits could be given carrots, fruit, hay or grain.

After arrival at the new animal housing the rabbits should be checked for any health problems and injuries, which could either be caused by the transport or which may have been acquired before the transport. The rabbits should be given 1-2 weeks of acclimatization time after the transport before they are used for research. This will depend on the degree of stress the animals were under at arrival, which in turn may depend on the nature of the journey, its duration and which parameter is being assessed during the recovery period (Swallow 1999). Olfert et al. (1993) state that newly acquired rabbits should be held in quarantine for at least three weeks, and examined regularly for disease. Animals dying during this period should be subjected to complete post-mortem examinations. Most laboratories we have contact with bring rabbits from suppliers of health defined rabbits bred behind a barrier. Often most of the rabbits come from the same supplier, thus there is no need for quarantine. However, when purchasing rabbits from different suppliers we recommend placing them in different animal rooms at least during an acclimatization period of up to three weeks.

There has been very limited research on the effect of transport on rabbits. Batchelor (1999) found considerable differences in body weight in rabbits that were housed in group pens compared to solitary in cages after transportation. The loss of weight in rabbits after transport is mostly due to a loss of gastrointestinal contents in of about 10% of the total body weight (Swallow 1999). These losses are probably maximal after about 15 h of transport, and similar to depriving animals of food and water during the same amount of time. However, the loss in live weight can take up to 7 days to recover (Swallow 1999). Our own research on the effect of providing male rabbits with or without hay and with a change or no change in the feed after a 10 h transport by truck, plane and truck showed that providing hay had a significant effect of reducing the occurrence of diarrhoea (Lidfors pers. com.).

4. HOUSING AND FEEDING

4.1 Housing in cages

Laboratory rabbits are traditionally housed singly in barren cages over periods of several weeks up to several years, depending on the research purpose. The cages for laboratory rabbits have changed from traditional wooden hutches to galvanized iron, aluminium, stainless steel and now plastic caging (Morton et al., 1993). Today three main types of cages can be found; wire, sheet metal with wire front and plastic with wire front (Stauffacher et al. 1994). Housing in small barren cages can lead to several abnormal behaviours and reduced welfare of rabbits (see section on abnormal behaviours). The cages with solid sides, back and top tend to isolate the animals, and prevent them from seeing the source of disturbance. This may cause the rabbits to be jumpy, and in breeding units lead to higher losses due to cannibalism. In order to allow rabbits to see their surroundings it is recommended that barred "windows" occupy 30-50 % of the total wall area (Stauffacher et al. 1994).

In order to allow rabbits to perform normal hopping movements and sit upright the regulations in several countries (Switzerland, United Kingdom) require that cages for rabbits are much bigger than the recommendations from the European Union and the World Rabbit Science Association (Table 10-1). The heights of the cages should be from 40 cm for rabbits <2 kg up to 60 cm for rabbits >5 kg in the Swiss Ordinance of Animal Protection 1991 (Stauffacher et al. 1994). In the suggestions for new rules in the Appendix A from the Council of Europe it is written that cages and pens for rabbits >10 weeks of age should be 3 500 cm^2 (45 cm height) for <3 kg rabbit, 4 200 cm^2

(45 cm height) for 3-5 kg rabbit and 5 400 cm² (60 cm height) for >5 kg rabbit. The weights are for the final body weight that any rabbit will reach in the housing. The floor area is minimum floor area for one or two socially harmonious animals, and the height is a minimum height. It is also stated that a raised area, i.e. shelf, must be provided within the cage, and if not providing a shelf the cage size must be 33% larger for a single rabbit and 60% larger for two rabbits. The suggested new rules in Sweden have been agreed since 1995, but so far no decision has been taken by the Agricultural Board.

Table 10-1. The recommendations by different authorities for minimum cage area (cm²) for rabbits

Weight (kg)	Swiss Ordinance of Animal Protection 1991	UK Home Office 1989	European Union Council Directive 1986	World Rabbit Science Assoc. 1992
< 2	3400	4000	1400	2000
2	4800	4000	2000	2000
3	-	4000	2500	-
3.5	7200	-	-	-
4	-	5400	3000	3000
> 5	9300	-	-	-
> 5.5	-	-	-	4000
> 6	-	6000	-	-

During the last 10 years there has been a development of new cage systems (Figure 10-3), using a cage with an increased floor area, a higher cage for upright sitting, a shelf for rabbits to hop up onto or hide under, racks to make hay feeding easier and flexible cage racks so that several cages can be built together. A raised shelf reduces restlessness, grooming, bar-gnawing, and timidity of being captured (Berthelsen and Hansen 1999). Wild rabbits rear when they are looking at their surroundings (Lockley 1961), and caged rabbits use boxes and shelves provided as sources of enrichment and lockout posts (Hansen and Berthelsen 2000). Mirrors as a form of enrichment do not provide an image of a companion, but probably stimulate activity by enriching the visual environment and increasing the amount of movement perceived by the rabbit (Jones and Phillips submitted). If rabbits are housed in a cage with a living area and darkened resting area, putting mirrors in the living area increases the time rabbits spend there, and especially the time that they spend investigating their environment and feeding (Jones and Phillips submitted).

Figure 10-3. A rabbit cage with a shelf, hay rack, food hopper and water bottle. It is possible to link two or more cages together and to move the shelf from one side to the other (photograph by Lena Lidfors 1996).

4.2 Housing in groups in floor pens

Rabbits can either be group housed in floor pens or in cages. In the latter case the limited area of a cage usually restrict this to pair-housing (Bigler and Oester 1994, Huls et al. 1991, Stauffacher 1993). This combines the benefits of cage housing, e.g. hygiene, experimental purposes, with animal welfare interest, and has been established in Switzerland, Germany, United Kingdom and Canada (Stauffacher et al. 1994).

Group housing of rabbits in floor pens has been introduced in many laboratories and countries during the last 10-15 years (Figure 10-4). This is beneficial for the rabbits because they can express social behaviours and exercise (Heath and Stott 1990, Batchelor 1991). Abnormal behaviours and physiological conditions caused by small barren environments are also reduced, but there is an increased risk of fighting between rabbits (Morton et al. 1993). The major factors that have to be considered when group housing rabbits are: compatibility of individual animals, size of pens, stocking density, husbandry practices and environmental enrichment (Morton et al. 1993).

Figure 10-4. Group housing in floor pens at the University of Guelph in Canada (photograph by Lena Lidfors 1996).

Rabbits which are not compatible will fight when placed together in a group, and the greatest problems occur when placing adult males together (Morton et al. 1993). There are strain differences in aggressiveness, i.e. Dutch rabbits are more aggressive than New Zealand White, whereas Lops are more docile (Morton et al. 1993). Some of the small strains of rabbits may show more aggression than the larger strains (Stauffacher et al. 1994). Individual animals may be highly aggressive, and fights can occur for unknown reasons, even in groups that have been stable for a long time (Morton et al. 1993). Therefore, groups of rabbits need to be carefully selected and regularly monitored. The best option is to keep litter mates which have been kept together from weaning (Zain 1988). Groups of intact, mature females not intended for breeding can be kept together (Morton et al. 1993). In periods of rest, does, bucks, and older young kept in mixed groups congregate and snuggle against each other or engage in mutual grooming (Stauffacher 1992). Does have showed a weak preference for a large, enriched, solitary pen over a group pen, but a strong preference for a group pen over a smaller, barren, solitary pen (Held et al. 1995). Fighting can still occur in groups of does, and a dominant female in oestrus can mount and damage the skin on the backs of other females and harass the group (Morton et al. 1993). The degree of compatibility of grouped rabbits will depend on factors such as strain, individual characteristics, sex, age and weight, size and structuring of pens, methods of husbandry and the interest and ability of the animal technicians (Bell and Bray 1984, Zain 1988, Morton et al. 1993,

Stauffacher 1993). Aggression may be seen in group pens even after the establishment of the dominance hierarchy.

From around 10 weeks of age it may be necessary to house males individually to avoid fighting (Morton et al. 1993). Groups of males kept in proximity of females tend to fight and urinate more frequently (Portsmouth 1987). Castration of males kept for longer periods in the laboratory may be one solution to be able to keep them in groups. The practical experience from castration of males is that aggression is reduced and stable for a long time afterwards (Gunn pers. com. in Morton et al. 1993, Lindberg pers. com. 2003). Castration should be carried out when the males reach sexual maturity at 50-80 days of age depending on strain and food composition (Stauffacher et al. 1994). It should be carried out only by well-trained persons and always before the males start to show aggressive behaviour, 3-4 weeks after weaning at the latest (Stauffacher et al. 1994, Morton et al. 1993). The testicles move down during sexual maturation, but then are withdrawn again. However, one has to consider what type of research the animals will be used for, as castration has an effect on the animal's physiology.

An alternative if castration is not advised is to place individually caged male rabbits in an exercise arena at regular intervals (Figure 10-5). This allows them to move around on a larger floor surface, to investigate enrichment objects and to get the smell from other males that have been exercised before them. This has not been evaluated in the laboratory units, maybe because it will take time for the animal keepers to take out and return the rabbits according to a routine schedule. However, in the case where castration is not possible it may introduce an important enrichment for the rabbits.

Rabbits placed in group housing should be of the same sex, of similar size and if possible, related and grouped when young, i.e. around the time of weaning (Morton et al. 1993). When establishing new groups of rabbits in floor pens the best option is to wean and mix at the same time around 6 weeks of age and to place 6-10 rabbits, preferably of the same sex, in one group (Morton et al. 1993). The ideal situation is when animals are kept in stable groups from birth (Zain 1988, Stauffacher 1993). Small groups may be most stable (Love and Hammond 1991). Some animals do not appear to settle well in groups, either because they are too dominant and bully the others or are too timid and prone to be bullied (Morton et al. 1993). When putting together rabbits that have been caged for six months or more it may be difficult to avoid fighting or self-inflicted injuries (Morton et al. 1993). However, individually-caged adult female rabbits of peaceful strains can be paired successfully, preferably in structured cages (Stauffacher 1993,

Morton pers. com. in Stauffacher et al. 1994). It is very important to provide refuge and hiding places for subordinate animals (Morton et al. 1993).

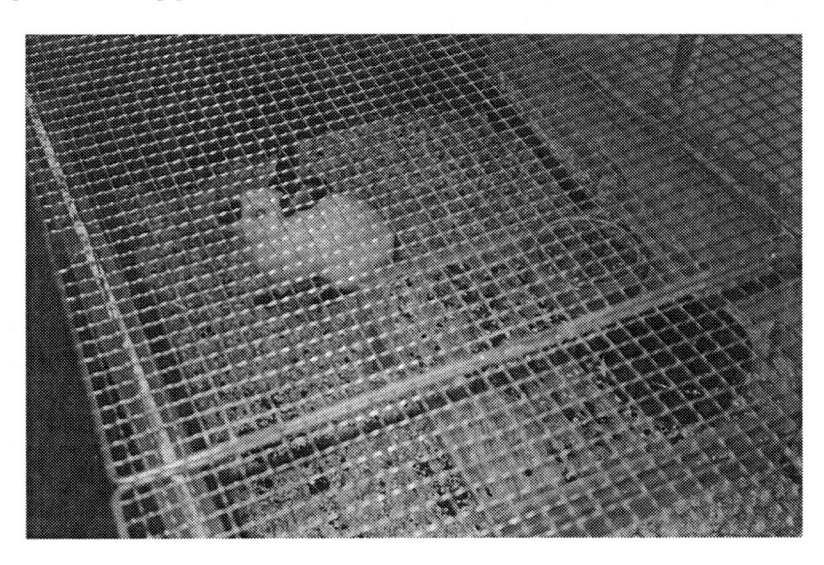

Figure 10-5. A male Russian rabbit placed in an exercise arena on the floor in a laboratory in Sweden (photograph by Lena Lidfors 1995).

The number of rabbits kept in each group pen should not exceed 6-8 mature animals (Morton et al. 1993, Stauffacher et al. 1994). This is recommended so that the rabbits can be adequately monitored for signs of bullying and ill-health. Groups of up to 20 laboratory rabbits have been successfully managed as stock and for the production of polyclonal antibodies (Stauffacher et al. 1994). However, more research is needed on optimum and maximum group size (Stauffacher et al. 1994).

When keeping breeding rabbits in groups they should be composed of 4-6 females, one male and their offspring until they are weaned at 30 days of age (Stauffacher et al. 1994). The group breeding housing system and management has been developed in Switzerland (Stauffacher 1989) and used in agricultural rabbit farming, but to our knowledge not being used for breeding of laboratory rabbits.

The minimum area of a floor pen should be large enough for the rabbits of a particular weight and size to be able to carry out normal behaviour, especially locomotion. One problem with using weight and size as a criterion of determining floor area is that young animals move more and might need more space to carry out play behaviours (Stauffacher et al. 1994). In the Swiss legislation each rabbit must be able to hop some steps or to jump up and down onto a shelf. This may help the rabbits to maintain a level of fitness and reduce the occurrence of disuse osteoporosis (Morton et al. 1993). The working group on "Refinement in rabbit husbandry" recommend that rabbits kept in groups have a clear area of 20 000 cm² with an overall minimum floor area of 6 000 – 8 000 cm² per rabbit for groups up to 6 rabbits. If more than 6 rabbits are kept in the groups an extra space of 2 500 cm² per rabbit is recommended (Morton et al. 1993). The height of floor pens should be 1.25 m., as rabbits can jump very high, and enrichment objects should be placed so that they can not be used for jumping over walls.

If European wild rabbits are used for research they should be housed in floor pens with sand or straw substrate so that they can hop around and exercise (Bell 1999). The reason for this is that wild rabbits kept in small cages have developed weak backs probably due to a lack of exercise. Young wild rabbits, between 4 and 8 weeks of age, fare better when they are kept in groups (Bell 1999).

4.3 Feeding and water

Rabbits are crepuscular and in the wild they usually graze during their active periods at dawn and dusk (Lockley 1961), or during early morning and at night (Cheeke 1987). However, feeding laboratory rabbits in cages means that they are almost invariably not fed the diet of grass for which they evolved to utilize. However, there is little evidence that they prefer a grass diet to one based on compound feed (Leslie et al. 2004). Despite this, it is often beneficial to supplement their ration of proprietary compound pellet with dietary enrichment, which as well as providing adequate nutrients (NRC 1966), in particular fibre (Lehmann 1990), which will increase the time usefully spent in procuring their food and reduce potentially damaging behaviours such as chewing their cage (Leslie et al. 2004). The visual stimulus of a varied diet is particularly important (Ruckebusch et al. 1971). A mixed diet is also a feature of natural herbivore feeding behaviour, due to their desire to sample regularly in case of the disappearance of one feed (Parsons et al 1994).

Several forms of dietary enrichments have been tried with a significant degree of success, including supplying fibrous food to reduce boredom: hay (Berthelsen and Hansen 1999), grass cubes or hay in a bottle (Lidfors 1997), and fresh grass (Leslie et al. 2004). The most useful enrichment is something to chew (Brummer 1975, Huls et al. 1991, Lidfors 1997, Berthelson and Hansen 1999) and high fibre objects are preferred, with hay or straw remaining an effective enrichment for long periods (Brummer 1975, Lidfors 1997). Hay and straw also cause less weight gain than proprietary fibre sticks or compressed grass cubes (Lidfors 1997). Abnormal maternal behaviours and trichophagia or fur-chewing (Brummer 1975, Mulder et al.

1992) are eliminated in caged rabbits when hay or straw is given. If the supplementary hay is ground, it is ineffective at reducing problem behaviours, demonstrating a need for long fibre (Mulder et al. 1992).

Water should always be available *ad libitum* to rabbits (Mader 1997). The need for water is 50-150 g per kg body weight, and rabbits raised on pellets need 50 g per kg body weight. More water is needed for growing animals and pregnant and lactating females. One rule of thumb is that the need for water per 24 h is 10 % of the body weight (Meredith 2000).

5. CARE AND HANDLING

It is important that all personnel that handle rabbits have learned how rabbits react to frightening sounds and handling in order to avoid injuries to both the rabbit and human handler. Rabbits that have had positive contact with humans and been handled with care previously will come to the front of the cage when opening the door and sit still when being lifted out. Similarly rabbits in floor housing will sit still when a person comes to pick them up. Rabbits that have been scared by humans or not been handled try to flee, and may injure themselves. The fear and distress may be communicated to other rabbits in the animal room (Beynen 1992).

Rabbits should be handled firm and gently, because if they sense insecurity they may struggle (Batchelor 1999). Lifting of rabbits of all ages should be by grasping the scruff of the neck in one hand while the other hand is placed under the rump to support the animal's weight (Batchelor 1999). Some authorities suggest that the rabbits ears should be grasped in addition to the scruff for greater control, but Batchelor (1999) have the experience that some rabbits react as if this was painful, which increase the likelihood of struggling in the rabbit. Rabbits should never be carried by the ears.

6. HEALTH AND DISEASE PROBLEMS

During the last thirty years the health situation of laboratory rabbits in general has moved from a position of disease problems that is still prevalent among many pet rabbits to a vastly improved level of health. The laboratory rabbits of today are bred, housed and cared for in such a manner that their overall health situation is much better than that of their ancestors a couple of decades ago.

The awareness among researchers and laboratory animal staff of the great importance of a good health situation for the animals has contributed to improvements in hygiene and in health monitoring systems, which in turn have led to a good general health situation among laboratory rabbits. The establishment of SPF breeding colonies in the late 1970s and early 1980s gave the rabbits an immensely improved health situation (Eveleigh et al. 1984). The first guidelines of FELASA (Federation of European Laboratory Animal Associations) for health monitoring of breeding colonies on a regular basis were also a milestone in the improvement of health for laboratory animals. At this time, in the mid-1980s, health discussions mainly dealt with hygienic aspects of health such as infectious diseases of different sorts and how the animals and the research results were affected by the infections.

6.1 General health

Clinically healthy rabbits have a well-groomed fur, alert and clear eyes, and quick reflexes of escape if threatened. When picked up and held by the caretaker a healthy rabbit accustomed to handling by human beings will give the impression of strong muscles in a resting state. When checking the health of laboratory rabbits on a daily basis the major points are the posture of the animal, look of eyes, ears and nose for signs of discharge and other abnormalities, the state of the fur (especially in group-housed rabbits since the effects of fighting might only show as a minor flaw in the fur), and the look of faeces and urine. Normal faeces should consist of dry fecal pellets of a uniform size and normal urine can vary in colour from yellow to dark red and is often cloudy due to excretion of calcium in the urine. See Meredith (1998) for more information.

FELASA has issued guidelines (Nicklas et al. 2002) on how to monitor the health status of laboratory rabbits and rodents. These guidelines comprise infectious agents, frequency of sampling, sample sizes and preferred methods of analysis for complete monitoring of the hygienic state of a breeding colony. The result of following the FELASA guidelines for health monitoring of laboratory rabbits and rodents can be called a Health Defined Rabbit. There are similar terms of earlier origin that are sometimes used for the same purpose, i.e. SPF (Specific Pathogen Free) or VAF (Virus Antibody Free). These different terms aim to give us information of the quality of health of the rabbits microbiologically.

A special health problem has been introduced along with the use of certain strains of rabbits with a genetically transmitted disease trait such as the WHHL (Watanabe Heritable Hyperlipidemic rabbit). This rabbit is characterized by the development of atherosclerotic lesions in arteries. Older rabbits tend to accumulate calcium deposits in various parts of the body, which of course affects their well-being.

Many diseases will not show clinical symptoms unless the rabbits are compromised by a number of different stressors at the same time, where all factors influencing the animals well-being positively and negatively add up (Nerem 1980).

6.2 Infectious diseases

Bacterial, viral or parasitic agents cause infectious diseases that rabbits can be affected by. Many of the infections of laboratory rabbits are subclinical, especially viral infections and may pass unnoticed, mainly affecting the results of studies performed. In the literature, there are plenty of data about diseases of the rabbits, the literature used here include: Harkness and Wagner 1995, Laber-Laird et al. 1996, and Hillyer and Quesenberry 1997.

Bacterial agents including *Pasteurella multocida* and *Bordetella bronchiseptica* are the main cause of respiratory inflammations, which may give symptoms such as sneezing, coughing, nasal discharge and lethargy. *Pasteurella multocida* and *Staphylococcus aureus* may also be involved in formation of abscesses in subcutaneous tissues, behind the eye bulb or in internal organs as well as inflammation of the mucus membranes of the eyes and in the middle ear. Other bacterial infections of rabbits are eye infections by *Moraxella catarrhalis*.

Infections of the gut by bacteria include mucoid enteritis that affects mainly young rabbits. The symptoms are depression, anorexia, diarrhoea and mucus in the stool and the cause is multifactorial with the bacterium *Clostridium spiroforme* being one of the major factors. *Escherichia coli* and other strains of *Clostridia* may also cause enteritis with diarrhoea as main symptom. Inflammations of the gut are aggravated by nutritional imbalance with a deficit of dietary fibres.

Viral infections comprise mainly viruses that affect the digestive tract. Rotavirus and rabbit enteric coronavirus may give the rabbits mild diarrhoea. Rabbit Viral Hemorrhagic Disease (RVHD) is a feared disease among pet rabbits but it is unlikely to infect laboratory rabbits. This disease affects many organs and the main symptoms are lethargy, anorexia, diarrhoea and haemorrhage from body openings such as the nose and urogenital opening.

Myxoma virus could be transferred to laboratory rabbits but this requires the presence of a vector, most often fleas but other insects may also act as vectors for myxoma virus. Myxomatosis is common in wild rabbits and could possibly infect laboratory rabbits in areas where wild rabbits are common.

Parasites that infect laboratory rabbits are mainly endoparasites. The largest problems are created by coccidiosis. This disease is caused by different strains of *Eimeria*. One strain, *E. stiedae*, infects the liver and causes different degrees of symptoms ranging from unapparent retardation of growth to fatal disease. Other *Eimeria* strains such as *E. perforans* and *E. magna* infect the intestine of rabbits. Symptoms depend on the amount of coccidia present in the gut and of the susceptibility of the rabbit. Most often, only the youngest animals show symptoms whereas in older rabbits coccidiosis is sub-clinical. The symptoms include weight loss and mild intermittent to severe diarrhoea.

Pinworms, *Passalurus ambiguus*, colonize the caecum and colon and the eggs are passed in the faeces. Pinworm infections are generally without symptoms.

Encephalitozoonosis or nosematosis is a disease common in pet rabbits and wild rabbit and regularly occurs in laboratory rabbits. It is caused by *Encephalitozoon cuniculi*, an intracellular protozoan that is transmitted from the urine of infected animals via the oral route to the intestine and tissues of susceptible rabbits. The parasite mainly damages the kidneys and the brain of infected animals but most often no symptoms are seen clinically.

Ectoparasites are uncommon in laboratory rabbits but a few should be mentioned since they are common in other rabbits. Ear mites, *Psoroptes cuniculi*, can cause considerable itching and wounds on and around the ears. *Cheyletiella parasitivorax* is the fur mite of rabbits. Mites, fleas and lice of rabbits cause considerable suffering since they produce anaemia and/or pruritus and result in generally poor condition in the rabbit.

6.3 Traumatic injuries

Fighting is the most common cause of traumatic injuries. Fighting can occur between all sexually mature males, between females that are not acquainted to another, between individuals in overcrowded pens if the feed hoppers and water bottles are not in sufficient numbers. Fighting males are very aggressive and may cause considerable damage to each other. Wounds may show readily but may also pass unnoticed, concealed by fur.

Fractures are uncommon and may result from handling cage-housed rabbits unaccustomed to being handled which cause them to struggle forcefully, and it mainly occurs in cage-housed rabbits with a weak skeletal structure in the vertebral column (Rothfritz 1992).

6.4 Diseases caused by housing, feeding and breeding

Housing and hygienic routines have a strong impact on the health of laboratory rabbits. Solid-bottom cages and pens need thorough cleaning and regular disinfections so as not to spread intestinal parasites and bacteria back to the rabbits. Cage-floors with perforation for droppings onto a pan reduce the number of coccidian spores etc. that the animals can ingest. "Sore hocks" is a condition seen in heavier breeds of rabbits kept on solid-bottom cages or pens with inappropriate hygiene and the symptoms are bleeding and infected wounds along the hind feet of the animals.

The diet of laboratory rabbits is crucial for the microbial balance in the gut and intestinal morphology (Yu and Chiou 1996) and thus the optimal rate of dietary fibres are essential to keep the animals from developing soft stool or diarrhoea.

Another type of health problem arises from the use of certain types of diets aiming to inflict metabolic changes in the animals, for example high cholesterol diets for the development of atherosclerosis. A side effect of using these diets in long-term studies is fatty liver and deposits of cholesterol in various organs.

7. EXPERIMENTAL TECHNIQUES

Rabbits are often used for collecting blood in immunisation studies. Blood can be taken from the marginal ear vein or by cardiac puncture. In the former case dilation of the ear vein facilitates the removal of blood, and with good blood flow up to 10 ml blood may be obtained (Batchelor 1999). Local anaesthetic facilitates the removal of blood from those animals which are distressed by the insertion of the needle into the ear vein. The cardiac puncture must be carried out under anaesthesia with the animal in dorsal recumbency. Rabbits have about 70 ml of blood per kg body weight, and a maximum of 7 ml/kg body weight should be taken at a single sample (Batchelor 1999). Others have suggested that up to 8 ml of blood could be taken per kg body weight at a single sample (Iwarsson et al. 1994)

The body temperature of rabbits should be taken by a suitable thermometer which has been lubricated for ease of introduction into the anus (Batchelor 1999). The rabbit should be gently restrained, the tail lifted and the thermometer gently inserted. The thermometer should never be forced into the anus, and if there are any obstructions to the passage it should be investigated. The thermometer should be left in situ for 1-2 minutes, whereafter the temperature is recorded. In rabbits the normal mean rectal temperature is 38.3°C with a range of 37.0-39.4 °C.

Rabbits may be given different substances by oral (p.o.), subcutaneous (s.c.), intravenous (i.v.), intramuscular (i.m.) or intraperitoneal (i.p.) administration. For more detailed description of these techniques see Batchelor 1999 and Iwarsson et al. 1994.

For anaesthesia of rabbits there are many different methods and substances (Svendsen 1994). Depending on the purpose of the procedure or operation a specific substance should be chosen, and then the recommended range of doses given in the literature should be followed (Olfert et al. 1993, Svendsen 1994, Batchelor 1999). Generally relief of pain should be given to all rabbits during operational procedures, if there is evidence that pain is present in the individual.

Postoperative care of rabbits includes placing them in a recovery cage, the box they were placed in before surgery or the home cage which has been lined with a tray liner (Batchelor 1999). The liner should be folded over the animal and a cotton surgical drape placed over it to minimise hypothermia. A 'Vet Bed', which is commercially available may also be used. The liner or 'Vet Bed' should be removed about 30 minutes after the animal has regained consciousness and is sitting up. When operating on several group housed animals the last animal operated on must have completely recovered consciousness before all the animals are returned to the pen simultaneously (Batchelor 1999). If animals are returned to a group pen while they recover consciousness they may be subjected to aggression. Incision sites should be covered with a clear plastic dressing spray, but rabbits may occasionally interfere with their stitches. A plastic collar can then be applied to the animal to restrict its access to the operation site after suturing (Batchelor 1999).

Euthanasia of rabbits is usually carried out by intravenous injection of an overdose of barbiturates, such as sodium pentobarbitone (Batchelor 1999). Physical dislocation of the neck is another option for rabbits up to one kg in body weight. However, it is recommended that this always is followed by heart puncture or cutting one of the main vessels so that the animal is drained of blood.

8. CONCLUSIONS

In order to ensure the welfare of laboratory rabbits there are several aspects to consider. First of all one has to make sure that the rabbits are purchased from a breeder producing health defined rabbits with a controlled genetic background, and with an enriched housing. The transport to the laboratory should be as short and stress-free as possible. Placing two rabbits in the transport box could reduce stress. At arrival to the laboratory the rabbits should be checked for any health problems and placed in its new cage. Acclimatization should be one to two weeks. Young rabbits could be housed pair-wise in cages, or in larger groups in floor pens. Cages should be large enough to give rabbits a place to perform hopping movements and to lie in a position fully stretched out. They should be provided with free access to hay and water, and fed at least once daily. Floor pens should be large enough to house the maximum number of rabbits planned for, at least one meter high to prevent escape and enriched with refuges, hay, gnawing sticks, etc. Regular cleaning of cages and floor pens are important to keep a good hygiene and healthy animals.

There has been relatively large amount of research on the welfare of laboratory rabbits over the past 20 years (Morton et al. 1993, Gunn 1994, Stauffacher et al. 1994, Hubrecht et al. 1999). Both improvements of cage systems and floor housing have been evaluated. There are several companies selling environmentally enriched cage systems and equipment for building floor pens and enrichment items for floor pens. Parallel with this development, new regulations have been planned and intensively discussed over the past years (Appendix A of the European Convention, regulations in Sweden), and some have been put into practice (regulations in Switzerland and UK). The most important welfare research to carry out for the future would be to verify if some of the behaviours rabbits perform in semi-natural environments, for example digging, is so important that it could be considered an essential behavioural need. Alternatives to permanent cage housing of mature bucks, as for example regular exercise in an enriched area, should also be scientifically evaluated. There is also a need for more controlled experiments on the effects of group housing and environmental enrichment on the results of laboratory research.

ACKNOWLEDGEMENTS

We would like to thank Clive Phillips for commenting on the first draft of this chapter and for adding important information to this text. Many thanks to AstraZeneca for taking the initiative to research into improving the housing of laboratory rabbits, which made it possible for Lena Lidfors to collect references and get knowledge about the welfare of rabbits. We also want to thank the Swedish Agricultural Board for taking the initiative to improve the regulations for laboratory animals and thus involving Lennart Lindberg and Lena Lidfors in giving suggestions on new regulations.

REFERENCES

Adams CE. The laboratory rabbit. In: The UFAW Handbook on the Care and Management of Laboratory Animals, Sixth edition. Poole TB ed, Longman Scientific & Technical, England, 1987; 415-435

Batchelor GR. Group housing on floor pens and environmental enrichment in sandy lop rabbits (I). Anim Technol 1991; 42:109-120

Batchelor GR. The laboratory rabbit. In The UFAW Handbook on the Care and Management of Laboratory Animals, 7th edition. Poole T ed, Blackwell Science, Oxford 1999; 395-408

Bell DJ. Aspects of the social behaviour of wild and domesticated rabbits *Oryctolagus cuniculus* L. Unpublished PhD thesis, University of Wales 1977

Bell DJ. Social olfaction in lagomorphs. Symposia of the Zoological Society of London, 1980; 45:141-164

Bell DJ. Mate choice in the European rabbit. In Mate Choice. PPG Bateson ed, Cambridge University Press, Cambridge 1983; 211-223

Bell DJ. The behaviour of rabbits: implications for their laboratory management. In Proceedings of UFAW/LASA Joint Symposium. Standards in Laboratory Animal Management, Part II. Potters Bar: Universities Federation for Animal Welfare 1984; 151-162

Bell DJ. The European wild rabbit. In The UFAW Handbook on the Care and Management of Laboratory Animals, 7th edition. Poole T ed, Blackwell Science Ltd, Oxford 1999; 389-394

Bell DJ, Bray GC. Effects of single- and mixed sex caging on post-weaning development in the rabbit. Lab Anim 1984; 18:267-270

Bell DJ, Webb NJ. Effects of climate on reproduction in the European wild rabbit *Oryctolagus cuniculus*. J Zoology 1991; 224:639-648

Bennett B. Storey´s Guide to Raising Rabbits. Storey books, Vermont 2001

Berthelsen H, Hansen LT. The effect of hay on the behaviour of caged rabbits (Oryctolagus cuniculus). Anim Welf 1999; 8:149-157

Beynen AC, Mulder A, Nieuwenkamp AE, van der Palen JGP, van Rooijen GH. Loose grass hay as a supplement to a pelleted diet reduces fur chewing in rabbits. J Anim Physiol Anim Nutr 1992; 68:226-234

Bigler L. Mutter-Kind-Beziehung beim Hauskaninchen. Lizentiatsarbeit University of Berne 1986

Bigler L, Lehmann M. Schlussbericht ueber die Pruefung der Tiergerechtheit eines Festwandkaefigs fuer Hauskaninchen-Zibben. Report Swiss Federal Veterinary Office, Bern 1991

Bigler L, Oester H. Paarhaltung nicht reproduzierender Zibben im Käfig. Berl Munch Tierarztl Wochenschr 1994; 107:202-205

British Rabbit Council. Breed Standards 1991-95. Newark 1991

Brooks DL. Nutrition and gastrointestinal physiology. In Ferrets, Rabbits and Rodents – Clinical Medicine and Surgery. Hillyer EW, Quesenberry KE eds, WB Saunders, London 1997; 169-175

Broom DM. The scientific assessment of animal welfare. Appl Anim Behav Sci 1988; 20:5-19

Brummer H. Symptome des Wohlbefindens und des Unwohlseins beim Kaninchen unter besonderer Beruecksichtigung der Ethopathien. In Wege zur Beurteilung Tiergerechter Haltung bei Labor-, Zoo- und Haustieren. Militzer K ed, Parey Schriften Versuchstierkunde 1986; 12:44-53

Brummer H. Trichophagia: a behavioural disorder in the domestic rabbit. Dtsch Tierarztl Wochenschr 1975; 82:350-351

Canali E, Ferrante V, Todeschini R, Verga M, Carenzi C. Rabbit nest construction and its relationship with litter development. Appl Anim Behav Sci 1991; 31:259-266

CFN (Centrala Försöksdjursnämnden). Använda försöksdjur i Sverige år 2002 - Statistik. Terenius L ed, CFN, Sweden 2002; 38-60

Cheeke PR. Rabbit Feeding and Nutrition. Academic Press, USA 1987; 160-175

Clarkson TB, Lehner NDM, Bullock BC. Arteriosclerosis research. In The Biology of the Laboratory Rabbit. Weisbroth SH, Flatt RE, Kraus AL eds, Academic Press, New York 1974; 155-165

Cohen C, Tissot RG. Serological genetics. In The Biology of the Laboratory Rabbit. Weisbroth SH, Flatt RE, Kraus AL eds, Academic Press, New York 1974; 167-177

Commission of the European Communities. Third report from the Commission to the Council and the European Parliament on the statistics on the number of animals used for experimental and other scientific purposes in the member states of the European Union, Brussels 22.01.2003, COM (2003) 19 final, 2003

Cowan DP. Aspects of the social organisation of the European wild rabbit (*Oryctolagus cuniculus*). Ethology 1987; 75:197-210

Cowan DP, Garson PJ. Variations in the social structure of rabbit populations: causes and demographic consequences. In Behavioural Ecology: the Ecological Consequences of Adaptive Behaviour. Sibly RM, Smith RH eds, Blackwell Sci Publ, Oxford 1985; 537-555

Cowan DP, Bell DJ. Leporid social behaviour and social organization. Mammal Rev 1986; 16:169-179

Daly JC. Effects of social organisation and environmental diversity on determining the genetic structure of a population of the wild rabbit (*Oryctolagus cuniculus*). Evolution 1981; 35:689-706

Donnelly TM. Basic anatomy, physiology and husbandry. In Ferrets, Rabbits and Rodents – Clinical Medicine and Surgery. Hillyer EW, Quesenberry KE eds, WB Saunders, London 1997; 147-159

Drescher B, Loeffler K. Einfluss unterschiedlicher Haltungsverfahren und Bewegungsmöglichkeiten auf die Kompakta der Röhrenknochen von Versuchs- und Fleischkaninchen. Tierarztl Umsch 1991a; 46:736-741

Drescher B, Loeffler K. Einfluss unterschiedlicher Haltungsverfahren und Bewegungsmöglichkeiten auf die Kompakta der Röhrenknochen von Mastkaninchen. Tierarztl Umsch 1991b; 47:175-179

Dunsmore JD. The rabbit in subalpine south-eastern Australia, 1. Population structure and productivity. Austr Wildl Res 1974; 1:1-16

Eveleigh JR, Taylor WTC, Cheeseman RF. The production of specific pathogen free rabbits. Anim Technol 1984; 35:1-12

Fraser KW. Emergence behaviour of rabbits, *Oryctolagus cuniculus*, in Central Otago, New Zealand. J Zoology, London, 1992; 228:615-623

Falkmer S, Waller T. Försöksdjursteknik- en praktisk handledning. Liber utbildning, Stockholm, 1984; 87-106

Gibb JA. Sociality, time and space in a sparse population of rabbits (*Oryctolagus cuniculus*). J Zoology, London, 1993; 229:581-607

Gibb JA, Ward CP, Ward GD. Natural control of a population of rabbit, *Oryctolagus cuniculus* L. for ten years in the Kourarau enclosure. New Zealand Department of Scientific and Industrial Research, Bulletin 1978; 223

González-Mariscal G, Díaz-Sánchez V, Melo AI, Beyer C, Rosenblatt JS. Maternal behaviour in New Zealand White rabbits: Quantification of somatic events, motor pattern, and steroid plasma levels. Physiol Behav 1994; 55:1081-1089

Gunn D. Evaluation on welfare in the husbandry of laboratory rabbits. PhD Thesis, University of Birmingham, England 1994

Gunn D, Morton DB. Rabbits. In Environmental Enrichment Information Resources for Laboratory Animals. Smith CP, Taylor V eds, AWIC Resource Series 1995; No 2:127-143

GV-SOLAS. Planung und Struktur von Versuchstierbereichen tierexperimentell tatiger Institutionen 4th edn. Biberach: Gesellschaft für Versuchstierkunde, Society for Laboratory Animal Science 1988

Hafez ESE. Sex drive in rabbits. Southwest Vet, 1960; 14:46-49

Hagen KW. Colony husbandry. In The Biology of the Laboratory Rabbit. Weisbroth SH, Flatt RE, Kraus AL eds, Academic Press Inc, New York 1974; 23-47

Hansen LT, Berthelsen H. The effect of environmental enrichment on the behaviour of caged rabbits *(Oryctolagus cuniculus)*. Appl Anim Behav Sci 2000; 68:163-178

Harcourt-Brown F. Textbook of Rabbit Medicine. Butterworth-Heineman, Oxford 2002

Harkness JE, Wagner JE: The Biology and Medicine of Rabbits and Rodents. William & Wilkins, Baltimore 1995

Hartman HA. The foetus in experimental teratology. In The Biology of the Laboratory Rabbit. Weisbroth SH, Flatt RE, Kraus AL eds, Academic Press, New York 1974; 92-153

Heath M, Stott E. Housing rabbits the unconventional way. Anim Technol 1990; 41:13-25

Held SDE, Turner RJ, Wootton RJ. Choices of laboratory rabbits for individual or group-housing. Appl Anim Behav Sci 1995; 46:81-91

Hillyer EV, Quesenberry KE. Ferrets, Rabbits and Rodents. Clinical Medicine and Surgery. W B Saunders Company, Philadelphia 1997

Hubrecht R, Beeston D, Cubitt S, Gunn-Dore D, Grey C, Hawkins P, Howard B, McBride A, Moore S, Ostle T, Wickens S, der Weduwen S, Wills T. Refining rabbit housing, husbandry and procedures: report of the 1998 UFAW/RSPCA rabbit behaviour and welfare group meeting. Anim Technol 1999; 50:155-164

Huls WL, Brooks DL, Bean-Knudsen D. Responses of adult New Zealand White rabbits to enrichment objects and paired housing. Lab Anim Sci 1991; 41:609-612

Iwarsson K, Lindberg L, Waller T. Common non-surgical techniques and procedures. In Handbook of Laboratory Animal Science. Svendsen P, Hau J eds, CRC Press, USA 1994; 229-272

Jackson G. Intestinal stasis and rupture in rabbits. Vet Rec 1991; 129:287-289

Jilge B. The rabbit: a diurnal or nocturnal animal? J Exp Anim Sci 1991; 34:170-183

Jones SE, Phillips CJC. Mirrors in rabbit cages: what does the rabbit see and how do they affect behaviour? Submitted

Kraft R. Vergleichende Verhaltensstudien an Wild- und Hauskaninchen. Z Tierz Zuechtungsbiol, 1979; 95:165-179

Kersten AM, Meijsser FM, Metz JH. Effects of early handling on later open-field behaviour in rabbits. Appl Anim Behav Sci 1989; 24:157-167

Kolb HH. The use of cover and burrows by a population of rabbits (Mammalia: *Oryctolagus cuniculus*) in eastern Scotland. J Zoology, London 1994; 233:9-17

Laber-Laird K, Swindle MM, Flecknell P. Handbook of Rodent and Rabbit Medicine. Elsevier Science Ltd, Oxford, UK 1996

Lehmann M, Wieser RV. Indikatoren für mangeinde Tiergerechtheit sonie Verhaltensstorungen bei Hauskaninchen. KTBL - Schrift 307 1985; 96-107

Lehmann M. Das Verhalten Junger Hauskaninchen unter Verschieden Umgebungsbedingungen. Dissertation, University of Bern, Switzerland 1989

Lehmann M. Activity requirement for young domestic rabbits: raw fibre consumption and animal welfare. Schweiz Archiv Tierheilkeld 1990; 132:375-381

Lehmann M. Social behaviour in young domestic rabbits under semi-natural conditions. Appl Anim Behav Sci 1991; 32:269-292

Leslie TK, Dalton L, Phillips CJC. The preference of domestic rabbits for pasture and concentrate feeds. Anim Welf 2004; 13:54-62

Lidfors L. Behavioural effects of environmental enrichment for individually caged rabbits, Appl Anim Behav Sci 1997; 52:157-169

Lloyd HG, McCowan D. Some observations on the breeding burrows of the wild rabbit on the island of Skokholm. J Zoology, London 1968; 156:540-549

Lockley RM. Social structure and stress in the rabbit warren. J Anim Ecol 1961; 30:385-423

Loeffler K, Drescher B, Schulze G. Einfluss unterschiedlicher Haltungsverfahren auf das Verhalten von Versuchs- und Fleischkaninchen. Tieraerztl Umsch 1991; 46:471-478

Love JA, Hammond K. Group-housing rabbits. Lab Anim 1991; 20:37-43

Mader DR. Basic approach to veterinary care. In Ferrets, Rabbits and Rodents – Clinical Medicine and Surgery. Hillyer EW, Quesenberry KE eds, WB Saunders, London 1997; 160-168

Meredith A. The Rabbit. In Online Information for Veterinary Students at Royal School of Veterinary Studies, University of Edinburgh 1998; http://www.aquavet.i12.com/Rabbit.htm

Meredith A. General biology and husbandry. In Manual of Rabbit Medicine and Surgery. Flecknell P ed, British Small Animal Veterinary Association, London 2000; 13-23

Metz JHM. Effects of early handling in the domestic rabbit. Appl Anim Ethol 1984; 11:71-87

Milligan SR, Sales GD, Khirnykh K. Sound levels in rooms housing laboratory animals: an uncontrolled daily variable. Physiol Behav 1993; 53:1067-1076

Morton DB, Jennings M, Batchelor GR, Bell D, Birke L, Davies K, Eveleigh JR, Gunn D, Heath M, Howard B, Koder P, Phillips J, Poole T, Sainsbury AW, Sales GD, Smith DJA, Stauffacher M, Turner RJ. Refinement in rabbit husbandry, Lab Anim 1993; 27:301-329

Mulder A, Nieuwenkamp AE, van der Palen JG, van Rooijen GH, Beynen AC. Supplementary hay reduces fur-chewing in rabbits. Tijdschrift fur Diergeneeskunde 1992; 117:655-658

Myers K. The effects of density on sociality and health in mammals. Proc Ecol Soc Aust 1966; 40-64

Myers K, Poole WE. A study of the biology of the wild rabbit, Oryctolagus cuniculus (L.), in confined populations II. The effects of season and population increase on behaviour. CSIRO Wildl Res 1961; 6:1-41

Myers K, Schneider EC. Observations on reproduction, mortality and behaviour in a small, free-living population of wild rabbits. CSIRO Wildl Res 1964; 9:138-143

Myers K, Parer I, Wood D, Cooke BD. The rabbit in Australia. In The European Rabbit: The History and Biology of a Successful Coloniser. Thompson HV, King CM eds, Oxford University Press, Oxford 1994; 108-157

Mykytowycz R. Social behaviour of an experimental colony of wild rabbits, Oryctolagus cuniculus (L.), III Second breeding season. CSIRO Wildl Res 1960; 5:1-20

Mykytowycz R. Territorial marking by rabbits. Scientific American 1968; 218:116-126

Mykytowycz R, Rowley I. Continuous observations of the activity of the wild rabbit, Oryctolagus cuniculus (L.) during 24-hour periods. CSIRO Wildl Res 1958; 3:26-31

Mykytowycz R, Fullager PJ. Effect of social environment on reproduction in the rabbit, Oryctolagus cuniculus (L.). J Reprod Fert 1973; 19:503-522

Mykytowycz R, Hesterman ER. An experimental study of aggression in captive European rabbits, Oryctolagus cuniculus (L.). Behaviour 1975; LII 1-2:104-123

National Research Council (NRC). Nutrient Requirements of Rabbits. National Academy of Sciences, Washington DC 1966

Nerem RM. Social environment as a factor in diet-induced atherosclerosis. Science 1980; 208:1475-1476

Nicklas W, Baneux P, Boot R, Decelle T, Deeny AA, Fumanelli M, Illgen-Wilcke B. FELASA recommendations for the health monitoring of rodent and rabbit colonies in breeding and experimental units. Lab Anim 2002; 36:20-42

Olfert ED, Cross BM, McWilliam AA. Rabbits. In Guide to the Care and use of Experimental Animals Vol. 2. Canadian Council on Animal Care 1993; 123-129.

Parer I. The population ecology of the wild rabbit, *Oryctolagus cuniculus* L., in a Mediterranean-type climate in New South Wales. Aust Wildl Res 1977; 4:171-205

Parer I. Dispersal of the wild rabbit, *Oryctolagus cuniculus*, at Urana in New South Wales. Aust Wildl Res 1982; 9:427-441

Parsons AJ, Newman JA, Penning PD, Harvey A, Orr RJ. Diet preference of sheep - Effects of recent diet, physiological state and species abundance. J Anim Ecol 1994; 63:465-478

Podberscek AL, Blackshaw JK, Beattie, AW. The behaviour of group penned and individually caged laboratory rabbits Appl Anim Behav Sci 1991; 28:353-363

Portsmouth J. Commercial Rabbit Keeping, 3rd Edition. Nimrod Press Ltd, Alton, UK 1987

Ross S, Denenberg VH, Frommer GP, Sawin PB. Genetic, physiological and behavioural background of reproduction in the rabbit. V. Nonretrieving of neonates. J Mammal, 1959; 40:91-96

Rothfritz P, Loeffler K, Drescher B. Einfluss unterschiedlicher Haltungsverfaren und Bewegungsmöglichkeiten auf die Spongiosastruktur der Rippen sowie Brust- und Lendenwirbel von Versuchs- und Fleischkaninchen. Tierarztl Umsch 1992; 47:758-768

Ruckebusch Y, Grivel ML, Fargeas MJ. Electrical activity of the intestine and feeding associated with a visual conditioning in the rabbit. Physiol Behav 1971; 6:359-365

SJVFS 2000:133. Statens jordbruksverks föreskrifter om transport av levande djur. Saknr L5, Sweden 2000

Southern HN. Sexual and aggressive behaviour in the wild rabbit. Behaviour 1948; 1:173-194

Spector W. Handbook of Biological Data. WB Saunders, Philadelphia 1956

Stauffacher M. Kaninchenhaltung in Zucht und Mastgruppen - ein neues tiergerechtes haltungskonzept fuer Hauskaninchen. Schweizer Tierschutz 1989; 116:20-35

Stauffacher M. Behaviour ontogeny and the development of abnormal behaviour. Proc. VII International Congress on Animal Hygiene Leipzig 1991; 3:1068-1073

Stauffacher M. Group housing and enrichment cages for breeding, fattening and laboratory rabbits. Anim Welf 1992; 1:105-125

Stauffacher M. Tierschutzorientierte Labortierethologie in der Tiermedizin und in der Versuchstierkunde – ein Beitrag zum Refinement bei der haltung und im Umgang mit Versuchstieren. In Ersatz- und Ergänzungsmethoden zu Tierversuchen. Schöffl H, Spielmann H, Gruber F, Koidl B, Reinhardt C eds, Springer, Wien 1993; 2:6-21

Stauffacher M, Bell DJ, Schulz K-D. Rabbits. In The accommodation of laboratory animals in accordance with animal welfare requirements. O'Donoghue PN ed, Proc Int Workshop Bundesgesundheitsamt, Berlin 17-19 May 1993, 1994; 15-30

Stodart E, Myers K. A comparison of behaviour, reproduction and mortality of wild and domestic rabbits in confined populations. CSIRO Wildl Res 1964; 9:144-159

Svendsen P. Laboratory animal anesthesia. In Handbook of Laboratory Animal Science. Svendsen P, Hau J eds, CRC Press, USA 1994; 311-337

Swallow JJ. Transporting animals. In The UFAW Handbook on the Care and Management of Laboratory Animals, 7th edition. Poole T ed, Blackwell Science, Oxford 1999; 171-187

Townsend GM. The grading of commercially bred laboratory animals. Vet Rec 1969; 85:225-226

Vastrade M. Spacing behaviour of free-ranging domestic rabbits, *Oryctolagus cuniculus* L. Appl Anim Behav Sci 1987; 18:185-195

Wagner JL, Hackel DB, Samsell AG. Spontaneous deaths in rabbits resulting from trichobezoars. Lab Anim Sci 1974; 24:826-830

Wallace J, Sanford J, Smith MW, Spencer KV. The assessment and control of the severity of scientific procedures on laboratory animals. Report of the Laboratory Animal Science Association Working Party. Lab Anim 1990; 24:97-130

Webb NJ, Ibrahim KM, Bell DJ, Hewitt GM. Natal dispersal and genetic structure in a population of the European wild rabbit (*Oryctolagus cuniculus*). Mol Ecol 1995; 4:239-247

Wieser RV. Funktionale Analyse des Verhalten als Grundlage zur Beurteilung der Tiergerechtheit. Eine Untersuchung zu Normalverhalten und Verhaltensstoerungen bei Hauskaninchen-Zibben. PhD Thesis, University of Berne 1986

Wood DH. The demography of a rabbit population in an arid region of New South Wales, Australia. J Anim Ecol 1980; 49:55-80

Wullschleger M. Nestbeschaeftigung bei saeugenden Hauskaninchenzibben. Revue Suisse de Zoologie 1987; 94:553-562

Yu B, Chiou WS. Effects of crude fibre level in the diet on the intestinal morphology of growing rabbits. Lab Anim 1996; 30:143-148

Zain K. Effects of early social environment on physical and behavioural development in the rabbit. PhD Thesis, University of East Anglia 1988

Zarrow MX, Denenberg VH, Anderson CO. Rabbit: frequency of suckling in the pup. Science 1965; 150:1835-1836

Chapter 11

THE WELFARE OF LABORATORY DOGS

Robert Hubrecht[1] and Anthony C Buckwell[2]
[1]*Universities Federation for Animal Welfare, Wheathampstead, Hertfordshire, and*
[2]*Division of Biomedical Services, University of Leicester, Leicester, United Kingdom*

1. INTRODUCTION

In this chapter we aim to provide a concise introduction to the care of dogs bred for and used in research, paying particular attention to the manner in which health, and welfare, may be enhanced and maintained, and to how an integrated management system should provide for the dogs' behavioural as well as physiological requirements.

It may be also be useful for readers to note that there are two recent, but at the time of writing, ongoing reviews of dog housing. The first, part of the revision of Appendix A of the European Convention ETS 123 (Draft proposals) is aimed at drawing up recommendations which, while in themselves not binding, are likely to have an impact on national legislation. As part of this process a review of the literature underlying the recommendations has been carried out and it is intended that this will be published at some stage (Group of experts (cats, dogs and ferrets), to be published). The second is the 8th publication in a series by the BVAAWF/FRAME/RSPCA/UFAW joint working group on refinement (Prescott et al. in preparation). This document comprehensively reviews refinements in all aspects of laboratory dog husbandry and use, drawing on scientific knowledge and experience with the aim of promoting the spread of best practice.

The function of laboratory housing is to keep animals in conditions that meet their needs, and to provide a practical, standardised environment to reduce experimental variability. It is probably the case that neither of these two aims has been met in the past. Although husbandry conditions are normally, although not always, held relatively constant within any one institution, there has always been variability in housing conditions between institutions. For example, within the UK, pen housing for pairs of dogs has been standard for many years, but cages as well as pens have continued to be used in the US and in some mainland European countries. For those interested in improving laboratory dog welfare, this diversity actually makes it easier to introduce improvements as given the fact that there are existing variations in structures of pens, it becomes less of an issue for other laboratories to consider making changes.

Housing systems for laboratory dogs are currently going through something of a revolution in the UK, and the advances that have been made in terms of social housing and enrichment are beginning to have an impact internationally. The sorts of barren enclosures described by Hubrecht (Hubrecht 1995a) are becoming rarer but old-style dog-housing is still in existence and this can cause problems for managers attempting to provide a good housing regime for their dogs. The lesson to be learnt here is that traditional methods of pen construction, using substantial materials, can make it difficult to introduce innovative ideas during the lifetime of the building. However, as old dog accommodation reaches the end of its life and new kennels are built it then becomes possible to implement real improvements in dog welfare. To do this without repeating the errors of the past, it is essential to fully consider the natural history of the dog, to understand its nature and biology and to take full account of current best practices. Moreover, because our knowledge of dogs' needs will continue to improve, their accommodation should be designed so that it is possible to incorporate new ideas without having to make expensive changes to the structure of the building. Flexible designs are the key to ensuring that this is possible.

2. BIOLOGY AND BEHAVIOURAL NEEDS

2.1 Origins and natural behaviour of the dog

The dog's biology and hence its needs have been moulded by both evolution and through natural selection. The archaeological evidence suggests that dogs have been domesticated for at least 14,000 years (Clutton-Brock 1995). However, analysis of mitochondrial DNA indicates that their origins might be much older (Vila et al. 1997). It has generally been accepted that dogs are derived from a lupine ancestor, most likely to be the Asiatic wolf *Canis lupus pallipes*, although Koler-Matznick (2002) takes a different view and argues that the dog lineage is separate to that of the wolf.

DNA analysis suggests that dogs may have originated from a number of founding events and that there has been occasional interbreeding with wolf populations. The result is that there are now in existence many breeds of dog showing enormous phenotypical differences, and also, to a certain extent, temperamental and behavioural differences (Hart 1995, Coppinger and Coppinger 1998).

Feral dogs live in loose and changeable social groupings that are strongly influenced by food distribution. Dogs differ from wolves in that they are not as cooperative, for example, in care of their young or while hunting. They do, however, seem to cooperate in territorial defence and interact as social groups (Boitani et al. 1995, MacDonald and Carr 1995). Wolf-like behaviours such as the tendency to follow a superior animal, male or female, (Mech 2000) and live in social groups (packs) (Mech 1970) may well have made the process of domestication easier (Bradshaw and Brown 1990). Moreover, these characteristics would have been favoured by human selection, and the dog remains an extremely social animal that is capable, given the correct rearing conditions, of behaving socially with humans as well as with conspecifics. Indeed, studies comparing dogs with other wild canids have shown that even under conditions of equivalent human socialisation, dogs show an increased dependence on human social contact (Fox 1978). Domestication may even have enhanced the dog's ability to communicate with humans (Hare et al. 2002, Miklósi et al. 2003).

It is now generally accepted that the period between 3 and 13 weeks of age, when the puppy is particularly sensitive to its physical and social environment, is most important with respect to its sociability, in later life, with humans or other conspecifics. Whilst experiences outside this range may also be influential, environmental complexity and social contacts during this period play a major role in determining the dogs' temperament and adaptability as a suitable subject for subsequent research.

2.2   Behaviour and senses

Although some wolf-like behaviours are evident, the process of domestication has modified the original behavioural repertoire, largely by a process of neotenisation (see Zimen 1981, Bradshaw and Brown 1990). The effect of this is that the maturation of some behaviours in the dog is delayed compared to the wolf. In some breeds (pointers and sheep dogs) the sequence of predation behaviour has been truncated so that hunting to the kill usually does not occur. In other cases certain communicatory behaviours have been lost (Goodwin et al. 1997). This loss of communicatory ability could be important in the management of these dogs in kennels because communication is essential to modulate social interactions. Hence, if a particular breed has partially lost the ability to communicate with other dogs, there is a potential for an increased risk of fighting or, at least, for higher aggression.

Barking is a potential problem in kennels. It was probably a behaviour that was, either deliberately or accidentally, selected for by humans at various stages in the dog's domestication as it would have been a particularly useful characteristic for guard or hunting dogs. As a result, it is likely that barking has now become partially divorced from communicatory function so that it in some circumstances the behaviour may only reflect a high state of arousal. Vocalisations, therefore, generally have to be interpreted in the light of the circumstances in which they occur. Nonetheless, dogs do, to some extent, use different vocalisations in different contexts. Howling, for example, which wolves use to communicate over long distances, may be similarly used by dogs when separated from their social group. However, other vocalisations such as barking and whining, can also be used in the same context (Lund and Jørgensen 1999).

One would expect that hearing and auditory signals would be optimised, and so it is not surprising that the dog's hearing is particularly sensitive to sound at the frequency levels where there is maximum energy in the bark, 500 Hz to 16 kHz. Dog hearing, generally, is both more sensitive and has a wider frequency range than humans. Dogs can hear sounds up to 50 kHz, however they are most sensitive to sounds between 1-20 kHz, and within some parts of this range are up to 4 times more sensitive to sounds than humans.

Dogs' vision is good, although perhaps not as good as that of man with respect to discrimination of form and pattern (Fox and Bekoff 1975). The common presumption that they see in black and white is incorrect. Dogs have dichromatic colour vision, are highly sensitive to movement and are capable of making discriminations on the basis of choice (Neitz et al. 1989).

All senses are important to animals but the sense of smell is particularly well developed in dogs and wolves and is used for communication and to locate prey. Fox and Bekoff (1975) and Thorne (1995) provide reviews of dog olfaction and taste and further information can be found in Alderton (1994). Unfortunately, olfaction is a poorly developed sense in humans, making it difficult for us to empathise with dogs and to adjust housing appropriately to either provide olfactory interest or to avoid removing odours which may be important to the dog. The importance of olfaction to dogs will be evident to anyone who has kept one as a companion animal. Dogs use pheromones to communicate sexual status (Sommerville and Broom 1998) as well as to mark territory and to communicate social status. In some instances visual signals such as scratching the ground or lifting the hind leg may accompany urine or faecal marking (Bekoff 1979). Here the function is presumably to draw attention to the scent mark, although scratching may also be used to scent mark from the interdigital scent glands (Simpson 1997).

Although the dog is a carnivore, it is a diet generalist capable of exploiting a wide range of food types. Moreover, it is often capable of surviving as a feral animal which indicates that, despite the process of domestication, some dogs at least retain most of their ancestral behavioural characteristics. It is clear that in order to be able to survive in a multitude of environments, cope with a complex social organisation and exploit novel foods, the dog has to be both curious and intelligent. These characteristics however may not serve it so well if it is housed in a barren environment where social contacts are limited. The aim of those involved in keeping dogs for experimental use should be to provide them with a captive environment that provides interest and diversity.

3. OPTIMAL ENVIRONMENT

The diversity of dog phenotypes makes it difficult to specify optimum conditions that would meet all the different breeds' needs. A wide variety of breeds are used in research, but by far the most common is the beagle probably because of its small to average size, its tractable nature and because it was traditionally housed in kennels. Over time, the body of work generated using the beagle has grown, so that considerations such as the presence of established laboratory breeding colonies and the ability to compare novel results with background data discourage the use of other breeds. It therefore makes sense to provide recommendations with the beagle in mind, with the caveat that should other breeds be used there may be differences in their requirements.

In the past the importance of an optimal environment has been underestimated in terms of both welfare and of improving the quality of the science. As an example of the importance of housing on scientific outcomes, Hubrecht et al. (1992) reported that one particular dog in laboratory housing was spending 51% of its time in abnormal repetitive behaviour (stereotypy). Staff reported that this dog required three times its normal daily food ration as a result of the locomotory stereotypy. The presence of dogs with such behavioural abnormalities is likely to lead to increased variation and data from them could be misleading. When considering an optimal environment for dogs it is particularly important to take into account early development and social as well as physical requirements.

3.1 Early development and the importance of integrating breeding and use

While the dog's needs are partly dictated by its nature arising from the selection processes of evolution and artificial selection, nurture also has a significant role. The dog's experiences throughout its life are extremely important in determining both its needs, and how it reacts to people, other dogs and novel events (eg Scott and Fuller 1965, Fox and Stelzner 1966). In common with other species, if dogs are reared in a deprived and barren environment then they are not as good at coping with change and novel experiences later in life.

It is not in the interests of scientists or technical staff to have dogs that are frightened of people, other dogs or of routine laboratory procedures. If dogs are frightened their welfare will be poor, their physiology will be abnormal and the quality of science will suffer. It is therefore of interest to note that a study on shelter dogs has shown that gentle handling and petting can produce smaller rise in cortisol levels of dogs subjected to serial blood samples (Hennessy et al. 1998). It is also worth remembering that early experiences have effects on brain development (see 4.2 in this chapter). Hence for good welfare and science it is important to ensure that dogs are exposed to a variable and stimulating environment throughout their life.

Whilst gentle and friendly interactions with dogs by staff are essential, dogs kept under laboratory conditions are there for a purpose, and it may not be in their best interests to treat them as companion animals. Therefore, laboratory dogs should be well socialised with humans and with conspecifics and the rearing regimen should provide opportunities to familiarise young dogs with routine laboratory procedures and environments. In the past, all too often the rearing environment bore only a limited similarity to the conditions under which the dogs would find themselves during procedure, and yet it is a relatively simple matter to train dogs to cooperate during routine procedures such as weighing. The dog learns that the procedure is non-threatening and also when it is appropriate to play and when it should be still. Because the dog is cooperating and not resisting the handler, it is likely to be less fearful and hence its welfare is improved. Life is made easier for the handler who does not have to restrain a boisterous or frightened dog, and because the dog is unstressed the quality of science arising from the study is also likely to be better. Much work has been recently done in developing best practice in this area by Heath, reported in Prescott et al. (in preparation). Best practice identified in this document includes ensuring that handling methods are standardised between the breeder and the user of the dogs. Other important factors are to ensure that handlers receive good quality training and that positive rather than negative reinforcement to obtain desired behaviours from the dog should be used.

3.2 The social environment

3.2.1 Social needs of conspecifics

The social nature of the dog is one of its most important characteristics with respect to providing it with an optimal environment. Hence the optimal laboratory environment must allow dogs to be housed in harmonious social groups. The presence of another dog provides continuous variety and interest within the captive environment, opportunities for play and the expression of other species typical behaviours. Dogs destined for use in laboratories would normally be raised in social groups, although they might be subjected to a number of regroupings from their natal group as they are sorted by weight gender and compatibility. It is only once they are allocated to an experimental procedure that there is an increased likelihood that they may be housed singly. It is generally accepted now that solitary housing represents a significant welfare insult to the dog and that it should be avoided wherever possible (Group of experts (cats, dogs and ferrets), to be published). Single housing for research purposes should always be justified in the welfare cost/benefit analysis for the proposed programme of work and ameliorated by maintaining visual, auditory and olfactory contact with conspecifics. Solitary housing requires exceptional justification.

The majority of laboratory dogs are used in toxicology studies in which traditionally dogs used to be housed singly to obtain food consumption data, and to avoid other problems such as the consumption of another dog's vomit. However, the industry in the UK has made enormous efforts to advance social housing and routinely houses dogs on study in pairs (Hubrecht 1995b). Dogs are usually only separated for short periods following feeding/dosing, and they remain in full visual and auditory contact with other dogs. Moreover, the period of separation has been progressively reduced. The Council of Europe draft proposals recommend that special justification is required for periods of separation greater than 4 hours (Draft proposals) whilst Prescott et al. (in preparation) states that separation can be for as little as 2 hours post dosing.

Dogs may also be singly housed in metabolism cages but again it is usually possible to place the cages so that the dogs remain in auditory and physical contact with other dogs. There may be some procedures that require single housing but the justification for separation always needs to be critically examined. For example, Prescott et al. (in preparation) identify a number of ways in which single housing can be avoided by using slightly different techniques (eg by replacing or refining surgical models that are vulnerable to damage by other dogs).

Dogs are sometimes housed singly for management reasons. Typical examples include aggressive dogs, stud dogs and pregnant bitches. Again, every effort should be made to avoid single housing or to reduce its duration to a minimum. For instance pregnant bitches, need not be housed singly until the bitch herself indicates that she no longer desires other canine contact.

3.2.2 Dog/human social needs

If puppies are socialised to humans between the ages of approximately 3-13 weeks, then they are much more likely to react in an appropriate social way to humans later in life. This is clearly beneficial in terms of producing animals that are not frightened of humans but it could potentially also be a welfare problem for the dogs if the match between the dogs requirements for human social contact and the resources available is not adequate. When dogs are kept as pets there is, in most cases, abundant human social contact, however, when the owners leave dogs in the house separation problems are a relatively frequent occurrence probably characterised by emotional states such as frustration and increased fear (Lund and Jørgensen 1999). It is possible that similar problems could also be an issue in some laboratory environments. Traditionally the task of animal carers has been to meet the physical needs of the animals in their care, and opportunities to socialise tended to be carried out in the small amount of spare time left to them. Hubrecht et al. (1992) drew attention to the extremely small amount of good quality human social contact time that dogs received in either rescue kennels or research kennels (0.04-0.67% of observed time). A survey of toxicology establishments in 1994 indicated that at that time handling or socialisation for enrichment varied from *none* through *if time permits or during procedures* to *once weekly* (Hubrecht 1995b*)*. However, the importance of socialisation for dogs has been increasingly recognised so that the Animal Procedures Committee (1998) concluded that "there must be sufficient staff time to allow social interaction with the animals". Also that "Unless contraindicated on scientific grounds dogs" (held singly or in small groups) "should be removed to a separate area and allowed to exercise, with other dogs where possible, and with staff supervision and interaction, ideally on a daily basis". It is now becoming increasingly clear that human socialisation can be integrated into the normal husbandry of the breeding unit to provide the appropriate degree of complexity, particularly during the 3 – 13 week period of "primary socialisation". Care should be taken, however, to ensure that negative handling experiences are not classified as social enrichment.

3.3 The physical environment

3.3.1 Temperature and humidity

The dog as a species is extremely adaptable, various breeds being capable of coping with climates ranging from the sub arctic to the tropical, but one should not loose sight of the fact that capacity for adaptation within individual breeds may be poor. Short coated breeds and those with extended or elongated bodies such as greyhounds may be less able to cope with cold stress than some other breeds, however there have been no studies to examine differences between breeds in their temperature requirements or preferences. Given the diversity of breeds and the lack of good scientific data, the Group of Experts for the preparation of draft proposals for revision of Appendix A to the European Convention ETS123 has noted that there is some variance between the temperature ranges recommended in the UK, Europe and the USA (Table 11-1) but that there have been no reports of any effect on welfare of dogs kept in the different ranges (Group of experts (cats, dogs and ferrets), to be published). They have therefore advocated a performance standard for temperature; that is, that there should be no adverse consequence of temperature in housing on welfare. In some cases it may be possible to provide variations in temperature in different locations within the dog enclosure. This can be done by providing localised heat sources (such as heat lamps, heated floor areas, etc) this then allows the dog to exert a degree of control and choice over its environment. It is important to remember that if one area of the enclosure is uncomfortable for the dog, then that area is effectively denied to it, so variations in temperature should not be too extreme. In some facilities the base of the sleeping area (which may be a raised step) in the enclosure is heated. While this seems to work in practice, there is a risk that the sleeping area could become too warm for the dog and for this reason it may be preferable that the heated area should be an adjacent wall so that the dog can control its temperature by either moving away or against the wall. Whilst under experiment there is a risk that variable temperatures could adversely affect the experimental response. As this would result in either invalid experiments and consequent wastage of animals, or the need for increased numbers, a more restricted temperature range should be used. In passing it is worth remembering that newborn puppies require higher temperatures (26-28°C) for the first week or so after birth (MacArthur Clark 1999).

Humidity is probably of relatively small importance to the welfare of the dog and the Draft proposals for the revision of Appendix A to the Council of Europe Convention ETS 123 note that *it is considered unnecessary to control relative humidity, as dogs can be exposed to wide fluctuations of ambient relative humidity without adverse effects.* Nonetheless, again in the interests of reducing experimental variation thought should be given as to whether it is necessary to control humidity for dogs under study.

Table 11-1. Temperature and humidity recommendations

	Temperature °C	Humidity %
UK	15-24	45-65
Council of Europe	15-21	
National Research Council	18-29	30-70

3.3.2 The olfactory, auditory and visual environment

There has been far too little scientific work carried out on providing an environment for dogs which is in harmony with their senses. The problem here is that the dog's senses differ so much from ours and this makes it difficult for us to imagine what dogs' needs in this area might be. The specialised adaptations to the nose and muzzle of the dog allow it to detect scents at concentrations at least 100 times less than those detectable by man. Nonetheless, little thought has been given to establishing whether certain scents are attractive or otherwise to dogs. It is certainly the case that, when housed socially, dogs retain a keen interest in their olfactory environment and spend more time sniffing the ground than dogs that are housed singly. This marking and investigation clearly takes place even when dogs are housed in groups of only two animals (Hubrecht et al. 1992). Overzealous removal of such odours might destabilise social groups, and the use of strongly scented cleaning agents could be aversive but unfortunately there has been little research work in this area.

Those that work with dogs should be aware of the fact that their own odours are also part of the dog's environment. Dogs normally greet a human with a sniff and it has been suggested that dogs may use odours from humans in different ways. Millot et al. (1987) studied the response of dogs to mannequins and suggested that anogenital sniffing is used by dogs to identify strange humans, whilst sniffing the face may be an attempt to determine the human's emotive state or behavioural intentions. Human odours include artificial scents and perfumes, so these should be used with caution (Fox 1986).

As described earlier, dog hearing is more sensitive than ours both in terms of frequency range and acuity. It is often the case that dog kennels as a result of their construction, linked with dogs, propensity to bark, are noisy places. It is therefore surprising that it is only comparatively recently that attention been drawn to the potential welfare problems (Hubrecht et al. 1997, Sales et al. 1997). These studies showed that at a number of different dog establishments, including various laboratory kennels, dogs were exposed to noise levels well in excess of levels that cause stress in other animals (Gamble 1982). High noise levels can have significant effects on the physiology and health of humans (Van Raaij and Oortgiesen 1996, Passchier Vermeer and Passhier 2000) and it is at least possible that there may be similar effects in dogs. Hence high noise levels might not only be a welfare issue, but there could also be serious implications regarding the quality of science carried out on dogs exposed to high sound levels. The noise in dog accommodation is often accentuated by the design. Hard materials are usually used to provide resistance to damage by the dogs themselves or by cleaning materials. However, walls made of hard materials reflect the sound back into the dog enclosure. Moreover, there may be multiple reflections thus increasing the dogs exposure to sound. The time taken for a probe sound to decay can be measured (reverberation time) and comparisons of two similar rooms both with acoustic ceilings, but one with the sound absorbent material painted over, demonstrated the importance of properly functioning sound absorption (Hubrecht et al. 1997). Sales et al. (1996) provide recommendations aimed at reducing noise in dog facilities (Table 11-2). It can be seen from the table that the control of noise is an example of how it is important to consider all aspects of husbandry and welfare before finalising building plans.

Table 11-2. Recommendations to control noise from Sales et al. (1996)

1) The nature of the site
- The siting should be carefully considered to reduce excess noise entering the building from the environment.
- If the egress of noise is likely to be a problem the nature of the sound baffling to reduce this egress should not be such as to reflect the sound back into the facility.

2) The design of the buildings
- Sound reducing measures should be considered at an early stage of the design process. These should include not only the avoidance of material that reflects sound but also incorporation of sound reducing material. In existing facilities absorbent materials could be added to the ceiling which should help to absorb some of the noise and prevent reflection and reverberation.

The layout of the building should be considered including
- the avoidance of long straight corridors between noisy areas and the areas housing the dogs.
- the use of double skinned walls and doors between noisy areas and the dogs.
- the separation of staff activity areas, including food preparation and dog training, from areas housing the dogs. For example, studies at some sites showed that personnel entering or leaving a room with a dog can stimulate barking. In new facilities it may be possible to design a service corridor that does not allow the other dogs to observe these events.
- In facilities where the dogs can see members of the public off site or outside

> working hours some form of solid fencing that would prevent the dogs seeing visitors to the site in the evening may reduce sound levels after the staff have gone home.
>
> 3) Husbandry procedures
> - Reducing the number of dogs in any one room would reduce the number of barking incidents and the social facilitation of barking and so should contribute to a reduction in noise levels.
> - Where husbandry systems involve letting the dogs into the outside run at night the benefit to the animals should be weighed against the problem of noise exposure.
> - Turning lights on in the morning closer to the time the staff arrive may reduce sound levels early in the day but trials would be needed to ensure that there were no adverse effects on welfare.

We know that human emotions are affected by colour and there is now some evidence to suggest that rodents show preferences for certain cage colours (Sherwin and Glenn, in press), however we have no idea what effect the colour of dog enclosures has on their mental state or welfare. Prescott et al. (in preparation) point out that many dog toys are designed to appeal to humans rather than dogs so that bright red toys which are eye catching to us might be much less so to dogs.

4. HOUSING, FEEDING AND CARE

4.1 The design of optimal housing

The enclosure which houses the dog is only part of the whole, and to ensure that welfare is maximised it is necessary to consider all aspects of the design and management of the complete facility. For example, human socialisation needs might be partially met by interactions during husbandry (as long as these are truly positive in nature) and this might lessen the need for other human socialisation periods. However, the primary enclosure which contains the dog is of critical importance, and as management practices are likely to change with time, it makes sense to incorporate the highest standards into the enclosure as possible. There are very few data available to allow precise definition of the space requirements of the domestic dog. We know from studies of feral dogs that they can roam over areas as large as 2850 hectares (Nesbitt 1975), and that dingoes, a closely related species, use even larger home ranges; but these figures should be interpreted with caution as home ranges are frequently determined by the availability of various resources such as food or mates. A recent study (Clubb and Mason, in press) has shown that pacing stereotypies are more common in carnivore species with large home ranges. This may indicate that these species have difficulty in adapting to restricted conditions, however it remains possible that the dog is a special case as the result of the changes produced by domestication. A more useful way of determining minimum enclosure dimensions is to consider the requirements of the dogs in terms of social housing, and functional use of space and then to consider the minimum space required to meet these needs.

4.1.1 Social grouping

As discussed earlier, the most important factor in dog housing is to ensure that each kennel has sufficient spaces and resources to permit social grouping. It is true that social housing carries some risk and that dogs may be injured or even killed as a result but, with appropriate management practices the risk can be minimised. To do this it is important to understand the causation of overt aggression.

Aggressive behaviour is a natural behavioural mechanism employed for a variety of useful purposes, including the establishment and maintenance of social hierarchies, but uncontrolled escalating aggression can lead to fighting and may consequently result in injury. Dogs fight when they contest limited resources (Wickens et al. 2001). There are also some important triggers of aggression such as excitement, and the presence of food or other valued resources. Recognising the situations that are most likely to precipitate fighting is essential. Examples include the introduction of an unfamiliar dog to an established group, situations in which the dog's physiological status is changing, particularly at sexual maturity and during pro-oestrus.

Specific actions that can be taken to avoid fighting include:
1. Ensuring dogs are well socialised if they are to be group-housed.
2. When introducing a previously unfamiliar dog to others, ensure that dogs are first housed within site, smell and sound of each other. It may also be helpful to run the dogs together initially in smaller groups where they have space to explore and interact freely.
3. Ensure that high value resources are not monopolised by providing sufficient enrichment for all the dogs.
4. Avoid presenting high value resources when the dogs' excitement is high. It is important, for instance, to avoid feeding dogs when they are likely to be particularly aroused such as during pen cleaning or immediately following exercise.

The argument is sometimes made that most laboratory facilities allow dogs to see and hear other dogs in neighbouring pens, however, there have been no studies to demonstrate whether this in any way meets social needs, or indeed whether it is frustrating (the dogs can see each other but not interact, for example, by smelling the inguinal region). The question remains then as to what constitutes an appropriate group size within each enclosure. In breeding facilities it is common to group stock dogs in numbers of 4-6, but the real issue is, what is the smallest group size that will still meet the social needs of dogs?. There has been very little research on this, but studies of dogs in a variety of housing types showed that dogs housed in pairs spent a similar proportion of their time budget in social interactions when compared with dogs housed in larger groups of 5-11 animals. This may indicate that the common practice of housing dogs in pairs is adequate to meet their social needs.

It follows that there must be sufficient space and resources within each enclosure for at least two animals. If this provision is written into legislation, or their associated codes of practice, (as is the case in the UK and also in the draft Council of Europe recommendations) it has the important effect of making social housing an attractive option, as single housing then becomes more expensive than social housing. As discussed earlier, dogs are often separated for dosing in toxicology studies, which means that the dogs are confined in half the normal space for a short period. Whilst this confinement may not be a significant welfare problem as the dogs remain in visual and auditory contact with their kennel mates, there is a problem in terms of enclosure design because it then becomes very much harder to achieve a design that allows functional separation of space and adequate enrichment.

4.1.2 Functional use of space

The natural environment is not uniform, and animals typically use different areas for different purposes. In the case of canids a home range will typically consist of a denning or breeding area, an area or areas for resting or sleeping and social interactions such as play. Other sections of the range will be used for hunting or foraging areas and the whole will be surrounded by sites which are scent marked. Some of these areas will be relatively fixed whilst the use of others will change over time as resources and needs fluctuate. It is clear that such an environment is likely to be both complex and variable. Out-of-date dog accommodation is often barren with no provision to allow the dogs to make different use of different areas. Good modern dog enclosures are sub-divided into varied areas, with, for example, a resting area and an exercise or play area. This sub division serves a number of useful practical purposes. Firstly it allows the dog choice to express certain behaviours in certain areas of its pen. For example Fox (1986) has argued that dogs housed in simple enclosures cannot satisfy their normal eliminative behaviours and are forced to defeacate where they eat and sleep. Providing separate areas that allow defecation away from the sleeping area, improves hygiene and allows normal behaviour patterns. It is true that some laboratory dogs will defaecate on sleeping areas, perhaps because of lack of options or because of natural variation, but this should not stop us from trying to provide an environment that stimulates more normal behaviour. Sub-divisions also allow socially housed dogs to control their social interactions. A single enclosure does not allow quarrelling animals to get out of sight of each other and this may lead to an escalation of aggression. Sub divisions provide an extra degree of complexity and interest, and allow the dog to express a small degree of choice about its location and visual field within its environment that would otherwise be impossible (see Poole and Stamp-Dawkins 1999). Such control may be particularly important for nervous dogs as it allows them to retreat to an area where they feel safe. Hubrecht et al. (1992), for example, found that an enclosed kennel within a pen was extensively used by young dogs as a refuge from other dogs and during play. Finally, and importantly, dog enclosures should also be large enough to allow easy staff entry and thus encourage staff interactions with the dog. Clearly then, cages can not be an appropriate method of housing dogs for use in experimental procedures.

4.1.3 Space allowances

Given these prerequisites, how big should a dog enclosure be to allow provision of these requirements? This is a difficult question to answer precisely, but practical experience does indicate that a dog pen of $4.5m^2$, as used in the UK) provides sufficient space for sub-division and enrichment to allow functional use of space for two adult beagle dogs. There have been studies that have examined US pen sizes with relation to dogs' needs for activity, but these have suffered from not including large enough enclosures in the studies (see Hubrecht 1995a). Studies of dogs in larger enclosures have shown differences dependent on enclosure size (Hubrecht et al 1992, Hetts et al. 1992). These indicate that larger enclosures are generally better than very small ones, but do not provide evidence regarding the minimum space that will still meet an animal's needs. It should be remembered that the space on its own does not provide adequate dog accommodation and that the whole is dependent on functional sub-division, enrichment, sight lines, social housing and management systems.

Some laboratory and breeder accommodation allow dogs access to outdoor areas that provide greater sensory opportunities, perhaps particularly olfactory ones. There is general agreement that such access can be enriching and is beneficial for the dogs. Unfortunately, such enclosures are not always possible due to concerns about animal activists. Where outdoor enclosures are provided, the dogs should always have access to a sheltered, and warm indoor area. Some establishments do not provide permanent access, but instead take the dogs to the outdoor enclosures for play and social interactions. There should be structures for play and enrichment items within the enclosure and a member of staff should be present to interact with the dogs and monitor for aggression.

It is important to ensure when designing new accommodation that the housing is easy to modify at any later stage and it should be possible to link enclosures together. Accommodation can then be updated cheaply on a much more regular basis than is possible with "bricks and mortar" types of internal structure and it becomes much easier to provide extra space and facilities to dogs when this is possible.

4.2 Housing enrichment

It is important that adequate enrichment should be provided within the dog enclosure (Hubrecht 1993, Hubrecht 1995c). The benefits are not only with respect to the welfare of the dog, but may also be important in terms of increased scientific validity of results. Dogs that are exposed to a stimulating environment early in life show a more rapid maturation of the brain (Fox and Stelzner 1966) and are also less likely to be afraid of novel experiences. Adult dogs also require adequate stimulation to avoid the risk of boredom and the development of abnormal behaviours.

Figure 11-1. Housing accommodation at Novo Nordisk, Denmark, designed to meet the behavioural needs of dogs. Note the pop hole giving access to an external run, the linked internal runs and enrichment including chews and a bed. (photograph copyright Novo Nordisk, used with permission)

One structural type of enrichment, which has been found to be particularly beneficial, is the use of a raised platform within the pen (Hubrecht 1993). The design used in this study was somewhat complex, with three stairs, but later designs based on this have successfully used ramps (Figure 11-1). Platforms such as these increase the overall space available to the dogs, they dramatically increase the choices that the dog has within the pen, and most importantly because dogs are curious about events outside their pen, they improve the sightlines out of the pen. Some facilities have installed cheaper and more replaceable box platforms within pens and these also improve visibility and choice (Figure 11-2).

Dogs are inquisitive animals which actively seek information about their surroundings. Pens should be designed so as to allow the dog some privacy from neighbouring dogs but also to allow good vision out of their pen. Platforms help with this but bars tend to limit visibility. It is possible to use glass creatively to improve the visibility within rooms (Figure 11-3).

Figure 11-2. Housing accommodation showing box platforms being used by dogs. (Photographs copyright GlaxoSmithKline, Department of Laboratory Animal Science, used with permission)

Dogs naturally chew objects, and in the absence of especially provided items will chew inappropriate structures within their enclosure such as plaster or even the steel wires controlling pop-holes. It is far better to provide them with purposely provided items that are safe and provide interest. Flavoured items seem to be preferred (Hubrecht 1993) and these can be suspended just above the floor to avoid: soiling, monopolisation by one or other of the dogs and blocking of drains. If they are suspended it can also be beneficial to use a spring or elastic cord at the top of the suspension to allow the dogs to place a paw on top of the chew item, bring it to the ground and gnaw it in a species-typical fashion (Hubrecht 1993) (Figure 11-4).

Figure 11-3. Housing accommodation showing glass partitions between pens. (Photograph copyright GlaxoSmithKline, Department of Laboratory Animal Science, used with permission)

Figure 11-4. Suspended chews. Note correct positioning a few cm above the floor of the pen. Inset shows spring allowing the dog to pull the chew to the floor allowing it to use it in a species-typical fashion. (Photograph copyright RC Hubrecht)

Dogs continue to use such chews over extended periods and for substantial proportions of their active time (Hubrecht 1993, Hubrecht 1995c) and they should be provided as standard unless there are overriding veterinary or scientific reasons not to do so. The use of enrichment items should not be unnecessarily curtailed just because an animal is under procedure. Chews with a certificate of analysis may be used in some toxicological studies, and it is even possible to provide inert chews for dogs in metabolism cages, which helps to ameliorate the barrenness of this particular environment. Other items that can be included in certain types of experimental institutions include water baths, ropes and pulls (Loveridge 1994, 1998), but there is a general need to validate the use of different types of enrichment and to determine the strength of the dogs' motivation to use them.

In many institutions background music is provided within the dog rooms. It is claimed that this masks background noise and is soothing for the dogs, but in many cases the music is tuned to a local pop station and appears to be more for the benefit of the staff than the dogs. The volume can sometimes be excessive as the volume is turned up to compensate for the noise of barking or pressure washing. There is only one study on the response of kennelled dogs to noise (Wells et al. 2002) and this was carried out in a shelter. The dogs in the study were somewhat calmer when classical music was played as opposed to rock music, but as the dogs were from unknown backgrounds, it is possible that previous exposure to music could have been a confounding factor.

Given the importance of olfaction for dogs it is surprising that so little has been done to investigate whether olfactory enrichment is a valuable idea. The most practical options so far are the chews that either have a natural odour such as rawhide or inert chews that have an odour impregnated in them. There is some evidence that housing dogs in groups may increase their interest in their environment (they spend more time sniffing the ground, Hubrecht et al. 1992) but we can not be certain that this is necessarily a positive enrichment. Frequent room cleaning results in the repeated removal of olfactory marks, which may result in a sub-optimal environment as far as the dog is concerned. It is now common in the UK for rooms to be pressure washed only once a week or fortnight, but with regular removal of faeces and damp sawdust from the floor as necessary. Such regimens may help to maintain more of the dog's olfactory environment.

4.3   Feeding

It is essential that dogs receive adequate nutrition to stay healthy but they should not be provided with so much that they become obese. There are well-established commercial diets which provide a well balanced nutrition, but the dogs will still need monitoring to ensure that each animal is meeting its needs. Guides to the daily requirement of the dog are available from a number of sources, eg MacArthur Clark (1999), and further reviews and advice are available in Prescott et al., in preparation). In the wild, a significant proportion of the canid's time is spent seeking and obtaining food, and were it possible to mimic these circumstances in the captive environment this would have the dual benefits that the animal would be forced to exercise and the activity would prevent boredom. This is obviously not the case in the laboratory. Food may either be provided *ad libitum* or restricted and there are various advantages and disadvantages associated with each of these options. For example *ad libitum* feeding can result in obesity and increased oral bacterial growth, while restricted feeding may increase competitive aggression amongst dogs housed in groups (see review in Prescott et al. (in preparation). On the other hand if food is provided *ad libitum* there may be an increase in the proportion of the time spent feeding which may help to offset boredom (Hubrecht et al. 1992). Generally, novelty is associated with increased palatability, and in this context it is worth remembering that diets in laboratories may not be as palatable as those provided to pet dogs. Variety is much more likely to be restricted, however it is possible that limited exposure to variety early in life could restrict preferences for food as an adult (Thorne 1995) thus reducing the importance of this variable. Dogs are able to associate the physiological value of a diet with its taste, and over time will tend to prefer diets that are more nutritious. Similarly diet aversion will occur to foods that produce negative physiological effects. It is, therefore, important to remember that diet aversion could occur if a diet became associated with a drug dose with unpleasant consequences.

4.4 Care

In the past the care of laboratory dogs was largely limited to providing them with adequate food, and keeping them physically healthy. Now it is generally accepted that the care given should be much more holistic, and that the needs of the dog for exercise, exploration, play and other social contacts both with humans and dogs should also be met as far as possible. Those responsible for management should set standards for dog care, and ensure that these standards are monitored and achieved. Each aspect needs to be planned taking into account other aspects of husbandry. For example, cleaning out of enclosures can be a negative experience for dogs, however with proper planning it can also be an opportunity for the dog to exercise out of its normal enclosure and to indulge in activities such as play and socialisation.

Staff who are engaged in day to day care of dogs should be empathetic to the dogs' needs. Managers should be aware of the stresses imposed on staff when they have to carry out experimental procedures or euthanasia and should ensure that the workload is not excessive. One of the problems associated with caring for dogs is that because of the familiarity of the dog to us as a domesticated animal, nearly everyone considers himself or herself an expert on dog behaviour. In fact, many dog signals are misunderstood or indeed missed altogether. Staff should understand the importance of body language and interpret it correctly. Body posture, when taken into account with the context in which it occurs, reflects the motivational state of the dog. Signs of appeasement, eg lipsmacking, paw lifting and tail wagging are often taken as signs of contentment when the dog may, in fact, be mildly stressed. Beerda et al. (1997, 1998, 2000) have examined the behavioural and physiological responses of dogs to acute and chronic stressors (Table 11-3). Acute stressors such as very loud noises or shocks resulted in very low posture in addition to increased salivary cortisol levels. Where the dog was able to predict the occurrence of a stimulus, the lowering of posture was more moderate but the dogs showed other signs such as shaking and oral behaviours (yawning, licking of muzzle, extrusion of tongue, etc). Chronic stress induced by suboptimal environments induced higher levels of activity, urination and paw lifting. The authors make the point that the assessment of chronic stress in dogs is difficult to reliably diagnose from behavioural measures alone and that other measures such as urinary cortisol analyses are needed in addition. Nonetheless, for day to day management, a screening system could be devised to look for behavioural indicators of stress, and then if these are found, physiological measures could be taken later.

Table 11-3. Indicators of stress

Indicators of chronic stress	Indicators of acute stress
Increased paw lifting	Body shaking
High levels of activity	Crouching
Increased change of activity	Oral behaviours (tongue out, licking muzzle, swallowing, etc.)
Body shaking	Restlessness
Yawning	Yawning
High Cortisol/creatinine ratio	Low Posture
Increased catecholamines	

In the husbandry of laboratory animals, the early detection and control of pain is particularly important for reducing suffering and staff should be familiar with signs associated with pain and distress. Pain may be indicated either by the presence of specific signs or the absence of some features of normal behaviour. The latter may be easier to miss, but in dogs are often more significant. Examples include inappetance, social withdrawal and failure to autogroom and listlessness. Literature relevant to the dog is reviewed in Prescott et al. (in preparation), see also Flecknell and Waterman-Pearson (2000). Following surgery, adequate analgesia should be provided.

5. HEALTH AND DISEASE PROBLEMS

Maintaining health and eliminating or controlling diseases are fundamental requirements for good animal welfare. A number of texts provide information on health and disease in dogs in a laboratory situation (e.g. Ringler and Peter 1984, Wolfensohn and Lloyd 1998, MacArthur Clark 1999). Many of the more common diseases recognised in laboratory dogs may be related to their particular lifestyle. Conjunctivitis, for instance, may be associated with exposure to dusty sawdust used for bedding material. The fact that laboratory dogs are frequently housed in indoor pens and the floor of the pen is hosed for cleaning may predispose to skin lesions if the dogs' coat is allowed to become wet during the routine cleaning process. All factors should be considered when investigating any incidence of disease among laboratory dogs and the emphasis should always be on prevention rather than cure in order to enhance welfare.

In laboratory dogs, the common infectious diseases may usually be eliminated or, if not, effectively controlled using a combination of vaccination and isolation. Laboratory dog colonies should either be maintained as "closed" colonies; where no stock from other outside sources is introduced or, more commonly, as a "semi-closed" colonies. In the later case, new stock are only introduced from one of a small number of known reliable sources and should be effectively quarantined for a period of time after arrival to ensure satisfactory health status before mixing them with any of the resident population.

Both breeders and users of laboratory dogs should have a routine prophylactic regime to control endoparasites and ectoparasites. The precise treatment regime used in any one colony should be prescribed and directed by the attending veterinarian who should be someone sufficiently familiar with the colony, including the sources of the dogs and their research purpose, to advise accordingly. Effective vaccines are now widely available against common infectious diseases of dogs. With appropriate vaccination programmes it should be possible to control such diseases as canine distemper, viral hepatitis, parvovirus, leptospirosis and respiratory disease caused by canine adenovirus and bordetellosis.

The high standards of hygiene now commonly applied in the management, housing, husbandry and care of laboratory dogs have, no doubt, contributed to reducing the incidence of disease amongst such colonies. Good hygiene is vital in promoting good welfare standards and, as a general rule, explains why in many cases the health status of laboratory dogs is found to be significantly higher than that in comparable boarding or rescue kennels.

All laboratory dogs should receive a regular clinical examination (Table 11-4) either carried out by the attending veterinarian or by suitably trained and competent staff working closely under his or her direction. This examination can frequently form part of an experimental protocol and should be designed to detect any deviation from normality that might be related to treatment as well as ill-health. At the very least it should comprise a periodic health check, sufficiently thorough to enable otherwise unapparent or relatively minor ailments to be detected and treated accordingly.

Table 11-4. A recommendation for health examination procedures

1. The general appearance and behaviour of the dog is first examined. The dog should adopt a normal posture and display normal movement with no signs or lameness. The normal, unprovoked, behaviour of the dog should be observed in the pen. Check that the dog shows no signs of any reluctance to move, and that the animal is not nervous or withdrawn and shows no reluctance, resistance, or resentment to being picked up or handled. The dog should show normal signs of social behaviour to both people and other dogs.
2. Next examine the dog's head to include an examination of the external appearance of the eyes, ears, nose and mouth (especially teeth and mucous membranes). The appearance of each should be normal; there should be no evidence of abnormal discharge(s). Examine carefully for excessive dental tartar accumulation, and the presence of papillomas on face, tongue, and the buccal and oral mucosae.
3. Check the skin for signs of normal elasticity. There should be no evidence of hair loss and no abnormal swelling(s). The animal should be in good bodily condition, well-muscled, and should not be too fat or too thin for its age and weight.
4. Check the respiration. Rate, rhythm and depth of respiration should be normal. Check that there is no evidence of laboured breathing and no evidence of a cough or other abnormal respiratory noises.
5. Gently palpate the abdomen. Look for evidence of abnormal swelling. The dog should not appear "tucked up" and there should be no evidence of pain or discomfort on abdominal palpation. Check that there are no umbilical or inguinal hernia(e). In males there should be two testicles, fully descended in the scrotum. Check that there is no abnormal discharge from either prepuce or vulva, as appropriate.
6. Check all females for signs of oestrus.
7. Check the legs, feet and tail. There should be no evidence of abnormal swellings, fractures, etc. Check the nails are not overgrown and trim as necessary. Check the footpads, there should be no evidence of cracking or abrasions.
8. Record any abnormalities detected.

Routine procedures such as nail trimming and teeth cleaning can easily be incorporated as part of such a regular health programme, all of which will contribute to the overall welfare of the dogs.

Although occurring relatively infrequently, the incidence of some health conditions appear to be more common in laboratory beagles, particularly polyarteritis and epilepsy. These conditions are examples of a variety of diseases known to have specific breed predisposition and although the heritability of many has yet to be confirmed, their association with specific breeds contributes to an assumption that their aetiology is likely to have an hereditary basis. For this reason they are of particular interest in breeding programmes and when detected in research dogs information should be passed back to the breeder. The list of some diseases and conditions associated with the Beagle includes:

Factor VII deficiency
Lymphocytic thyroiditis
Cleft palate
Brachury
Umbilical hernia
Ectasia Syndrome
Pulmonic stenosis

Unilateral kidney aplasia
Renal hypoplasia
Intervertebral disc disease
Atopic dermatitis
Cataract
Retinal dysplasia

The existence of certain breed associated conditions underlines the importance of appropriate health screening within laboratory dog breeding colonies and of selective breeding within the management of these colonies. The basic requirement is to breed for "soundness"; a term which, when applied to laboratory dogs, may be defined in terms physical conformation, freedom from hereditary disease, or propensity to disease, and also importantly in terms of temperament. The overall strategy should be to select positively, thus enhancing the expression of desirable characteristics. Within a closed, outbred colony the temptation to indiscriminately eliminate breed stock, and thereby limit the genetic pool within the population, simply because they may be carriers of undesirable recessive genes should be resisted. The importance of a suitable temperament in dogs that are to be used for research cannot be underestimated. It is essential that the laboratory dog easily adapts to its kennel environment and that it quickly integrates into a peer-group hierarchy, particularly if group-housed. The dog has to become accustomed to the routine in its research environment and must be able to cope with the requirements of the research protocol. Unlike pet dogs, laboratory dogs rarely have a single carer. In the laboratory situation, whilst the animal technical staff are likely to be responsible for routine care and maintenance and will consequently have most contact with the dog, others such as research staff will have need to handle the animals in order to carry out a variety of potentially stressful procedures. Laboratory dogs should be calm with staff and other dogs, they should be willing to be handled and restrained, and show no tendency towards nervousness or aggression. Dogs that are unwilling to interact calmly with a variety of staff and other dogs, or that resist minimal restraint soon become very difficult to work with and it is more likely that they will become stressed and their welfare jeopardised.

6. EXPERIMENTAL TECHNIQUES

A wide variety of experimental procedures, including cardiovascular work and other general biological or biomedical studies are carried out on dogs, but by far the majority of dogs are used in toxicology. This is not the place to go into the refinement of these procedures in detail, but it is worth considering some broader issues.

There is a need to critically re-evaluate procedures that require single or solitary housing for any length of time. As discussed earlier there has been a progressive refinement of the duration of single housing for dogs in toxicology studies, and it is now routine to house dogs in metabolism cages within sight of each other. Other reasons given for single housing include surgical preparations or implants that might be vulnerable to damage if the dogs fought. Here, the risk of aggression should be realistically assessed, and consideration given to devising preparations or implants that are more suitable. Dogs are also sometimes housed singly during telemetry studies. Here, advances in technology should allow the radio signal to be read in larger pens and to separate signals from several dogs within one unit. It is the responsibility of scientists carrying out these procedures to request these advances from the manufacturers of the equipment. Restraint should always be minimal, as it is a stressor for dogs. Dogs should be habituated to handling, trained to cooperate with dosing procedures, and consideration should be given to breeding dogs that are not unduly stressed by such techniques (see above). Dosing should always be carried out humanely, using the most appropriate route and volume of substance. The aim should be to minimise or avoid adverse effects (Morton et al. 2001). Procedures should always be carried out in a humane way. It is important to continually critically reappraise research techniques to identify further refinements or better ways of achieving the same outcome and continually reassessed to identify refinements to avoid unnecessary pain and distress.

7. CONCLUSIONS

From the above it can be seen that the dogs needs can be summarised as follows. Dogs should be housed socially in a complex enriched environment, and the space requirements must be sufficient to allow this, except under extraordinary conditions such as when the animals are temporarily housed in metabolism cages. Dogs should be adequately socialised, habituated to complex environments and where appropriate trained for procedures. Dogs should be purpose bred for laboratory use. Their use should be reduced to a minimum, through reduction and replacement with non-sentient alternatives. Procedures should be refined and housing and management systems should be optimised so that it is not necessary to compromise on any aspect of welfare. A useful checklist has been devised by the BVAAWF/FRAME/RSPCA/UFAW Joint Working Group on Refinement report (Table 11-5), and to this should be added systems for staff training.

Table 11-5. Factors to consider for good dog housing (Prescott et al. in preparation)

The following list pre-supposes that the housing meets the dogs' physiological needs and allows easy husbandry. <u>Other factors will need to be considered also, but the most important are emphasised here:</u>

1. Housing in socially compatible groups
2. Sufficient space to allow essential enrichment of the pen
3. Housing should provide interest and choice. (In some circumstances this could be provided through the management practices of the entire facility. In many cases it will also be necessary to provide choice within the pen environment to provide interest, allow different behaviours in different areas, and allow the dog to control its social interactions.)
4. Choice of microenvironment within the pen (e.g. light, temperature, noise)
5. A comfortable resting/sleeping area
6. Good visibility of the room outside the pen whilst still allowing a semi-enclosed area for retreat
7. Sufficient depth of the pen to allow retreat
8. Objects to chew, presented so that the dog can hold them in a species specific fashion
9. Pen design should allow flexibility in running pens together
10. Pens should be designed and constructed so as to facilitate refurbishment and remodelling as knowledge develops and improves our understanding of the housing needs of dogs
11. Noise control through the use of good management practices, adequate socialisation, well-designed housing, enrichment, and noise absorbent ceilings, upper walls and/or baffles
12. The numbers of dogs in a room should be limited to reduce allelomimetic barking (around 20)
13. Bitches should be provided with a secluded area for littering out of sight of other dogs and free from disturbance

REFERENCES

Alderton D. Foxes, Wolves and Wild Dogs of the World. Blandford, London 1994

Animal Procedures Committee. Report of the Animal Procedures Committee for 1998. The Stationery Office, London 1998

Beerda B, Schilder MBH, van Hooff JARAM, de Vries HW. Manifestations of chronic and acute stress in dogs. Appl Anim Behav Sci 1997; 52:307-319

Beerda B, Schilder MBH, van Hooff JARAM, de Vries, HW, Mol JA. Behavioural, saliva cortisol and heart rate responses to different types of stimuli in dogs. Appl Anim Behav Sci 1998; 58: 365-381

Beerda B, Schilder MBH, van Hooff JARAM, de Vries HW, Mol JA. Behavioural and hormonal indicators of enduring environmental stress in dogs. Anim Welf 2000; 9:49-62

Bekoff M. Scent marking by free ranging domestic dogs: olfactory and visual components. Biol Behav 1979; 4:123-129

Boitani L, Francisci F, Ciucci P, Anreoli G. Population biology and ecology of feral dogs in central Italy. In The Domestic Dog: Evolution, Behaviour, and Interactions with People. Serpell J ed, Cambridge University Press, Cambridge 1995; 217-244

Bradshaw JWS, Brown S. Behavioural adaptations of dogs to domestication. Waltham Symposium 1990; 20:18-24

Clubb R, Mason G. Pacing polar bears and stoical sheep: testing ecological and evolutionary hypotheses about animal welfare. Anim Welf in press

Clutton-Brock J. Origins of the dog: domestication and early history. In The Domestic Dog: Evolution, Behaviour, and Interactions with People. Serpell J ed, Cambridge University Press, Cambridge 1995; 7-20

Coppinger, R. and Coppinger, L. Differences in the behavior of dog breeds. In Genetics and the Behavior of Domestic Animals. Grandin T ed, Academic Press, San Diego 1998; 167-202

Draft proposals for revision of Appendix A to the Council of Europe Convention ETS 123

Flecknell PA, Waterman-Pearson A. Pain Management in Animals. W.B. Saunders, London 2000

Fox MW, Stelzner D. Behavioural effects of differential early experience in the dog. Anim Behav 1966; 14:273-281

Fox MW, Bekoff M. The behaviour of dogs. In The Behaviour of Domestic Animals, 3rd Edition. Hafez ESE ed, Baillière Tindall, London 1975; 370-409

Fox MW. The Dog: Its Domestication and Behaviour. Garland STPM Press, New York 1978

Fox MW. Laboratory Animal Husbandry: Ethology, Welfare and Experimental Variables. State University of New York Press, Albany 1986

Gamble MR. Sound and its significance for laboratory animals. Biol Rev 1982; 57:395-421

Goodwin D, Bradshaw JWS, Wickens SM. Paedomorphosis affects agonistic visual signals of domestic dogs. Anim Behav 1997; 53:297-304

Group of Experts (cats, dogs and ferrets) for the preparation of draft proposals for revision of Appendix A to the European Convention ETS 123. To be published

Hare B, Brown M, Williamson C, Tomasello M. The domestication of social cognition in dogs. Science 2002; 298:1634-1636

Hart BL. Analysing breed and gender differences in behaviour. In The Domestic Dog: Evolution, Behaviour, and Interactions with People. Serpell J ed, Cambridge University Press, Cambridge 1995; 65-77

Hennessy MB, Williams MT, Miller DD, Douglas CW, Voith VL. Influence of male and female petters on plasma cortisol and behaviour: can human interaction reduce the stress of dogs in a public animal shelter? Appl Anim Behav Sci 1998; 61:63-77

Hetts S, Clark JD, Calpin JP, Arnold CE, Mateo J.M. Influence of housing conditions on beagle behaviour. Appl Anim Behav Sci 1992; 34:137-155

Hubrecht RC. A comparison of social and environmental enrichment methods for laboratory housed dogs. Appl Anim Behav Sci 1993; 37:345-361

Hubrecht RC. Dog welfare. In The Domestic Dog: Evolution, Behaviour, and Interactions with People. Serpell J ed, Cambridge University Press, Cambridge 1995a; 179-198

Hubrecht R ed. Housing, Husbandry and Welfare Provision for Animals Used in Toxicology Studies: Results of a UK Questionnaire on Current Practice 1994. A Report by the Toxicology and Welfare Working Group, UFAW, Wheathampstead, Hertfordshire 1995b

Hubrecht RC. Enrichment in puppyhood and its effects on later behavior of dogs. Lab Anim Sci 1995c; 45:70-75

Hubrecht RC, Serpell JA, Poole TB. Correlates of pen size and housing conditions on the behaviour of kennelled dogs. Appl Anim Behav Sci 1992; 34:365-383

Hubrecht R, Sales G, Peyvandi A, Milligan S, Shield B. Noise in dog kennels, effects of design and husbandry. Animal Alternatives, Welfare, and Ethics. van Zutphen LFM, Balls M eds, Dev Anim Vet Sci 1997; 27:215-220

Koler-Matznick J. The origin of the dog revisited. Anthrozoos 2002; 15:98-118

Loveridge G. Provision of environmentally enriched housing for dogs. Anim Technol 1994; 45:1-19

Loveridge GG. Environmentally enriched dog housing. Appl Anim Behav Sci 1998; 59:101-113

Lund JD, Jørgensen MC. Behaviour patterns and time course of activity in dogs with separation problems. Appl Anim Behav Sci 1999; 63:219-236

MacArthur Clark JA. The dog. In The UFAW Handbook on the Care and Management of Laboratory Animals, Volume 1 - Terrestrial Vertebrates, 7th Edition. Poole T ed, Blackwell Science Ltd, Oxford 1999; 423-444

Macdonald DW, Carr GM. Variation in dog society: between resource dispersion and social flux. In The Domestic Dog: Evolution, Behaviour, and Interactions with People. Serpell J ed, Cambridge University Press, Cambridge, 1995, 199-216

Mech LD. The Wolf: The Ecology and Behaviour of an Endangered Species. The Natural History Press, Garden City, New York, USA 1970

Mech LD. Leadership in wolf, *Canis lupus*, packs. Canad Field - Nat 2000; 114:259-263

Miklósi A, Kubinyi E, Topál J, Gácsi M, Virányi Z, Csányi V. A simple reason for a big difference: Wolves do not look back at humans, but dogs do. Current Biology 2003; 13:763-766

Millot JL, Filiatre JC, Eckerli A, Gagno AC, Montagner H. Olfactory cues in the relations between children and their pet dogs. Appl Anim Behav Sci 1987; 19:189-195

Morton DB, Jennings M, Buckwell A, Ewbank R, Godfrey C, Holgate B, Inglis I, James R, Page C, Sharman I, Verschoyle R, Westall L, Wilson AB. Refining procedures for the administration of substances. Lab Anim 2001; 35:1-41

Neitz J, Geist T, Jacob GH. Color vision in the dog. Visual Neurosci 1989; 3:119-125

Nesbitt WH. Ecology of a feral dog pack on a wildlife refuge. In The Wild Canids. Fox MW ed, Van Nostrand Reinhold, New York 1975; 391-395

Passchier Vermeer W, Passchier WF. Noise exposure and public health. Environ Health Perspect 2000; 108:123-131

Poole T, Stamp-Dawkins MS. Environmental enrichment for vertebrates. In The UFAW Handbook on the Care and Management of Laboratory Animals, Volume 1- Terrestrial Vertebrates, 7th Edition. Poole T ed, Blackwell Science Ltd, Oxford 1999; 13-20

Prescott MJ, Morton DB, Anderson A, Buckwell AC, Heath SE, Hubrecht RC, Jennings M, Robb D, Ruane R, Swallow J, Thompson P. Refining dog husbandry and care: Eighth report of the BVAAWF/FRAME/RSPCA/UFAW Joint Working Group on Refinement. In preparation

Ringler DH, Peter GK. Dogs and cats as laboratory animals. In Laboratory Animal Medicine. Fox JG, Cohen BJ, Loew FM eds, Academic Press, Orlando, Florida 1984; 241-271

Sales G, Hubrecht R, Peyvandi A, Milligan S, Shield B. Noise in dog kennelling: a survey of noise levels and the causes of noise in animal shelters, training establishments and research institutions. Animal Research Report No 9. UFAW, Wheathampstead, Hertfordshire 1996

Sales G, Hubrecht R, Peyvandi A, Milligan S, Shield B. Noise in dog kennelling: is barking a welfare problem for dogs? Appl Anim Behav Sci 1997; 52:321-329

Sherwin CM, Glenn E. Cage colour preferences and effects of home-cage colour on anxiety in laboratory mice. Anim Behav in press

Scott JP, Fuller JL. Genetics and the Social Behaviour of the Dog. University of Chicago Press, Chicago 1965

Simpson BS. Canine communication. Vet Clin North Am Small Anim Pract 1997; 27: 445-464

Sommerville BA, Broom DM. Olfactory awareness. Appl Anim Behav Sci 1998; 57:269-286

Thorne C. Feeding behaviour of domestic dogs and the role of experience. In The Domestic Dog: Evolution, Behaviour, and Interactions with People. Serpell J ed, Cambridge University Press, Cambridge 1995; 103-114

Van Raaij MTM, Oortgiesen M. Noise stress and airway toxicity: a prospect for experimental analysis. Food Chem Toxicol 1996; 34:1159-1161

Vila C, Savolainen P, Maldonada JE, Amorim IR, Rice JE, Honeycutt RL, Crandall KA, Lundeberg J, Wayne RK. Multiple and ancient origins of the domestic dog. Science 1997; 276:1687-1689

Wells DL, Graham L, Hepper PG. The influence of auditory stimulation on the behaviour of dogs housed in a rescue shelter. Anim Welf 2002; 11:385-393

Wickens S, Hubrecht,R, Buckwell T, Gregory D, Robb D, Willson M, Rochlitz I. Report of the 2000 UFAW/RSPCA Carnivore Welfare group meeting. Anim Technol 2001; 52:43-56

Wolfensohn S, Lloyd M. Handbook of Laboratory Animal Management and Welfare, 2nd Edition. Blackwell Science, Oxford 1998; 218-226

Zimen E. The Wolf: His Place in the Natural World. Souvenir Press, London 1981

Chapter 12

THE WELFARE OF PIGS AND MINIPIGS

Peter Bollen and Merel Ritskes-Hoitinga
[1]*Biomedical Laboratory, University of Southern Denmark, Odense, Denmark*

1. INTRODUCTION

The welfare of pigs is usually discussed in relation to animal production. Nearly all pigs are bred for the purpose of food production, often compromising welfare in favour of economical considerations (Schrøder-Petersen and Simonsen 2001). However, during the last two decades, pigs have increasingly been used as research animals, mainly because of prospects of xenotransplantation, but also as alternative non-rodent models in regulatory toxicology (Bollen and Ellegaard 1997). In Denmark, the use of pigs in biomedical research has increased by more than 500% from 1980 to 2001. Nevertheless, overall pig use is about two percent of the total number of animals used in biomedical research (Danish Animal Experiments Inspectorate 2001), which is insignificant compared to numbers of pigs in food production. Similar figures are seen in European statistics. In 1991, 48,420 (0.41%) pigs were used in biomedical research in the member states of the European Union, whereas this had increased to 6,749 (0.56%) in 1996 and 66,131 (0.67%) in 1999 (Commission of the European Communities 2002).

Animal welfare is a topic of concern in biomedical research. Apart from moral concerns and legislative obligations to secure good welfare for animals kept at the laboratory, a reduced animal welfare makes the results from biomedical research with animals unreliable (Van Zutphen et al. 1993). Moreover, when pigs are going to be used as organ donors, an optimal organ quality can only be guaranteed from a healthy, non-stressed pig (Olsson 2000).

We regard the welfare of a pig, as of any animal, a the state in which the animal is able to cope with its environment (Broom 1986). Welfare can vary from very poor to very good and scientific and objective measurements of welfare can be made. The state of an animal includes health and physiological functioning, as well as the feelings of the individual animal. Suffering is one of the most important aspects of poor welfare and, beside health and physiological functioning, feelings should be included wherever possible when trying to assess welfare (Dawkins 1990). In order to understand pig welfare, a good knowledge of the natural behaviour of pigs and differences between domesticated breeds is a necessary first step.

2. PIG BREEDS AND BEHAVIOUR

The wild boar, *Sus scrofa*, is the ancestor of all modern breeds of pigs. The first evidence of domestication in Europe is some 3,500 years old, although the cradle of the domesticated pig is claimed to have been in China, about 10,000 years ago (Porter 1993). *Sus scrofa* is a member of the pig family (Suidae), which includes species of non-ruminant, even-toed ungulates. Suidae are omnivores, and are divided into five genera, with a total of 13 species.

Through selection, numerous breeds of pigs have been developed. Existing breeds are relatively young in comparison to the history of domestication of pigs. Local Celtic pigs with lop-ears were widespread in European countries in the 18th and 19th centuries, and are considered to be the primogenitor of the Landrace. Crosses with Chinese breeds in the early 1800s, and with English breeds in the late 1800s, established the Danish Landrace, which was widely used for the derivation of other national Landrace breeds. The Yorkshire or Large White, a breed with pointed, standing ears, originated in England, and was exported to many countries around the world during the last half of the 19th century (Porter 1993). Pigs are numerous worldwide. The total number of pigs produced in the 15 countries of the European Union was 124 million in 2002 (FAO 2002).

Pure-bred pigs, such as the Landrace, Yorkshire, Hampshire and Duroc, are almost exclusively found in breeding herds. Modern production herds, however, utilize cross-breeding to produce hybrid grower pigs. Hybrid breeds are very common nowadays, and consequently, pigs obtained for the laboratory from an agricultural source will most likely be of a mixed breed. Hybrid breeds are selected for favourable production traits, and regarding animal welfare, hybrids have the advantage of being less sensitive to environmental stress than pure-bred pigs. Certain pure-bred breeds, such as the Pietrain, are particularly susceptible to stress (McGlone et al. 1998).

Miniature pigs are especially bred for research purposes. Many of the present breeds of miniature pigs have their origin in the Minnesota (Hormel) minipig, which was developed in the early 1950s at the Hormel Institute in Austin, USA. Miniature pigs derived from this population are the Göttingen minipig and the Sinclair minipig. Other breeds are the closely related Hanford and Pitman-Moore miniature pigs. A Mexican feral pig was introduced into the laboratory in 1960 and later referred to as the Yucatan minipig. A smaller sub-line of the Yucatan minipig was selected to develop the Yucatan micropig. Several other breeds have been established, among others the Ohmini and Clawn minipig in Japan (Bollen et al. 2000).

With regards to physiology and behaviour, there is no clear difference between large pigs and minipigs, and thus welfare aspects of both types of pigs are similar. However, normal genetic variation occurs, as it does between different breeds (McGlone et al. 1998). Since pigs have been selected for rapid growth, unlike minipigs, particular large differences are found between the growth curves of pigs and minipigs (Bollen 2001).

An important reason why pigs became the subject of domestication is their behaviour. Their feeding, sexual and social behaviour, besides their general adaptability, favoured domestication. Compared to wild pigs, domesticated pigs are calmer, quieter and less active, although they still share many behavioural characteristics with wild pigs, such as social, feeding and explorative behaviour. Wild pigs live in small social groups formed by sows and their offspring. Young males live in bachelor groups, whereas adult boars tend to be solitary. In groups of females a social dominance order is quickly established. Generally the level of aggression, mainly expressed by butting or biting the neck and ears, soon subsides in stable social groups. Sexually mature males may fight fiercely, especially in the presence of females. Newborn pigs establish a social order, the teat order, within a few days after birth. As soon as a teat order has been established, each piglet consistently suckles the same teat. Dominant piglets generally occupy the more productive anterior teats. If young pigs from different litters are mixed after weaning, a new dominance hierarchy will be established. Ranking is the result of aggressive interaction, but may also take place without overt aggression. It exists as long as a group is together. Subordinates which have been separated from a group will be attacked after reintroduction, whereas a dominant animal may be separated and reintroduced without any trouble (Holtz and Bollen 1999).

Wild pigs spend most of the day rooting and collecting food. Their feeding and exploratory behaviour are closely linked. During a day, pigs eat many small food portions. They are omnivorous and diurnal, with elevated activity during the late afternoon. In the laboratory, activity of pigs is stronger related to the presence and activity of humans than the light-dark cycle (Bollen et al. 2000). Rooting seems to be an important behaviour to pigs, since pigs in a semi-natural enclosure spend 6-8 hours searching for food, even when full rations of commercial diet is available. Other important foraging behaviours are grazing and browsing (Wood-Gush et al. 1990). The importance of foraging behaviours can be demonstrated in preference tests.

3. THE GENERAL ANATOMY OF PIGS

Pigs are used as large animal models in biomedical research because their anatomy and physiology has large similarities with human anatomy and physiology, but there are also differences. Moreover, specific anatomical structures can explain welfare problems in pigs, e.g. the sparse hair layer of pigs makes this animal particular sensitive to temperature induced stress, and the relatively small heart in relation to body size poses the risk of cardiovascular shock during stress. Therefore, a brief overview of the anatomy of pigs is given in this section.

The skin of pigs has anatomical and physical similarities with human skin, including hair layer, pigmentation and thickness of the dermal layer. The subdermal layer is much thicker than in humans. Pigs have short, sturdy bones. The number of ribs varies greatly, from 13 to 17, but most commonly 14 or 15 ribs are present. The bony structure of the thorax is considerably smaller than the external dimensions suggest. The skull is extremely sturdy, and has a largely extended frontal sinus. Pigs have an extensive dentition, with 44 teeth. For this reason they are often used in dental research. The formula for the permanent teeth is i3/3, c1/1, p4/4, m3/3. The canine teeth or tusks can be found in both boars and sows, but are most developed in boars. The tusks of boars grow throughout the animal's life. Pigs are born with 8 teeth, the needle teeth (i1/1, c1/1). After two months, two more deciduous incisors and three of the deciduous premolars have erupted (i3/3, c1/1, p3/3). At 6 months, the first permanent teeth erupt. The pig is at least 18 months old before all permanent teeth have erupted.

Pigs are monogastric, and the stomach has a large fundus and a diventriculum. The area where the oesophagus enters the stomach has non-glandular mucosa, and is prone to peptic ulcers. The bile duct and pancreatic duct enter the duodenum. Similar to man, pigs have an unbranched pancreatic duct. The jejunum and ileum contain numerous patches of lymph nodes. The large intestines are coiled and voluminous. Because of the structure and large volume, the caecum lies to the left.

The liver has much fibrous tissue, giving the surface a net-like appearance. The pancreas has two lobes, which are difficult to identify. The kidneys have multiple papillae protruding into a central renal pelvis via the calyces. In structure and size they are similar to human kidneys. Lymph nodes are located along all major arteries in the abdomen. The uterine horns in sows are very long, and the cervix is tightly closed with mucosal ridges. The boar has large testes and various accessory genital glands. The ejaculate is voluminous, 200-500 ml. The penis is fibrous, and has a corkscrewed tip. The preputial diverticulum produces a strong odour.

The heart of the pig is small in relation to body size. The coronary distribution is similar to that of humans, but the left azygous vein ends directly in the heart, instead of in the caval vein. Pigs develop atherosclerosis of the major arteries after ingestion of a lipid-rich diet over a relatively short period. Pigs have a tracheal bronchus to ventilate the right cranial lobe. The lymph nodes of pigs are inverted, with the follicular centres in the medulla. The parathyroids are not located near the thyroids, but are related to the thymus. The right adrenal lies in close contact with the caudal caval vein (Bollen et al. 2000)

Detailed biological, haematological, clinical chemical, and respiratory and cardiovascular parameters can be found in Pond and Mersmann (2001).

4. THE ASSESSMENT OF WELFARE IN PIGS

To our knowledge no specific information on the assessment of welfare in laboratory pigs is available. Therefore it is necessary to rely on information on the assessment of welfare in pigs from agricultural sources. This can be justified because pigs kept at the laboratory often originate from an agricultural source. An exemption on this are minipigs, which are purposely bred for laboratory purposes. However, until now not much attention has been paid to the assessment of welfare in minipigs.

The major indicators of welfare of an animal are related to behavioural and physiological parameters, injuries, other aspects of health, growth, reproductive parameters and life expectancy. Behavioural and physiological measures have been used for years to evaluate the ability of domestic animals to cope in the environment, but do not include measures of suffering. Therefore it is often necessary to evaluate several parameters in the assessment of welfare. These could include combinations of behavioural measurements, such as preference tests, aversion tests, motivation tests and abnormal behaviour, physiological parameters, such as glucocorticoid and catecholamine measurements, heart rate and immune response, as well as injury and diseases, growth, reproduction and life expectancy. Based on combined measurements, a conclusion can be drawn on negative feelings, such as suffering (Von Borell et al. 1997).

Changes in behaviour patterns are often first seen when animals first perceive changes in their environment. A wide variety of behavioural parameters have been used to assess welfare. Abnormal behaviours are those which differ in pattern, frequency or context from those which are shown by most members of the species in conditions which allow a full range of behaviour (Broom and Johnson 1993). There can be abnormalities of feeding, grooming and sexual behaviour. One category of abnormal behaviour is the stereotypy. Stereotypies are repeated, relatively invariate sequences of movements which have no obvious purpose (Fraser and Broom 1990). Stereotypies develop when the animal is severely or chronically frustrated. Hence their development indicates that the animal is having difficulty in coping and its welfare is compromised. Examples of stereotypies in pigs are bar biting, tail biting and sham chewing, which occur when pigs cannot carry out foraging behaviours. Other abnormal behaviours include those which are directly connected with physical restrictions and those which are responses to cope with problems. Restrained animals may be unable to show certain movements. Activity, behaviour and responsiveness may be much lower in depressed individuals than in those which are not depressed. Social responses may be exaggerated or misdirected (Von Borell et al. 1997).

Injuries caused by other pigs, by humans or by contact with the immediate physical environment can be quantified in animal welfare studies. A typical injury which is often quantified in pig welfare studies is tail injury caused by biting. Tail biting is not necessarily a sign of low welfare, but it poses a risk for the animal with the injury, causing infections and reduced welfare for the victim. Methods for assessing disease in pigs are of particular importance in welfare studies, since all disease results in poorer welfare, some disease conditions being worse than others. The importance of the disease depends not only on the incidence or risk but also on the duration of the disease and on the intensity of pain and discomfort which a diseased animal experiences. Production-related diseases are of great relevance in considering welfare in relation to housing conditions and management practices. Among the major diseases in this area are clinical conditions of the legs and feet which result in lameness, urinary tract infections, reproductive system disorders, mastitis and other conditions affecting lactation and cardiovascular disorders. In each case, clinical analysis of severity is possible and a combined measure of frequency and severity can be used (Von Borell et al. 1997).

For practical reasons, five basic needs are often taken as a measure of welfare. These include the freedom from hunger, thirst and malnutrition; appropriate comfort and shelter; the prevention, or rapid diagnosis and treatment of, injury, disease or infestation; freedom from fear, and freedom to display most normal patterns of behaviour (DEFRA 2003).

5. THE WELFARE OF PIGS DURING HOUSING

In an omnivorous species such as the pig, exploration is expected to be closely linked to foraging behaviour. Exploration develops early under natural conditions and forms a substantial part of the activity of free-ranging domestic pigs (Petersen 1994). Pigs may be motivated to explore even if there are no obvious novel stimuli (Wood-Gush and Vestergaard 1993). Rooting, grazing and browsing are the prominent exploration and foraging behaviours (Wood-Gush et al. 1990). Therefore, particular attention should be paid to flooring and bedding material of pig stalls, to satisfy these behaviours and guarantee good welfare.

Bedding material such as straw has several aspects which are attractive for pigs, such as odour, rooting and chewing material, nutritive value, insulation and nest building material. Pigs that are provided with bedding are reported to be more active and exhibit increased rooting and exploratory behaviour when compared to pigs housed on bare flooring without any bedding (Arey and Franklin 1995, Beattie et al. 1995). The effects of straw on the well-being of pigs include improved comfort, thermal insulation and dietary effects (Fraser 1985). Oral behaviours of pigs in barren environments often become re-directed towards pen fittings and other pigs (McKinnon et al. 1989). Therefore, a major function of straw is to provide a stimulus and outlet for rooting and chewing, resulting in a reduction of such activities directed at pen-mates (Fraser et al. 1991). Destructive behaviours like tail biting or oral stereotypic activities (i.e. bar biting) are reported to be reduced by straw (Fraser 1985, Spoolder et al. 1995). The provision of straw does not reduce fighting between newly-mixed growing pigs (Arey and Franklin 1995). Also the provision of soft chewable toys reduced excitability in pigs (Grandin and Curtis 1984). Amongst different toys offered, a knotted nylon rope, a rubber hose, a chain and an hourglass-shaped, soft, pliable rubber dog toy, the latter was suggested for enrichment of confined pigs. Pigs in small enclosures played more with toys than pigs in larger pens (Apple and Craigh 1992).

Social grouping is also important for pig welfare. Pigs are social animals which live in stable maternal family groups, and should therefore preferably be housed in small groups. Adult males, which can be solitary under natural conditions, can be housed individually. Where individual housing is necessary for experimental purposes, pigs should have visual, auditory and olfactory contact with other pigs, whenever possible. The optimal group size is 2 to 6 sows (Graves 1984).

Pigs are highly sensitive to environmental temperature and place a high behavioural priority on thermoregulation. Their lying area should be maintained within the thermoneutral range. Pigs may be kept in a uniform, temperature-controlled environment, in which case the whole room should be maintained within the thermoneutral zone. Alternatively, they may be kept in a pen with different microclimates, by providing local heating or insulating bedding material. In these circumstances, only the lying area needs to be maintained within the thermoneutral zone, whilst the main room may be kept at a lower temperature. The thermoneutral zone for minipigs is well documented (Georgiev et al. 1977). Energy conversion and rectal temperature measurements at different environmental temperature and relative humidity (RH) showed that the optimum temperature differs at different ages (Table 12-1).

Table 12-1. The optimal temperature of Göttingen minipigs of different age and at different RH.

	Thermoneutral zone RH = 50, 70, 90 %	Optimal temperature RH = 50 %	Optimal temperature RH = 90 %
Piglet (6-8 weeks old)	20-30 °C	29.0 °C	27. 8 °C
Adolescent (14-16 weeks old)	15 - 25 °C	24.0 °C	23.0 °C
Adult (34-36 weeks old)	15-25 °C	17.3 °C	17.8 °C

There are no experimental studies known to us which define the optimal range of relative humidity for pigs. Under practical conditions, pigs can be exposed to a large variation in relative humidity without leading to adverse effects. A very low level of humidity may predispose to irritation of the respiratory tract, whilst an excessively high relative humidity may impair thermoregulation. On the basis of practical experience, it is recommended that relative humidity is maintained in the range between 50 and 80 %.

The activity of pigs peaks typically during the morning and late afternoon. Changes in lighting can be used to reduce adverse behaviour of pigs in intensive housing systems. Tail biting was found to decrease greatly when pigs were housed in low light and aggression among unfamiliar regrouped individuals was also greatly reduced when they were kept in darkness (Barnett et al. 1994).

There is limited information on which to base recommendations on noise levels. Continuous loud noise (85 dBA) can adversely affect behaviour, whilst high frequency noise (500-8000 Hz) can induce physiological and behavioural alarm responses. Pigs have a region of best sensitivity from 250 Hz to 16000 Hz (Heffner and Heffner 1990). Generally, the level of background noise from animals and ventilation systems should be kept to a minimum and sudden and unpredictable noises should be avoided. High levels of noise are potential stressors to pigs. Noise may originate from the animals themselves and from the environment. Noise from animals is intensified by a high number of animals housed in the same stable, especially at feeding times when the pigs are particularly excited. High intensity sounds can be also produced by ventilation systems, which are often working continuously.

Space allowances for pigs are usually based on best practical experience. Crowding of pigs may occur in an agricultural setting, but is not an issue for laboratory pigs. Beside pen dimensions, the area necessary for different functional behaviours should be considered. In addition to space for lying, pigs require space for exercise, recreation and excretion. Where pigs are housed individually or kept in small groups for chronic experiments, greater space allowances per animal are required than for short experiments. The total area for animals housed in small groups should never be less than that for singly housed animals.

Proper feeding is essential for maintaining a good level of welfare. Although nutrition requirements of pigs have been investigated thoroughly (ARC 1981, NRC 1998), no scientific investigations into the nutrient requirements of minipigs had been carried out, which was why this was subject of research at our laboratory (Bollen 2001). Specific recommendations for the composition of minipig diets for breeding and maintenance exist (GV-Solas 1993), but the scientific evidence is not available. Established nutrient levels for large pigs were recommended as a guide for miniature pigs until experimental data on the nutrient needs of miniature pigs became available (Cunha 1966, Detmers 1968). But since nutrient requirements for pigs are based on *ad libitum* feeding in order to obtain maximum growth, the question was raised whether these recommendations were appropriate for minipigs (Ritskes-Hoitings and Bollen 1997). Restricted feeding is often necessary for minipigs, to prevent obesity. However, restricted feeding could impose a reduced welfare, with a continuous feeling of hunger. The introduction of straw, as a substrate to manipulate after restricted feeding, could relieve this adverse effect (Von Borell and Hurnik 1991).

6. THE WELFARE OF PIGS IN BIOMEDICAL RESEARCH

Pigs are used as model animals in various research areas (Bustad et al. 1966, Mount and Ingram 1971, Stanton and Mersmann 1986, Tumbleson 1986, Swindle 1992, Tumbleson and Schook 1996). It is important to avoid unnecessary pain, suffering, distress and lasting harm in order to prevent a reduced animal welfare during experimentation. If suffering is detected, this has to be eliminated as quickly as possible. Pigs are also established model animals in operative surgery training. Traditionally, dogs were used in teaching programmes, but their procurement has been increasingly difficult, and they are also much more expensive than pigs (Swindle 1983). In countries where it is permitted to use animals in surgical training programmes, recovery from anaesthesia is prohibited, in order to prevent further suffering.

Acclimatisation to the laboratory environment is indispensable for pigs used in biomedical research. Especially handling for experimental procedures is considered stressful in unacclimatised pigs. In minipigs, it was shown that the animals became rapidly docile when positive contact, such as talking to the animal and stroking its back, was made during daily caretaking. Over 80 % of newly acquired minipigs became used to being handled within 4 weeks after arrival at the laboratory, compared to 10 months in minipigs lacking positive contact during daily caretaking. Therefore, acclimatisation is considered important for reducing stress upon handling (Tsutsumi et al. 2001).

7. THE WELFARE OF GENETICALLY ENGINEERED PIGS

Since 1985 gene insertion in pigs has been possible, but the first reports of adverse effects of gene modification in pigs were published in 1989, when growth hormone releasing factor and insulin-like growth factor had been inserted into pigs (Pursel et al. 1989a, Pursel et al. 1989b). Most of these transgenic pigs did not survive long enough to study welfare aspects, but in 1990 pigs with inserted growth hormones were born. These grew much more rapidly and efficiently than controls, but several adverse effects occurred, such as lameness, ulcers, cardiomegaly, lethargy, renal disease, susceptibility to stress and infertility (Pursel et al. 1990).

The example above had a more efficient animal production as its goal. The ability to modify the mammalian genome by genetic engineering has allowed researchers to contemplate xenotransplantation as a possible solution to the shortage of human donor organs. Until recently, genetic engineering of pigs was limited to microinjection of up to three foreign genes. In 2000 the first specific pigs were produced, using a combination of gene targeting and embryo stem cell technique to knockout galactosyltransferase, making xenotransplantation much more likely, although the efficiency still is less than 1% (Nottle et al. 2002). In minipigs, transgenic techniques also lead to a low efficiency. One study reported that five out of 68 offspring were found to be transgenic after microinjection of 402 oocytes from 171 donors. Of 23 synchronised recipients, sixteen maintained pregnancy, giving an overall success rate of 1.24% (Uchida 2001). The low efficiency of transgenic techniques may be overcome by cloning by somatic nucleus transfer, which makes it possible to create identical copies of founder transgenic pigs. But cloning still suffers from low efficiencies of around 3% (Giles and Knight 2003). Even when embryos do successfully implant in the uterus after nucleus transfer, pregnancies are often interrupted or animals die shortly after birth because of developmental abnormalities (Pennisi and Vogel 2000). Even though the first pig clones have been born, no long term experience exists on topics such as health, welfare and ageing. Do cloned pigs have good health and welfare, and what is their life expectancy? Due to the lack of experience, these questions remain unanswered for the moment. However, in February 2003, Dolly the sheep, the first mammal to be cloned, died at six years of age. Dolly was euthanized because of a severe viral lung infection. Post mortem examination also revealed arthritis, but no other abnormalities were found (Giles and Knight 2003). However, Dolly was a non-genetically engineered clone. It is difficult to foresee what effects genetical modification have on animal welfare. Genetic modification is often necessary for the purpose of immunomodulation during xenotransplantation, and pathological changes may be expected as a result of changed physiological traits (Olsson 2000). However, studies with mice have shown that techniques used for gene modification as such are not necessarily associated with a compromised animal welfare (Van der Meer 2001).

8. THE WELFARE OF PIGS USED FOR XENOTRANSPLANTATION

The shortage of human organs for transplantation has drawn attention to alternative sources of organs. Although xenotransplantation, the transplantation of organs and tissues between different species, is still in an early phase, the pig is the primary candidate as an organ donor. Anatomical and functional similarity to humans, size, and, above all, low cost, make the pig more favourable than primates for this purpose. However, the prospect use of pig organs in human medicine raises ethical questions (Cortesini 1998). One of these questions addresses the acceptability of this form of animal use, and the modification of pigs through transgenic techniques. Whereas, authorities are quick in issuing guidelines for dealing with aspects of patient safety, such as zoonosis (Daar 2000), animal issues are often not discussed. Nevertheless, doctors are advocating rapid introduction of xenotransplantation into clinical practice (Editorial, The Lancet 1999).

There are special reasons why animal welfare issues should be of concern in connection with xenotransplantation. Animals have an interest and right to have a good animal life, and, moreover, patients have the right to know that the source animal was raised with a minimum of stress, suffering and discomfort and that the transplant was harvested in a humane manner (Olsson 2000). Although it is possible to breed, rear and house pigs in such a way as to ensure animal welfare, quality assurance of donor organs and the control of zoonosis, including biosecurity facilities, microbiological screening and routine examinations (Fishman 2000), may compromise animal welfare substantially. Certain facilities, presented as high welfare bioexlusion facilities, try to counterbalance the negative effects which biosecurity facilities generally have on animal welfare, by providing pigs with separated sleeping areas, feeding areas and loitering areas. Sterile objects, such as sterile water bags, can be provided as enrichment while housed in gnotobiotic isolators, whereas isolated rearing facilities, designed according to the Nurtinger principle, addressing the behavioural needs for growing pigs (Tucker at al. 2002a). Pigs raised in high welfare bioexlusion facilities did not develop pathological changes, although white blood cell count is reduced, probably as a result of the limited exposure to pathogens (Tucker et al. 2002b).

9. CONCLUSION

With respect to animal welfare only minor differences appear to exist between pigs and minipigs, due to genetic variation between breeds. The most important differences between pigs and minipigs are size and feed intake. Pigs have been selected for rapid lean growth during *ad libitum* feeding, whereas minipigs have been selected for small size. However, minipigs become grossly obese when fed *ad libitum*, and therefore restricted feeding is often applied. Restricted feeding may impose a risk on animal welfare, with a continuous feeling of hunger.

Acclimatisation to handling is necessary to reduce stress reactions during experimental procedures. Acclimatisation can be accelerated by daily positive contact during caretaking.

In general, welfare principles from pig farming can be transferred to the laboratory. Housing, social grouping, environmental conditions and feeding should all ensure the animal's capability to cope with its environment. However, research may cause an additional risk of reducing animal welfare, and new fields of research may result in yet unknown changes in the level of animal welfare. For this reason it is particularly important to monitor the level of welfare in pigs used for research purposes continuously.

REFERENCES

Apple JK, Craig JV. The influence of pen size on toy preference of growing pigs. Appl Anim Behav Sci 1992; 35:149-155

ARC, Agricultural Research Council. The Nutrient Requirements of Pigs. Commonwealth Agricultural Bureaux, Slough 1981

Arey DS, Franklin MF. Efffects of straw and unfamiliarity on fighting between newly mixed growing pigs. Appl Anim Behav Sci 1995; 45:23-30

Barnett JL, Cronin GM, McCallum TH, Newman EA. Effects of food and time of day on aggression when grouping unfamiliar adult pigs. Appl Anim Behav Sci 1994; 39:339-347

Beattie VE, Walker N, Sneddon IA. Effects of environmental enrichment on behaviour and productivity of growing pigs. Anim Welf 1995; 4:207-220

Bollen PJA. Nutrition of Göttingen minipigs, a study of the influence of ad libitum and restricted feeding on the physiology of the Göttingen minipig. Thesis, Biomedical Laboratory, University of Southern Denmark 2001

Bollen P, Ellegaard L. The Göttingen minipig in pharmacology and toxicology. Pharmacol Toxicol 1997; 80(Suppl 2):3-5

Bollen PJA, Hansen AK, Rasmussen HJ. The Laboratory Swine. CRC Press, Boca Raton 2000

Broom DM. Indicators of poor welfare. Br Vet J 1986; 142:524-526

Broom DM, Johnson KG. Stress and Animal Welfare. Chapman and Hall, London 1993

Bustad LK, McClellan RO, Burns MP. Swine in Biomedical Research. Frayn & McClellan, Seattle 1966

Commission of the European Communities. Third report from the commission to the council and the European parliament on the statistics on the number of animals used for experimental and other scientific purposes in the member states of the European Union. Brussels 2002

Cortesini R. Ethical aspects in xenotransplantation. Transplant Proc 1998; 30:2463-2464

Cunha TJ. Nutritional requirements of the pig. In Swine in Biomedical Research. Bustad LK, McClellan RO, Burns MP eds, Frayn & McClellan, Seattle 1966; 681-695

Daar AS. Xenotransplantation and cloning: working with the World Health Organization to develop ethical guiding principles. Transplant Proc 2000; 32:1549-1550

Danish Animal Experiments Inspectorate. Annual report 2001. Copenhagen 2001

Dawkins MS. From an animals point of view: motivation, fitness, and animal welfare. Behav Brain Sci 1990; 13:1-61

DEFRA, Department for Environment, Food and Rural Affairs. Codes of Recommendations for the Welfare of Pigs 2003; http://www.defra.gov.uk/animalh/welfare/farmed/pigs/

Dettmers A. On nutrition of the miniature pigs at the Hormel Institute. Lab Anim Care 1968; 18(1):116-119

Editorial. Xenotransplantation, time to leave the laboratory. The Lancet 1999; 354 (9191):1657

FAO, Food and Agricultural Organization. FAO statistical database, agriculture 2002; http://apps.fao.org

Fishman JA. Xenotransplantation from swine, making a list, checking it twice. Xenotranplantation 2000; 7:93-95

Fraser D. Selection of bedded and unbedded areas by pigs in relation to environmental temperature and behaviour. Appl Anim Behav Sci 1985; 14:117-126

Fraser AF, Broom D. Farm Animal Behaviour and Welfare. Wallingford, CAB International 1990

Fraser D, Bernon DE, Ball RO. Enhanced attraction to blood by pigs with inadequate dietary protein supplementation. Can J Anim Sci 1991; 71:611-619

Georgiev J, Georgieva R, Kehrer A, Weil S. Relationship between environmental temperature, humidity and energy conversion in the Gottinger miniature pig. Berliner und Münchener Tierärztlicher Wochenschrift 1977; 90(20):392-396

Giles J, Knight J. Dolly's death leaves researchers woolly on clone ageing issue. Nature 2003; 421:776

Grandin T, Curtis SE. Toy preferences in young pigs. J Anim Sci 1984; 59(Suppl 1):85

Graves HB. Behaviour and ecology of wild and feral swine (Sus Scrofa). J Anim Sci 1984; 58:482-492

GV-Solas, Gesellschaft fuer Versuchstierkunde/Society for Laboratory Animal Science. Futtercomponente und ihr Einsatz bei Versuchstiere. Mitteilungen des Ausschusses fuer Ernaehrung der Versuchstiere, Berlin 1993

Heffner RS, Heffner HE. Hearing in domestic pigs (Sus scrofa) and goats (Capra hircus). Hear Res 1990; 48(3):231-240

Holtz W, Bollen P. Pigs and minipigs. In The UFAW Handbook on the Care and Management of Laboratory Animals. Poole TB ed, Blackwell Scientific, London 1999; 464-488

McGlone JJ, Désaultés C, Morméde P and Heup P. Genetics of behaviour. In The Genetics of the Pig. Rothschild MF, Ruvinsky A eds, CAB International, Wallingford 1998; 295-311

McKinnon AJ, Edwards SA, Stephens DB, Walters DE. Behaviour of groups of weaner pigs in three different housing systems. Br Vet J 1989; 145:367-392

Mount LE, Ingram DL. The Pig as a Laboratory Animal. Academic Press, London 1971

Nottle MB, d'Aprice AJF, Cowan PJ, Boquest AC, Harrison SJ, Grupen CG. Transgenic perspectives in xenotransplantation. Xenotransplantation 2002; 9 (5): 305-308

NRC, National Research Council. Nutrient Requirements of Swine. National Academy Press, Washington 1998

Olsson K. Xenotransplantation and animal welfare. Transplant Proc 2000; 32:1172-1173

Pennisi E, Vogel G. Clones, a hard act to follow. Science 2000; 288:1722-1727

Petersen V. The development of feeding and investigatory behaviour in free-ranging domestic pigs during their first 18 weeks of life. Appl Anim Behav Sci 1994; 42:87-98

Pond WG, Mersmann HJ. Biology of the Domestic Pig. Ithaca, Cornell University Press 2001

Porter V. Pigs, a Handbook of the Breeds of the World. Helm Information, Mountfield 1993

Pursel VG, Miller FK, Bolt DJ, Pinkert CA, Hammer RE, Palmiter RD, Brintster RI. Insertion of growth hormone genes into pig embryos. In Biotechnology of Growth Regulation. Heap RB, Prosser CG, Lamming GE eds, Butterworths, London 1989a; 181-188

Pursel VG, Pinkert CA, Miller FK, Bolt DJ, Campbell RG, Palmiter RD, Brintster RI, Hammer RE. Genetic engineering of livestock. Science 1989b; 244:1282-1288

Pursel VG, Hammer RE, Bolt DJ, Palmiter RD, Brintster RI. Integration, expression and germline transmission of growth related genes. J Reprod Fertil Suppl 1990; 41:77-87

Ritskes-Hoitinga J, Bollen P. Nutrition of (Göttingen) minipigs: facts, assumptions and mysteries. Pharmacol Toxicol 1997; 80(Suppl 2):5-9

Schrøder-Petersen DL, Simonsen HB. Tail biting in pigs. Vet J 2001; 162:196-210

Spoolder HAM, Burbidge, JA, Edwards SA, Simmins PH, Lawrence AB. Provision of straw as a foraging substrate reduces the development of excessive chain and bar manipulation in food restricted sows. Appl Anim Behav Sci 1995; 43:249-262

Stanton HC, Mersmann HJ. Swine in Cardiovascular Research, vol. 1-2. CRC Press, Boca Raton 1986

Swindle MM. Basic Surgical Exercises using Swine. Praeger, New York 1983

Swindle MM. Swine as Models in Biomedical Research. Iowa State University Press, Ames 1992

Tsutsumi H, Morikawa R, Niki R, Tanigawa M. Acclimatisation and response of minipigs toward humans. Lab Anim 2001; 35:236-242

Tucker A, Belcher C, Moloo B, Bell J, Mazzulli T, Humar A, Hughes A, McArdle P, Talbot A. The production of transgenic pigs for potential use in clinical xenotransplantation, microbiological evaluation. Xenotransplantation 2002a; 9:191-202

Tucker A, Belcher C, Moloo B, Bell J, Mazzulli T, Humar A, Hughes A, McArdle P, Talbot A. The production of transgenic pigs for potential use in clinical xenotransplantation, baseline clinical pathology and organ size studies. Xenotransplantation 2002b; 9:203-208

Tumbleson ME. Swine in Biomedical Research, vol. 1-3. Plenum Press, New York 1986

Tumbleson ME, Schook LB. Advances in Swine in Biomedical Research, vol. 1-2. Plenum Press, New York 1996

Uchida M, Shimatsu Y, Onoe K, Matsuyama N, Niki R, Ikeda JE, Imai H. Production of transgenic miniature pigs by pronuclear microinjection. Transgenic Res 2001; 10 (6): 577-582

Van der Meer M. Transgenesis and animal welfare. Thesis, Department of Laboratory Animal Science, Utrecht University 2001

Van Zutphen LFM, Baumans V, Beynen AC. Principles of Laboratory Animal Science. Elsevier, Amsterdam 1993

Von Borell E, Hurnik JF. Understanding stereotypies. Pig Internat 1991; 4:27-29

Von Borell E, Broom DM, Csermeley D, Dijkhuizen AA, Edwards SA, Jensen P, Madec Cneva F, Stamataris C. The Welfare of Intensively Kept Pigs. Report of the Scientific Veterinary Committee, European Commission, Doc XXIV/B3/ScVC/0005/1997 1997

Wood-Gush DGM, Jensen P, Algers B. Behaviour of pigs in a novel semi-natural environment. Biol Behav 1990; 15:62-73

Wood-Gush DGM, Vestergaard K. Inquisitive Exploration in Pigs. Anim Behav 1993; 45:185-187

Chapter 13

THE WELFARE OF NON-HUMAN PRIMATES

Jann Hau[1] and Steven J. Schapiro[2]
[1]Division of Comparative Medicine, Department of Neuroscience, University of Uppsala, Sweden; [2]Department of Veterinary Sciences, The University of Texas M. D. Anderson Cancer Center, Bastrop, TX, USA

1. INTRODUCTION

Although non-human primates (in this chapter called primates) account for less than a fraction of one percent of the animals used for biomedical research, they remain vital and presently irreplaceable as models for certain types of research (Hau et al. 2000). The most common areas of research using primates are microbiology (including HIV/AIDS), neuroscience and biochemistry/chemistry. *Chlorocebus aethiops*, *Macaca mulatta*, *M. fascicularis* and *Papio* spp. are the most commonly used species (Carlsson et al. 2003), so this chapter will focus on these Old World monkey species.

The use of primates in biomedical research is associated with a number of problems of an ethical nature, and for this reason the continued use of primates has been questioned and debated in recent years (see *e.g.*, Balls 2000, Hau et al. 2000). Some propose a complete ban on the use of primates for research (Balls 2000). In countries like the UK, the Netherlands and Sweden, the use of Great Apes for biomedical research is now prohibited. To single out and ban these few species from study may represent both unscientific thinking and speciesism. One wonders why advocates for this type of policy do not begin by protecting *Homo sapiens*, rather than starting at a slightly lower taxonomic level. In this context, it seems important to discuss the actual use of these primates and whether they are participating in trivial or harmful (causing pain, suffering or distress) experimentation.

Instead of proposing a ban on certain species, it would be more logical to ensure that procedures associated with pain and suffering would not be allowed for the animals of the highest sentience and self-awareness. By applying this principle as part of the ethical review process, as is, in fact, done in many countries, scientific and medical progress may continue to develop without a legal ban on certain species.

Among the problems associated with the use of primates in biomedical research, the facts that primates are phylogenetically high-ranking and are likely to possess a degree of sentience similar to that of *Homo sapiens* are major considerations. This means that primates, second only to humans, are probably the animals with the highest degree of self-awareness, consciousness and capacity for suffering. In addition to this, they are also basically wild, undomesticated animals that have spent relatively few generations in captivity. This is in contrast to the vast majority of laboratory animal species who all have a long history of coexistence with humans, either as pets, pests (mice and rats) or farm animals subjected to generations of selection for docility and agreeableness.

The use of animals by humans is always associated with responsibility for the welfare of the animals concerned. However, when animals are used for research and subjected to procedures that may cause them pain, distress, suffering or lasting harm, there are additional obligations to refine techniques so as to cause only the minimum of pain and distress. Here it is important to distinguish between the pain and distress that may be an unavoidable component of the research and that which is avoidable through, for example, better experimental technique or husbandry. Russell and Burch, as long ago as 1959, made this distinction between direct (necessary) and contingent (avoidable) inhumanity. It is generally agreed that any animal suffering not required to achieve the scientific objective can only confound the experimental results.

The EU's Scientific Committee on Animal Health and Animal Welfare recently published a detailed document on the welfare of non-human primates used in research (European Commission 2002) that should be read by all readers of the current chapter. Welfare can be defined in different ways, as described in the introductory chapter to this book and in some parts of the world the term well-being is frequently used as if synonymous with welfare. This however, is not necessarily the case. Well-being may be seen as the quality of life of an animal, primarily in the short term (DeGrazia 1996), whereas welfare is a longer term - whole life - average. For example, in order to increase the welfare of an animal, it may be necessary to temporarily reduce its well-being, for example during restraint for vaccination. However, this animal's welfare is increased because of a reduced risk of serious disease resulting in a longer life of better quality.

Like animal well-being, animal welfare can be either low or high and should always be assessed by those responsible for the care of animals in captivity. The World Health Organization defines health as complete mental and physical well-being and not merely the absence of disease or infirmity, yet most scientists regard animal health as a sub-set of welfare and not the other way round. Animals can be perfectly healthy but still have a poor quality of life, if singly housed in a small cage. While all captive animals deserve to live in appropriately enriched conditions, when animals possessing the highest levels of sentience and cognitive abilities are kept in captivity, it seems reasonable to be especially conscientious about maintaining their psychological well-being.

Unlike humans, non-human primates cannot tell you when they are in pain, distress, or experiencing discomfort. The same applies to human infants, and to patients who are conscious, but unable to communicate. Responsible staff therefore, must be able to recognise the signs that indicate when animals are suffering. This places an obligation on scientists and animal technical staff to learn the signs indicative of suffering. Only when they can recognize pain and distress in an animal, can they do something about it, such as changing housing and/or husbandry practices, providing pain relief, withdrawing the animal from the study or humanely killing it. Although the animals cannot communicate verbally with us, we can make certain predictions that hold good for all primates, (including ourselves, e.g., we are all able to experience pain, frustration, pleasure, etc.). A reasonable start might be to ask whether a human experiencing a specific condition would be uncomfortable or experience pain or distress. But we need to be careful and apply what has been termed a 'critical anthropomorphism' (Morton et al. 1990). This means that one has to take into account the biological characteristics of the species of animal and also, if possible, to have some idea of what an individual animal has experienced in its life. For example, signs of pain and distress will vary by species, and primates that have been raised in the wild may respond very differently from those who have always been in a captive environment. Animals that have already been subjected or conditioned to experimental procedures, such as injections or blood removal, behave differently from those that are naïve or unconditioned. Moreover, the experience an animal has had with a particular human being (e.g., the one performing the injection) will undoubtedly influence its behaviour and physiological homeostasis. In some facilities, where the technical staff and the veterinary staff wear lab coats of different colour, it is very obvious that the animals behave differently depending on what colour lab coat they see. Publications examining the interactions between humans and animals have clearly demonstrated major positive effects of handling and socializing (Davis and Balfour 1992,

Hemsworth and Coleman 1998), a matter of considerable importance when animals are subjected to scientific procedures. When animals are properly conditioned to human handling, pain thresholds are likely to increase and trivial procedures, like restraint and injections, are less likely to be responded to as if they were stressful.

2. HOUSING, HUSBANDRY, WELFARE AND THE FIVE FREEDOMS

Laboratory primates are kept in confinement for their entire life span (if not caught from the wild) and some are moved to retirement facilities, rather than sacrificed at the end of an experiment. Long term experiments (*e.g.*, repeated scanning in Positron Emission Tomography cameras) and re-use in multiple experiments (*e.g.*, repeated infections with different infectious agents) are not uncommon, and it is important to consider the burdens we place on the animals for the whole of their lives, not just the impact of individual scientific procedures. Since many primates are likely to spend long periods in captivity, high quality housing conditions and husbandry practices should be among the top priorities for those responsible for these animals.

Primates are social animals who, in their natural environments, have evolved to live in groups. It is generally recognized that the social needs of primates are so important that they should always be housed with member(s) of the same species, unless veterinary or scientific reasons require that they be temporarily singly housed. Non-human primates show severely disturbed behaviour when raised in isolation and the opportunity to perform social behaviours, like grooming, must be seen as an essential need for these animals. If prevented from grooming socially, female primates have been observed to "groom" the ground (a redirection, in which the thwarted activity is performed on a substitute object; Bastock et al 1953), but a more common occurrence for many primates is an increase in self-directed grooming (Schapiro et al. 1996). In fact, many singly housed primates engage in abnormally high levels of self-directed behaviours (Lutz et al. 2003), sometimes with adverse consequences (as in the case of self-injurious behaviour; Novak 2003).

Although rhesus macaques can exhibit significant aggression and be difficult to maintain in harmonious groups (Augustsson and Hau 1999), quite a few laboratories and breeding colonies have used group-housing strategies very successfully (Bernstein 1991, Goo and Fugate 1984, Schapiro et al. 1994). It is, however, a challenge at many research facilities to establish groups that display long-term social harmony. This is particularly difficult among groups comprised of "experimental" animals, where group composition may change frequently as a function of study manipulations (treatments, sampling, euthanasia). The initial period following new group formation is the most problematic, but by managing the age and sex composition of the group, aggression-related injuries and social distress can be minimized (Schapiro et al. 1994). Application of the best care and husbandry techniques and use of competent and motivated technical staff and professional primate ethologists can ensure that individuals are not subjected to unreasonable aggression from cage mates. It is important to note that aggression among individuals in a primate social group is natural and only becomes undesirable when it adversely affects the animals' welfare and/or utility as a research subject.

Continuous behavioural observations, using relevant ethograms (see e.g., Primate Info Net 2003), can be used to establish time budgets for individual groups of primates. Time budgets provide a tool for the analysis of how primates spend their time and will reveal whether initiatives to increase desirable behaviours (e.g., play, grooming, foraging) or decrease undesirable behaviours (e.g., self-injurious behaviour, stereotypies) are successful. Both simple (Augustsson and Hau 1999, Nystrom et al. 2000) and more complex (Schapiro et al. 1996) observational systems have been published to facilitate the routine monitoring of behaviour in group-housed primates.

Group housing is often not a feasible option when animals are on experiment and pair housing may often be a practical alternative for laboratory primates. Pair-housing has numerous advantages over single housing (Schapiro et al. 1996, 2000) and has been used successfully with a variety of primate species (Crockett et al. 1994, Majolo et al. 2003, Reinhardt 1989a,b, 1994a,b). Pair-housing may be problematic however, if one member of the pair enters or finishes an experiment before the other. It can then be difficult for the remaining animal to adjust to a new partner, even if one is indeed available within the project or colony. Similar problems may occur if a group is diminishing in size due to participation of group members in research projects. Randomization procedures can also make pairing difficult, however we suggest that facility managers consider forming pairs during the quarantine or conditioning period, and then have study directors randomize pairs rather than individuals. This should decrease the likelihood of disruption of compatible pairs to begin an experiment without compromising research objectives.

Protocols for the safe formation of compatible pairs of previously single-caged adults of the same sex have been documented for numerous species, including chimpanzees (Fritz and Fritz 1979), baboons (Jerome and Szostak, 1987), rhesus macaques (Reinhardt 1989a,b, McMillan et al. 2003), stump-tailed macaques (Reinhardt 1994a,b), long-tailed macaques (Lynch 1998), and pig-tailed macaques (Byrum and St. Claire 1998). Adult animals need to be carefully familiarized before being brought together in the same cage, but little danger seems to be associated with pair formation of juvenile primates with strange single-housed adults or by pairing unfamiliar juveniles (Reinhardt 1994a,b). Schapiro and Bushong (1994) demonstrated that when single-caged individuals were brought together in compatible pairs their behavioural and physical health improved, evidenced primarily through decreased susceptibility to stress-related diseases (Schapiro et al. 2000). It has also been shown that the companion may serve as a stress buffer in experiments requiring restraint or similar fear-provoking situations (Mason 1960, Coe et al. 1982, Coelho et al. 1991).

Housing primates in pairs as compared with single housing does not interfere with many husbandry procedures, health management practices or research protocols (Reinhardt et al. 1989a,b, Schapiro and Bushong 1994, Schapiro 2002; Shively 2001). If an animal has to be kept alone for scientific procedures or for veterinary reasons, alternative social opportunities to minimize the stress resulting from social isolation (Coelho et al. 1991, Mahoney 1992, Lindburg and Coe 1995) should be provided, including grooming-contact bars (Crockett et al. 1997) or visual access to at least one compatible conspecific.

The minimum specified cage sizes for primates are small (see e.g., Council of Europe 1986, Convention ETS 123, Appendix A). There is, however, a welcome trend toward increasing the minimum acceptable size and it seems likely that future guidelines and requirements of regulatory authorities will stipulate larger cages for primates. In the Uppsala University rhesus monkey facility (Guhad et al. 1998), the animals are group housed in large rooms, and have access to larger outdoor enclosures that are used even during the Scandinavian winter (Figures 13-1 and 13-2).

Providing large cages for animals experimentally infected with potentially zoonotic microorganisms is rarely simple. In the new facility of the Swedish Institute for Infectious Disease Control, the cages (sections) are one square metre in floor area and two metres in height (Figures 13-3 and 13-4). Two or more cages can be combined and in fact, two connected cages are required for a pair of primates up to 5 kg in body weight. Individual housing is not allowed and animals not infected with dangerous zoonotic diseases are group housed in this facility.

THE WELFARE OF NON-HUMAN PRIMATES 297

Figure 13-1. Group housed Rhesus monkeys, Uppsala University, Sweden (photo by courtesy of Mr. O. Rosendahl)

Figure 13-2. Group housed Rhesus monkeys, Uppsala University, Sweden (photo by courtesy of Mr. O. Rosendahl)

However, a balance between welfare concerns and study requirements must often be achieved for primate subjects in experimental procedures. In such circumstances, one may have to settle for less than perfect solutions with respect to animal welfare in order not to subject the animals to unnecessary stress associated with experimental procedures. An example of this view can be found on the home page of the European Primate Resources Network (2003): "In some pharmacological or toxicological investigations, there may be a need to administer test substances, collect blood samples and monitor aspects of metabolism over days. The type of procedures involved and the degree of human intervention required might be expected to require fundamentally different housing conditions from that appropriate to a long term study designed to investigate hierarchical behaviour in large social groups, with minimal human interaction. Although intuitively one might feel that it is always desirable to provide the largest cage which resources permit, it may well be that the impact of catching and restraining animals for interactive procedures, such as blood sampling, is more marked in animals housed in larger cages or social groups." If sufficiently competent staff are engaged in training (using positive reinforcement strategies) the animals however, primates can be trained to cooperate and move to single cages attached to large group cages for administration of substances and blood sampling.

Figure 13-3. Large primate cages at the Swedish Institute for Infectious Disease Control in Stockholm, Sweden (photo by courtesy of Drs Spangberg and Westlund)

THE WELFARE OF NON-HUMAN PRIMATES 299

Figure 13-4. Training section of large primate cage at the Swedish Institute for Infectious Disease Control In Stockholm, Sweden (photo by courtesy of Drs Spangberg and Westlund)

2.1 The Five Freedoms

The Five Freedoms framework as described in Chapter 1.1 was first described in 1965 and is derived from its use in farming. The five freedoms, in the wording of Webster (1986) are:

Freedom from hunger and thirst
Freedom from physical discomfort and pain
Freedom from injury and disease
Freedom from fear and distress
Freedom to conform to essential behaviour patterns.

The framework is generally applicable to all animals kept in captivity and may provide a practical system for evaluating conditions for laboratory primates. The five freedoms will be dealt with individually:

2.1.1 Freedom from hunger and thirst

Access to food and water and to a diet that will maintain health and vigour is normally not a welfare issue for laboratory animals. The recent report on primate nutrition from the National Academy of Science provides extensive information concerning the nutrient requirements of most nonhuman primate species commonly used in research (Ad hoc committee on nonhuman primate nutrition 2003). However, both dental problems and obesity occur in the captive setting. Group housing may result in dominant animals consuming too many calories; one of several reasons that support the use of behavioural management techniques, such as positive reinforcement training (cooperative feeding; Bloomsmith et al. 1994, Schapiro et al. 2003) and the scattering of food in straw, wood chips or similar floor coverings (Chamove et al. 1982).

2.1.2 Freedom from physical discomfort and pain

Freedom from discomfort is achieved by providing a suitable environment, including shelter and comfortable resting areas. Chronic discomfort may be inevitably associated with certain experimental protocols and disease models.

Freedom from pain is normally one area where therapeutic intervention can easily be employed. For example after surgery, post-operative pain is not a desired scientific outcome, but rather an unwanted side effect. The elimination of post-operative pain not only relieves the animal's discomfort, but also enhances the value of the research by potentially reducing variance in the data. Other sources of pain may be more difficult to relieve, since the administration of analgesic drugs may interfere with the achievement of the research objectives. In these cases, however, the withholding of analgesics must be completely scientifically justified.

2.1.3 Freedom from injury and disease

Housing adult primates together may result in injuries from fights. This problem can be minimised by careful observation and alert technical staff trained to recognise relevant behaviours. When a primate has been injured in a group, it must be decided whether the injury needs treatment and/or whether the animal needs to be isolated for its own protection. It can be difficult to reintroduce a primate who has been isolated from its group for some time, and in these cases it may be advantageous for the veterinary staff to seek the advice of the technical and ethological staff.

Occasionally, primates will suffer from either an induced or spontaneous disease while they are serving as an experimental subject. When these animals require veterinary attention or the alleviation of pain or discomfort, the situation is not different than if a non-subject member of the colony was similarly afflicted. The only potential exception would be if the intervention has been shown to compromise the experimental data. At this stage in the evolution of the management of captive primates, preventative medical practices and the rapid diagnosis and treatment of ailments have significantly increased the likelihood that study animals can enjoy considerable freedom from most diseases.

2.1.4 Freedom from fear and distress

Good housing and husbandry practices help ensure that captive conditions minimise fear and distress among subjects. Both freedom from fear and distress and freedom from pain are critical for the welfare of laboratory primates and present opportunities for all staff in contact with the animals to improve conditions by actively implementing numerous refinement initiatives. It is well recognised that animals that are attempting to handle severe stress by means of dramatic physiological and/or behavioural coping strategies are unsuitable as laboratory subjects. These animals' adaptation mechanisms may well diminish the usefulness of the results generated; the added variance adversely affecting the validity of the data and its interpretation. There are therefore, scientific, as well as humanitarian, reasons for eliminating significant stressors for research primates.

Fear that stimulates an animal to flee or fight is a short-term stress that involves the adrenal medulla, where catecholamines such as adrenalin, noradrenalin, and dopamine are released. Fear helps physiologically prepare the animal for whatever action it decides to take, and there is a redistribution of blood to vital organs, such as muscle; increased heart rate and blood pressure; and increased acuity of the sense organs for hearing and sight. Fearfulness, as in anticipating an intervention or simply being removed from a familiar environment to a strange one, may end up causing pain to an animal. Thus, seemingly trivial procedures such as handling, sexing and weighing an animal, or giving it a subcutaneous injection, may be associated with fear of the unknown and the perception of pain in a timid animal not accustomed to human contact. In contrast, animals confident with their surroundings and staff are less affected by simple procedures and will often welcome the contact with the staff member, in spite of being dosed or receiving an injection. It is thus imperative that those staff members who will eventually handle and restrain an animal take the necessary time (perhaps as little as a few minutes every day for a week or two) to systematically habituate or condition the primate to the target circumstance, so it can adjust and gain in confidence. Finally, other routine husbandry procedures, such as weaning and identification methods, may be perceived as routine for the humans, but they are one off events for the animals. An animal may therefore come to experience (and expect) that every time it is handled, something unpleasant happens, inevitably causing fearfulness and mental distress. The impact of these adverse states on an animal is probably best assessed using a combination of behavioural analyses and clinical measurements, focusing on deviations from normal responses, rather than attempting to quantify physiological responses and so-called stress markers.

Mental distress is the product of an animal experiencing and responding to an unpleasant environment and is influenced by the animal's early experiences and current psychological state. The relative weights of feelings such as boredom, frustration, anxiety, contentment and happiness determine the degree of the animal's mental distress. Primates kept in inadequate environments frequently manifest stereotyped behaviours (Mason 1993, Wemelsfelder 1993), a fairly observable indication of their distress. For both standard (*Macaca* spp., *Chlorocebus*, *Papio* spp.) and unusual primate species, detailed knowledge of the natural habitat and of natural behaviour will provide considerable information to help design suitable captive environments. Methods to assess the suitability of captive environments include 1) measuring the number of different behaviours from the species-typical repertoire expressed by the animals and 2) determining the amount of effort that animals are willing to expend to gain access to particular components of the environment (Lawrence and Rushen 1993). Animals should be provided with adequate opportunities to work to perform those behaviours that help them maintain their "behavioural homeostasis" (Ashby 1952). Providing opportunities for captive primates to engage in social grooming would be one way to address this issue.

2.1.5 Freedom to express normal behaviour

The benefit to animals of being housed in an environment that functionally simulates natural conditions, thereby promoting the expression of species typical behaviours, has been increasingly acknowledged during the past decade. Unfortunately, laboratory animals have often been kept in barren environments that result in the performance of abnormal behaviours. Perhaps the most illustrative example of this is the self-injurious behaviour observed in captive primates (Novak 2003). In line with the comments above, the revised Appendix A to the Council of Europe Convention ETS 123 (1986) has, as a general rule, that laboratory animals should be kept in an enriched (enhanced in some way to be more than simply a barren space) environment and that they should be housed in harmonious social groups whenever possible. Being housed with another – or other – member(s) of the same species is recognised as the single most important enrichment practice for non-human primates. There is thus every good reason to house animals in compatible groups when possible. Some of the intricacies of housing research primates socially have been discussed above.

There are many practical options for ensuring that the environment experienced by captive nonhuman primates is more than simply barren space. A variety of techniques are available to functionally simulate the physical, foraging and sensory components of primates' natural habitats in captivity. There is a considerable body of published data that describes the effects of many of these enrichment options. The reader is directed to Segal (1989) and Novak and Petto (1991), two edited volumes containing many reports on enrichment procedures. A brief discussion of a few of the techniques that have been employed to stimulate natural behaviour patterns for primates housed in research settings follows. In general, the entire volume of the cage/enclosure should be fully usable by the primates for locomotion, play, rest, escape and other natural behaviours. This can be accomplished by providing enhancements to the physical environment, including such items as trees, perches, swings, nest boxes, visual barriers, swimming pools and other cage furnishings (Reinhardt 1989b, Anderson et al. 1994). Cages should be tall enough to allow captive primates to get above human eye level. Primates in the wild spend a considerable portion of their day foraging for, processing, and eating their food, and numerous strategies are available to encourage captive primates to do the same. These include distributing food in the substrate on the floor (Chamove et al. 1982), providing food in puzzles or devices (Bayne et al. 1992) and providing novel or 'time consuming to process' foods (Bloomsmith et al. 1988). The sensory environment of captive primates can be enhanced to more accurately simulate natural conditions via the presentation of interesting smells, sights, and/or sounds (Bloomsmith and Lambeth 2000). In the natural setting, primates can control many, but not all, of their interactions with the environment and their social partners. In captivity, much of this control is taken from the animals. A final effective strategy for promoting normal behaviour is to provide captive primates with the ability to control some of their interactions with the environment and/or conspecifics (Fragaszy and Adam-Curtis 1991, Lambeth et al. 2001).

3. EXPERIMENTAL PROCEDURES

Primates may carry diseases, which can be harmful and, in some cases, fatal to humans. The best known examples are Marburg virus in African green monkeys and *Herpesvirus simiae* in macaques. The Marburg virus resulted in fatalities in the 1970s and *H. simiae* transmission from rhesus monkeys to humans has also resulted in human deaths, most recently a few years ago in the United States. Although rhesus colonies of Specific Pathogen Free quality have been established (Schapiro et al. 1994, Schapiro 2002), it will be many years before rhesus monkey *H. simiae* will cease to be a health hazard for humans in contact with these animals. Consequently, in order to minimise the risk of human exposure to *H. simiae* infection, primates have to be handled carefully and appropriate personal protective equipment has to be worn by those in contact with the animals. This unfortunately complicates handling and sampling procedures and many simple procedures are often performed on animals anaesthetised solely to reduce the risk of infection.

Primates, however, are intelligent and can readily learn (using positive reinforcement training techniques) to cooperate with personnel during common procedures. Reinhardt (2002), Schapiro and colleagues (2003) and Laule and colleagues (2003) provide a number of general examples of training primates to cooperate for husbandry, veterinary and research procedures. Additional examples of specific behaviours that can be trained include: transfer to a holding area (Goodwin 1997, Bloomsmith et al. 1998), capture from the home cage (Reinhardt 1992a,b), capture from the group (Reinhardt and Cowley 1990, Kessel-Davenport and Gutierrez 1994, Mendoza 1999, White et al. 2000), blood collection (McGinnis and Kraemer 1979, Reinhardt 1991, Laule et al. 1996, Moore and Suedmeyer 1997), blood pressure measurement (Smith and Ansevin 1957, Mitchell et al. 1980, Turkkan 1990), systemic injection (Spragg 1940, Levison et al. 1964, Byrd 1977, Priest 1991, Reinhardt 1992b), urine collection (Kelley and Bramblett 1981, Ziegler et al. 1987, Bond 1991, Anzenberger and Gossweiler 1993, Shideler et al. 1994, Lambeth et al. 2000), saliva collection (Bettinger et al. 1998), topical drug application (Reinhardt and Cowley 1990, Segerson and Laule 1995), oral drug application (Turkkan et al. 1989), semen collection (Brown and Loskutoff 1998, Perlman et al. 2003), insemination (Desmond et al. 1987), vaginal swabbing (Bunyak et al. 1982, Hernándes-López et al. 1998) and veterinary examination (Brown 1998, Perlman et al. 2001). The time invested in training an animal is small when compared to the time and resources saved if capture, restraint and anaesthesia are no longer necessary in order to obtain a sample or administer an injection.

The rapid development of sophisticated imaging technologies (Geyer et al. 2000, Scott and Wise 2003) and the use of implantable telemetry devices (Reinhart et al. 2000, Crofts et al. 2001, Murphy et al. 2001) are likely to have increased the demand for primates in biomedical research. Although these new tools may have increased demand, they have also made many invasive procedures and chronic catheterisation experiments obsolete. Research questions that previously could only be addressed using major invasive surgery and catheterisation techniques can now be addressed using less invasive imaging and/or telemetry techniques. There are many aspects of these advancements that are encouraging from an animal welfare point of view.

4. PRIMATE SPECIFIC WELFARE PROBLEMS

Unlike for most laboratory rodent species, the origin or source of the individual primates used in the laboratory poses interesting welfare and management problems (National Research Council 2003). While laboratory primates can be obtained from breeding establishments in user countries, the current demand for primates in the US and Europe has overwhelmed the ability of domestic producers to satisfy that demand. Consequently, many primates are still imported from facilities in Asia and Africa. There are obvious scientific advantages to the maintenance of reliable, well-defined, domestic sources of research primates, however primates breed relatively slowly and production has not kept pace with demand (National Research Council 2003). Therefore, domestic supplies must be supplemented with purpose-bred animals from breeding facilities (often island colonies) in source countries. Even this has not satisfied the demand, and wild-caught animals are sometimes introduced into breeding colonies or used for biomedical research. Every effort should be made to establish productive captive breeding facilities to minimise the need for bringing in wild breeding stock.

In addition, wild primates are occasionally trapped and relocated in combination with conservation and pest management programs. Capture and confinement of wild primates is known to be stressful to the animals (Johnson et al. 1973, Else 1985, Suleman 1998) and often results in high morbidity and mortality (Uno et al. 1989, Suleman et al. 1999, 2000, 2001). Initiatives are underway to improve capture, confinement and translocation methods (Moinde et al. 2003).

Weaning is probably the most stressful life event for most primates in captivity and primates separated from their mothers at an early age show abnormal behaviour, sleep disturbance and long-lasting physiological changes (Harlow and Harlow 1962, Sackett and Terao 1992). Those breeding primates for commercial purposes tend to wean infants at an age that is significantly younger than mothers would typically wean their young in natural environments. Early weaning is practiced by captive managers for several reasons. Among breeders of macaques, early weaning is thought to minimise the risk of mothers infecting their offspring with *H. simiae*. Among breeders of many primate species, early weaning is thought to maximise productivity, increasing the number of infants that can be produced per female per year. There is scant evidence, however, that early weaning (3 months) results in increased productivity compared with weaning at 6 months of age. One study suggests that later weaning (8-12 months) may marginally reduce reproductive rates (Goo and Fugate 1984), however.

Transportation of long duration is a special problem for a number of reasons. Following a period of quarantine in the home country, the primates are normally transported singly in small cages (Figure 13-5). Since few airlines will transport primates, they may have to travel and/or wait for many hours under varying climatic conditions. When they finally arrive at the destination facility, they are again quarantined – oftentimes in single cages. Establishment of local breeding facilities at, or close to, user laboratories or establishment of user laboratories at, or close to, existing breeding facilities would eliminate many of the stressful events associated with long distance transport.

The longevity of primates and their widespread reuse in multiple experiments can be viewed as problems with the use of these animals in biomedical research. Reuse of animals in multiple experiments is generally not permitted in Europe, but dispensation is often granted for primates. In the United States, apes are not euthanised when they are no longer suitable for research, but are housed in retirement facilities, presumably until they die of natural causes. Considering that primates are often subjected to transportations of long duration at an early age, followed by a number of experimental protocols during a number of years, lifetime welfare assessment schemes should be in routine use at all research facilities.

Figure 13-5. Cynomolgus monkey transport cage, Indonesia (photo by courtesy of Dr HE Carlsson)

5. RECOGNITION AND ASSESSMENT OF ANIMAL WELFARE STATE

In virtually all biomedical research using primates, pain is an undesirable side effect of certain procedures and must be relieved by what is termed 'pre-emptive analgesia'. If pain is not relieved, neurons adjacent to the activated ones in the dorsal horn of the spinal cord will also be sensitised and the painful area will extend laterally. Consequently the animal will perceive more pain (hyperalgesia), and stimuli that are normally not painful become painful (allodynia). Behavioural studies are important for assessment of potential pain perception by the animals. Quantification of endorphin levels or nerve activity may be useful, but cannot substitute for careful behavioural observation, and, in any event, may not reflect the actual degree of pain perceived by the animal. The behavioural changes exhibited will depend on the site affected (e.g., immobility and posture when abdominal muscles have been traumatised; lameness after surgery or arthritis; depth of respiration after thoracotomy, and vocalisation when the affected site is touched) as will other quantifiable clinical signs, such as food and water consumption, body weight, vomiting, diarrhoea, lameness, respiration rate and temperature.

Anxiety is difficult to quantify, but fear can be assessed by behavioural studies combined with measurement of 1) urination and defecation frequencies, 2) organ functions such as heart rate, 3) muscle tone, or 4) quantification of catecholamines in the circulation.

When animals are exposed to environmental stressors such as extremes of temperatures, or when they perceive a threat by virtue of their senses, the hypothalamic pituitary adrenal axis is activated, resulting in the release of adrenocorticotrophic hormone. This, in turn, will stimulate the adrenal cortex to synthesise and release corticosteroids; predominantly cortisol in primates. Thus, if high levels of corticosteroids are present in the circulation and persist for long periods of time, despite their short half-lives, it would be reasonable to conclude that the animal was being subjected to a severe stress - 'bad stress', also termed 'dystress' (Dystress is created from the Greek word 'dus' having an implication of bad, ie bad stress.) (Morton and Hau 2003). If the animal does not adapt, then other activities of the hypothalamic axis are affected and this may influence nearly all the homeostatic mechanisms of the body (Sapolsky 1994) – with potentially adverse impacts on the scientific objective. Such animals show poor growth rates and sometimes even body weight loss, as well as disturbances of the endocrine system to the point where the animal is no longer 'fit' in a biological sense. It may not be able to breed and reproduce, and it may show considerable pathology, including gastric ulcer and hippocampal damage (Uno et al. 1989, Sapolsky 1994). The animal becomes more susceptible to infectious and non-infectious diseases due to suppression of the immune system. Dystress is principally an unconscious phenomenon and can be assessed by measuring circulating ACTH and corticosteroids, the response of the adrenal gland to ACTH, signs of pathological lesions at autopsy, as well as the impaired function of whole body systems (e.g., endocrine, immune and reproductive systems).

6. CONCLUSIONS

Many trends and developments in refinement and improvement of primate welfare deserve recognition; including the desirable introduction of relatively non-invasive imaging and telemetry techniques. Similarly, the use of detailed behavioural analyses and the search for immunochemical and neurological markers of stress in secretions like saliva, urine and faeces are additional advancements (Carver and Hau 2000, Stavisky et al. 2001). The present chapter is meant to give a brief update on the welfare of primates and the many welcome improvements that have been implemented in recent years. The increasing recognition of the importance of group housing in reasonably complex and interesting environments for these gregarious and intelligent animals bodes well for the future.

In conclusion, primate husbandry, care and use have been much improved, but they are areas that must be subjected to continuous refinement. Every primate must have an individual health record giving all relevant clinical and pathological observations and veterinary procedures and analyses performed. Health records have often been restricted to a history of only physical health, but also should include a psychosocial profile. Such a profile should list information on the animals' housing history, including social partners, dominance rank in the pair/group, compatibility with other animals, affiliative behaviour, etc. When relevant, the file should also include the animal's response to training and human contact. Of course, all of the experimental procedures that the animal has participated in during its lifetime should also be included in the record. The file should follow the animal throughout life and accompany the animal if and when it is moved between facilities. Detailed individual files of this kind will provide an important tool to help ensure the best possible treatment and care of non-human primates in research facilities and for implementing refinement at the level of the individual primate.

ACKNOWLEDGEMENTS

The authors gratefully acknowledge the constructive criticism of Professors David Morton and Clive Phillips. Mr Rosendahl and Drs Spangberg, Westlund and Carlsson very kindly allowed us to use the photographs illustrating this chapter.

REFERENCES

Ad hoc committee on nonhuman primate nutrition. Nutrient Requirements of Nonhuman Primates. Second Revised Edition. National Academies Press, Washington, DC, 2003

Anderson JR, Rortais A, Guillemein S. Diving and underwater swimming as enrichment activities for captive rhesus macaques (*Macaca mulatta*). Anim Welf 1994; 3:275-283

Anzenberger G, Gossweiler H. How to obtain individual urine samples from undisturbed marmoset families. Am J Primatol 1993; 31:223-230

Ashby WR. Design for a Brain. Chapman and Hall, London, UK, 1952

Augustsson H, Hau J. A simple ethological monitoring system to assess social stress in group-housed laboratory rhesus macaques. J Med Primatol 1999; 28(2):84-90

Bastock M, Morris D, Moynihan M. Some comments on conflict and thwarting in animals. Behav 1953; 6:66-84

Balls M. Proponent's statement: Moving toward the zero option for the use of non-human primates as laboratory animals. In Progress in the Reduction, Refinement and Replacement of Animal Experimentation. Balls M, van Zeller AM, Halder ME eds, Elsevier Publishers 2000; 1587-1591

Bayne KA, Dexter S, Mainzer H, McCully C, Campbell G, Yamada F. The use of artificial turf as a foraging substrate for individually housed rhesus monkeys (Macaca mulatta). Zoo Biology 1992; 1:39-53

Bernstein IS. Social housing of monkeys and apes: Group formations. Lab Anim Sci 1991; 41: 329-399

Bettinger T, Kuhar C, Sironen A, Laudenslager M. Behavior and salivary cortisol in gorillas housed in an all male group. American Zoo and Aquarium Association (AZA) Annual Conference Proceedings, Tulsa, Oklahoma, USA, 1998; 242-246

Bloomsmith MA, Lambeth SP. Videotapes as enrichment for captive chimpanzees (*Pan troglodytes*). Zoo Biology 2000; 19:541-551

Bloomsmith MA, Alford PL, Maple TL. Successful feeding enrichment for captive chimpanzees. Am J Primatol 1988; 16:155-164

Bloomsmith MA, Laule GE, Alford PL, Thurston RH. Using training to moderate chimpanzee aggression during feeding. Zoo Biology 1994; 13:557-566

Bloomsmith MA, Stone AM, Laule GE. Positive reinforcement training to enhance the voluntary movement of group-housed chimpanzees within their enclosure. Zoo Biology 1998; 17:333-341

Bond M. How to collect urine from a gorilla. Gorilla Gazette 1991; 5:12-13

Brown CS. A Training Program for Semen Collection in Gorillas (Videotape with commentary). Omaha's Henry Doorly Zoo, Omaha, USA, 1998

Brown CS, Loskutoff NM. A training program for noninvasive semen collection in captive western lowland gorillas *(Gorilla gorilla gorilla)*. Zoo Biology 1998; 17:143-151

Bunyak SC, Harvey NC, Rhine RJ, Wilson MI. Venipuncture and vaginal swabbing in an enclosure occupied by a mixed-sex group of stumptailed macaques *(Macaca arctoides)*. Am J Primatol 1982; 2:201-204

Byrd LD. Introduction: Chimpanzees as biomedical models. In Progress in Ape Research. Bourne GH ed, Academic Press, New York, NY, 1977; 161-165

Byrum R, St.Claire M. Pairing female *Macaca nemestrina*. Lab Primate Newsletter 1998; 37:1

Carlsson HE, Schapiro SJ, Farah IO, Hau J. The present use of primates in research: a global overview 2001. Submitted for publication 2003

Carver JFA, Hau J. Development of a saliva sampling method and a rocket immunoelectrophoretic assay for quantification of salivary IgA in the squirrel monkey (*Saimiri sciureus*). In Vivo 2000; 14:735-739

Chamove AS, Anderson JR, Morgan-Jones SC, Jones SP. Deep woodchip litter: Hygiene, feeding, and behavioural enhancement in eight primate species. Int J Study Anim Problems 1982; 3:308-318

Coe CL, Franklin D, Smith ER, Levine S. Hormonal responses accompanying fear and agitation in the squirrel monkey. Physiol Behav 1982; 29:1051-1057

Coelho AM, Carey KD, Shade RE. Assessing the effects of social environment on blood pressure and heart rates of baboon. Am J Primatol 1991; 23:257-267

Council of Europe. European Convention for the Protection of Vertebrate Animals used for Experimental and other Scientific Purposes (ETS 123). Council of Europe, Strasbourg 1986

Crockett CM, Bowers CL, Bowden DM, Sackett GP. Sex-differences in compatibility of pair-housed adult longtailed macaques. Am J Primatol 1994; 32:73-94

Crockett CM, Bellanca RU, Bowers CL, Bowden DM. Grooming-contact bars provide social contact for individually caged laboratory macaques. Cont Top Lab Anim Sci 1997; 36:53-60

Crofts HS, Wilson S, Muggleton NG, Nutt DJ, Scott EAM, Pearce PC. Investigation of the sleep electrocorticogram of the common marmoset (Callithrix jacchus) using radiotelemetry. Clin Neurophysiol 2001; 112:2265-2273

Desmond T, Laule GM, McNary J. Training to enhance socialization and reproduction in drills. American Zoo and Aquarium Association (AZA) Regional Conference Proceedings, Wheeling, WV, USA, 1987; 352-358

Davis H, Balfour D. The Inevitable Bond: Examining Scientist-animal Interactions. Cambridge University Press, Cambridge, UK, 1992

DeGrazia, D. Taking Animals Seriously: Mental Life and Moral Status. Univ. Cambridge Press, 1996

Else JG. Captive propagation of vervet monkeys (*Cercopithecus aethiops*) in harems. Lab Anim Sci 1985; 35:373-375

European Commission. The welfare of non-human primates used in research, Report of the Scientific Committee of Animal Health and Animal Welfare, Adapted on 17 December 2002; http://europa.eu.int/comm/food/fs/sc/scah/out83_en.pdf

European Primate Resources Network 2003, http://www.dpz.gwdg.de/eupren/eupren.htm

Fragaszy DM, Adams-Curtis LE. Environmental challenges in groups of capuchins. In Primate Responses to Environmental Change. Box HO ed, Chapman and Hall, New York 1991; 237-264

Fritz P, Fritz J. Resocialization of chimpanzees. J Med Primatol 1979; 8:202-221

Geyer S, Matelli M, Luppino G, Zilles K. Functional neuroanatomy of the primate isocortical motor system. Anat Embryol 2000; 202:443-474

Goo GP, Fugate JK. Effects of weaning age on maternal reproduction and offspring health in rhesus monkeys (*Macaca mulatta*). Lab Anim Sci 1984; 34:66-69

Goodwin J. The application, use, and effects of training and enrichment variables with Japanese snow macaques (*Macaca fuscata*) at the Central Park Wildlife Center. American Zoo and Aquarium Association (AZA) Regional Conference Proceedings 1997; 510-515

Guhad F, Augustsson H, Hau J. The Torneby Primate Facility: Optimisation of housing conditions for rhesus macaques in Sweden. Scand J Lab Anim Sci 1998; 25:173-176

Harlow HF, Harlow MK. Social deprivation in monkeys. Scientific American 1962; 207:137-146

Hau J, Farah IO, Carlsson HE, Hagelin J. Opponents' statement: Non-human primates must remain accessible for vital biomedical research. In Progress in the Reduction, Refinement and Replacement of Animal Experimentation. Balls M, van Zeller AM, Halder ME eds, Elsevier Publishers 2000; 1593-1601

Hemsworth PH, Coleman GJ. Human-Livestock Interactions: the Stockperson and the Productivity and Welfare of Intensively Farmed Animals. CAB International, 1998: 152

Hernándes-López L, Mayagoitia L, Esquivel-Lacroix C, Rojas-Maya S, Mondragón-Ceballos R. The menstrual cycle of the spider monkey *(Ateles geoffroyi)*. Am J Primatol 1998; 44:183-195

Jerome CP, Szostak L. Environmental enrichment for adult, female baboons (*Papio anubis*). Lab Anim Sci 1987; 37:508-509

Johnson PT, Valerio DA, Thompson GE. Breeding the African green monkey, *Cercopithecus aethiops*, in a laboratory environment. Lab Anim Sci 1973; 23:355-359

Kelley TM, Bramblett CA. Urine collection from vervet monkeys by instrumental conditioning. Am J Primatol 1981; 1:95-97

Kessel-Davenport AL, Gutierrez T. Training captive chimpanzees for movement in a ransport box. The Newsletter 1994; 6:1-2

Lambeth SP, Perlman JE, Schapiro SJ. Positive reinforcement training paired with videotape exposure decreases training time investment for a complicated task in female chimpanzees. Am J Primatol 2000; 51(Suppl 1):79-80

Lambeth S, Bloomsmith M, Baker K, Perlman J, Hook M, Schapiro S. Control over videotape enrichment for socially housed chimpanzees: Subsequent challenge tests. Am J Primatol 2001; 54(Suppl 1):62-63

Lawrence AB, Rushen J eds. Stereotypic Animal Behaviour; Fundamentals and Applications to Welfare. CAB International, Wallingford, Oxon, UK 1993; Chapters 2 and 4

Laule GE, Thurston RH, Alford PL, Bloomsmith MA. Training to reliably obtain blood and urine samples from a diabetic chimpanzee (*Pan troglodytes*). Zoo Biology 1996; 15:587-591

Laule GE, Bloomsmith MA, Schapiro SJ. The use of positive reinforcement training techniques to enhance the care, management, and welfare of primates in the laboratory. J Appl Anim Welf Sci 2003; 6:163-173

Levison PK, Fester CB, Nieman WH, Findley JD. A method for training unrestrained primates to receive drug injection. J Exp Anal Behav 1964; 7:253-254

Lindburg DG, Coe J. Ark design update: Primate needs and requirements. In Conservation of Endangered Species in Captivity. Gibbons EF, Durrant BS, Demarest AJ eds, SUNY Press, Albany, NY, 1995; 553-570

Lutz C, Well A, Novak M. Stereotypic and self-injurious behavior in rhesus macaques: A survey and retrospective analysis of environment and early experience. Am J Primatol 2003; 60:1-15

Lynch R. Successful pair-housing of male macaques (*Macaca fascicularis*). Lab Prim Newsl 1998; 37:4-5

Mahoney CJ. Some thoughts on psychological enrichment. Lab Anim 1992; 21(5):27,29,32-37

Majolo B, Buchanan-Smith HM, Morris K. Factors affecting the successful pairing of unfamiliar common marmoset (*Callithrix jacchus*) females: Preliminary results. Anim Welf 2003; 12:327-337

Mason GJ. Forms of stereotypic behaviour. In Stereotypic Animal Behaviour; Fundamentals and Applications to Welfare. Lawrence AB, Rushen J eds, CAB International, Wallingford, Oxon, UK 1993; Chapter 2

Mason WA. Socially mediated reduction in emotional responses of young rhesus monkeys. J Abnorm Soc Psychol 1960; 60:100-110

McGinnis PR, Kraemer HC. The Stanford outdoor primate facility. In Comfortable Quarters for Laboratory Animals, Seventh Edition. Animal Welfare Institute, Washington, DC 1979; 20-27

McMillan J, Maier A, Tully L, Coleman K. The effects of temperament on pairing success in female rhesus macaques. Am J Primatol 2003; 60(Suppl 1):95

Mendoza SP. Squirrel Monkeys. In The UFAW Handbook on the Care and Management of Laboratory Animals, Seventh Edition. Poole T, English P eds, UFAW, Blackwell Science, Oxford, UK, 1999; 591-600

Mitchell DS, Wigodsky HS, Peel HH, McCaffrey TA. Operant conditioning permits voluntary, noninvasive measurement of blood pressure in conscious, unrestrained baboons *(Papio cynocephalus)*. Behav Res Methods Instrumen 1980; 12:492-498

Moinde NN, Higashi H, Suleman MA, Hau J. Habituation, capture and relocation of endangered arboreal non human primates: Experience from relocation of Sykes monkeys (*Cercopithecus mitis albotorquatus)* on the coast of Kenya. Anim Welf In Press

Moore BA, Suedmeyer K. Blood sampling in 0.2 Bornean orangutans at the Kansas City Zoological Gardens. Animal Keepers' Forum 1997; 24:537-540

Morton DB, Hau J. Welfare Assessment and Humane Endpoints. In: Handbook in Laboratory Animal Science Volume I: Essential Principles and Practices, Second Edition. Hau J, Van Hoosier G eds, CRC Press, Boca Raton, Florida, USA, 2003; 457-486

Morton DB, Burghardt G, Smith JA. Critical Anthropomorphism, Animal Suffering and the Ecological Context. Hasting's Center Report Spring Issue on Animals, Science and Ethics 1990; 20(3):13-19

Murphy DJ, Renninger JP, Coatney RW. A novel method for chronic measurement of respiratory function in the conscious monkey. J Pharmacol Toxicol 2001; 46:13-20

National Research Council. International perspectives: The future of nonhuman primate resources. Proceedings of the workshop held April 17-19, 2002. National Academies Press, Washington, DC, 2003

Novak MA Self-injurious behaviour in rhesus monkeys: New insights into its etiology, physiology, and treatment. Am J Primatol 2003: 59:3-19

Novak MA, Petto, AJ. Through the Looking Glass: Issues of Psychological Well-being in Captive Nonhuman Primates. American Psychological Association, Washington, DC, 1991

Nystrom P, Schapiro SJ, Hau J. Accumulated means analysis: a novel method to determine reliability of behavioral studies using continuous focal sampling. In Vivo 2000; 15:29-34

Perlman JE, Bowsher TR, Braccini SN, Kuehl TJ, Schapiro SJ. Using positive reinforcement training techniques to facilitate the collection of semen in chimpanzees (*Pan troglodytes*). Am J Primatol 2003; 60 (Suppl 1):77-78

Perlman J, Guhad FA, Lambeth S, Fleming T, Lee D, Martino M, Schapiro, S Using positive reinforcement training techniques to facilitate the assessment of parasites in captive chimpanzees. Am J Primatol 2001; 54(Suppl 1):56

Priest GM. Training a diabetic drill (*Mandrillus leucophaeus*) to accept insulin injections and venipuncture. Lab Prim Newsl 1991; 30:1-4

Primate Info Net 2003; http://www.primate.wisc.edu/pin/

Reinhardt V. Behavioral responses of unrelated adult male rhesus monkeys familiarized and paired for the purpose of environmental enrichment. Am J Primatol 1989a; 17:243-248

Reinhardt V. Evaluation of the long-term effectiveness of two environmental enrichment objects for singly caged rhesus macaques. Lab Animal 1989b; 18:31-33

Reinhardt V. Training adult male rhesus monkeys to actively cooperate during in-homecage venipuncture. Anim Technol 1991; 42:11-17

Reinhardt V. Transport-cage training of caged rhesus macaques. Anim Technol 1992a; 43:57-61

Reinhardt V. Improved handling of experimental rhesus monkeys. In The Inevitable Bond. Examining Scientist-Animal Interactions. Davis H, Balfour AD eds, Cambridge University Press, Cambridge, UK, 1992b; 171-177

Reinhardt V. Pair-housing rather than single-housing for laboratory Rhesus macaques. J Med Primatol 1994a; 23:426-431

Reinhardt V. Social enrichment for previously single-caged stumptail macaques. Anim Technol 1994b; 5:37-41

Reinhardt V. Comfortable quarters for primates in research institutions. In Comfortable quarters for Laboratory Animals, 9th Edition. V Reinhardt, A Reinhardt, eds, Animal Welfare Institute 2002, http://www.awionline.org/pubs/cq02/cqindex.html

Reinhardt V, Cowley D. Training stumptailed monkeys to cooperate during in-homecage treatment. Lab Prim Newsl 1990; 29: 9-10

Reinhart GA, Preusser LC, Opgenorth TJ, Wegner CD, Cox BF. Endothelin and ETA receptors in long-term arterial pressure homeostasis in conscious nonhuman primates. Am J Physiol-Regul Integr Comp Physiol 2000; 279:R1701-R1706

Russell WMS, Burch RL. The Principles of Humane Experimental Technique. UFAW Special Edition 1992, Publrs UFAW 1959

Sackett GP, Terao K. Separation stress in mother and nursery reared pigtail monkey infants (M. nemestrina). XIV Congress of the International Primatology Society, Strasbourg 1992, Abstract No 0462

Sapolsky RM. Why Zebras Don't Get Ulcers. WH Freeman & Co., New York, 1994

Schapiro SJ. Effects of social manipulations and environmental enrichment on behavior and cell-mediated immune responses in rhesus macaques. Pharmacol Biochem Behav 2002; 73:271-278

Schapiro SJ, Bushong D. Effects of enrichment on veterinary treatment of laboratory rhesus macaques (*Macaca mulatta*). Anim Welf 1994; 3:25-36

Schapiro SJ, Lee-Parritz DE, Taylor LL, Watson L, Bloomsmith MA, Petto A. Behavioral management of specific pathogen-free rhesus macaques: group formation, reproduction, and parental competence. Lab Anim Sci 1994; 44:229-34

Schapiro SJ, Bloomsmith MA, Porter LM, Suarez SA. Enrichment effects on rhesus monkeys successively housed singly, in pairs, and in groups. Appl Anim Behav Sci 1996; 48:159-172

Schapiro SJ, Nehete PN, Perlman JE, Sastry KJ. A comparison of cell-mediated immune responses in rhesus macaques housed singly, in pairs, or in groups. Appl Anim Behav Sci 2000; 68:67-84

Schapiro SJ, Bloomsmith MA, Laule GE. Positive reinforcement training as a technique to alter nonhuman primate behaviour: Quantitative assessments of effectiveness. J Appl Anim Welfare Sci 2003; 6:175-187

Scott SK, Wise RJS. PET and MRI studies of the neural basis of speech perception Speech Commun 2003; 41:23-34

Segal, EF. Housing, Care and Psychological Wellbeing of Captive and Laboratory Primates. Noyes Publications, Park Ridge, NJ, 1989

Segerson L, Laule GE. Initiating a training program with gorillas at the North Carolina Zoological Park. American Zoo and Aquarium Association (AZA) Annual Conference Proceedings 1995; 488-489

Shideler SE, Savage A, Ortuño AM, Moorman EA, Lasley BL. Monitoring female reproductive function by measurement of fecal estrogen and progesterone metabolites in the white-faced saki *(Pithecia pithecia)*. Am J Primatol 1994; 32:95-108

Shively CA. Psychological Well-Being of Laboratory Primates at Oregon Regional Primate Research Center. Willamette Week, Portland, OR, 2001, March 21, http://www.wweek.com/html2/shivreport.html

Smith CC, Ansevin A. Blood pressure of the normal rhesus monkey. Proc Soc Exp Biol Med 1957; 96:428-432

Spragg SDS. Morphine addiction in chimpanzees. Comp Psychol Monogrs 1940; 15:1-132

Stavisky RC, Whitten PL, Hammett DH, Kaplan JR. Lake pigments facilitate analysis of fecal cortisol and behavior in group-housed macaques. Am J Phys Anthropol 2001; 116:51-58

Suleman MA. Studies on stress in African green monkeys (*Cercopithecus aethiops*), Ph.D. Thesis, Uppsala University 1998

Suleman MA, Yole D, Wango E, Sapolsky R, Kithome K, Carlsson HE, Hau J. Peripheral blood lymphocyte immunocompetence in wild African green monkeys (*Cercopithecus aethiops*) and the effects of capture and confinement. In Vivo 1999; 13:25-27

Suleman MA, Wango E, Farah I.O, Hau J. Adrenal cortex and stomach lesions associated with stress in wild male African green monkeys (*Cercopithecus aethiops*) in the post-capture period. J Med Primatol 2000; 29:338-342

Suleman AM, Wahungu GM, Muoria PK, Karere GM, Oguge NO, Moinde NN. Tana River Primate Census and Habitat Evaluation. Report to Kenya Wildlife Services and the World Bank, 2001

Turkkan JS. New methodology for measuring blood pressure in awake baboons with use of behavioral training techniques. J Med Primatol 1990; 19:455-466

Turkkan JS, Ator NA, Brady JV, Craven KA. Beyond chronic catheterization in laboratory primates. In Housing, Care and Psychological Wellbeing of Captive and Laboratory Primates. Segal EF ed, Noyes Publications, Park Ridge, NJ 1989, 305-322

Uno H, Tarara R, Else JG, Suleman MA, Sapolsky RM. Hippocampal damage associated with prolonged and fatal stress in primates. J Neurosci 1989; 9:1705-1711

Webster J. Health and welfare of animals in modern husbandry systems – dairy cattle. In Pract 1986; 8:85-89

Wemelsfelder F. Animal Boredom : Towards an Empirical Approach of Animal Subjectivity. Dissertation, Leiden Rijksuniv., The Netherlands 1993

White G, Hill W, Speigel G, Valentine B, Weigant J, Wallis J. Conversion of canine runs to group social housing for juvenile baboons. AALAS 51st National Meeting, Official Program, 2000; 126

Ziegler TE, Bridson WE, Snowdon CT, Eman S. Urinary gonadotropin and estrogen excretion during the postpartum estrus, conception, and pregnancy in the cotton-top tamarin (*Saguinus oedipus oedipus*). Am J Primatol 1987; 12:127-140

Chapter 14

ANIMAL WELFARE ISSUES UNDER LABORATORY CONSTRAINTS, AN ETHOLOGICAL PERSPECTIVE: RODENTS AND MARMOSETS

Augusto Vitale, Francesca Cirulli, Francesca Capone and Enrico Alleva
Department of Cell Biology and Neurosciences, Istituto Superiore di Sanità, Rome, Italy

1. INTRODUCTION

Over the last decade both the scientific community and the wider public have become increasingly aware of the issues surrounding laboratory animal welfare. However the questions posed remain controversial: on the one hand, the validity of the very concept of animal welfare as a coherent topic for debate is still not universally acknowledged; on the other, there exists a diverse array of definitions and methodologies proposed to evaluate and monitor it (see Chapter 1 in this book).

The aim of this chapter is to consider some issues concerning the welfare of laboratory animals, in particular rodents and non-human primates, applying an ethological perspective, and to suggest ways of ameliorating the welfare of laboratory animals, without compromising the quality of experimental data. Achieving a satisfactory balance between welfare requirements on the one hand and the need for reliable, good quality data on the other is clearly not an easy task. While not attempting to give simple answers to difficult questions, we propose that a thorough ethological knowledge of the study species can play a vital role in significantly improving its captive conditions and the carrying out of the experimental protocols, and that the experimental animals' quality of life and the quality of the data collected will inevitably be linked (Vitale and Alleva 1999).

We will focus on rodents and non-human primates: the first ones being the most used animal models; the second ones are less used, but have a highly evocative power. Among these, special attention will be given to marmosets, because these are among the non-human primates that, at the moment, are increasingly found in biomedical laboratories.

We will consider the possibility, or need, to offer the laboratory animals a captive environment able to mimic the stimuli the animals can encounter in the natural environment. Then, we will discuss the need to prevent and assess instances of aggressive behaviour in captive situations that, if not routinely checked, can lead to the disruption of the social environment and, consequently, can severely decrease the level of welfare of individuals. This point is particularly important nowadays, when the need to keep animals in captivity in an adequate social environment is widely recognised. Finally, our attention will focus on the possibility to assess and limit nociception in experimental protocols involving painful experiences. For all of these aspects the ethological point of view will be the common methodological and exploratory framework.

2. DESIGNING NATURALISTIC ENVIRONMENTS UNDER LABORATORY CONSTRAINTS

2.1 Rodents

A variety of rodent studies have indicated that the use of an enriched environment is a successful strategy to increase the welfare of captive animals. Environmental physical enrichment consists of increasing the complexity of captive environments, adding features like wheels, tunnels, platforms and toys. Although the term enrichment should imply that modifications are beneficial for animals, housing rodents in an enriched environment does not necessarily lead to an increase in their welfare (Nevison et al. 1999). Indeed the impact of environmental enrichment on animal well-being strictly depends on its efficacy in catering for their ethological needs and not upon the anthropomorphic whims of the caretakers (Newberry 1995). Differences in enrichment, objects and cage structure, appearing subtle to humans, may substantially address the needs of animal for performing specific naturally-selected responses, which remain in laboratory animals as an "evolutionary legacy" (Dawkins 1988). Providing nesting material, for example, represents an easy strategy to increase welfare of laboratory rodents, since they show a strong motivation to build nests (Sherwin 1997), as they do in commensal habitats (Estep et al. 1975). Even in the presence of pre-fabricated shelters (e.g. plastic tubes), currently used in several laboratories, isolated laboratory rodents continue performing nest-building activities, using these artificial nests for sleeping only if sawdust is removed (Galef and Sorge 2000; Sherwin 1996). We have investigated the utilization of a physically-enriched environment in juvenile and adult CD-1 mice of both sexes. This cage, already used in a previous study (Caston et al. 1999) and commonly sold in pet-shops, consisted of an assembly of four differently-shaped and -coloured plastic compartments. Data from this study indicate that elected features of a physically-enriched environment have a distinct relevance from the mouse point of view and therefore may exert a different impact on animal well-being. As an example, providing a standard cage with a tubular circular structure, rather than with a running wheel, does not seem to add any comfort to laboratory mice of both ages and sexes, while a small nest-like compartment may selectively fulfil the needs of adult females (Pietropaolo et al. submitted).

In most enrichment studies physical and social enrichment are used at the same time, so it is often difficult to disentangle the exact contribution of each component in determining a specific outcome. We have investigated whether changing either the social or the physical complexity of the periadolescent environment may produce long-term effects on behavioural and neuroendocrine responses of male CD-1 mice (Pietropaolo et al. submitted). Social or physical enrichment during periadolescence exerts distinct long-term effects on behavioural responses. Subjects exposed to a physically-enriched environment show decreased exploratory activity, reduced interactions with the unfamiliar object and low levels of aggressiveness. Social enrichment does not affect exploratory behaviour, but appears sufficient to modify the quality of the agonistic response. Mice kept in groups during adolescence show a more appropriate agonistic strategy, suggesting that social experiences at critical developmental stages are more relevant than physical complexity for the development of adequate strategies to cope with social challenges. These results may be of relevance in order to ameliorate current husbandry standards of the most commonly used laboratory rodent species, the mouse. Housing subjects in a socially complex unit at periadolescence seems to be an important and convenient strategy to preserve an intact social agonistic behaviour, thus improving the welfare of laboratory animals by giving them the opportunity to develop their natural behavioural patterns.

The features of the social environment are very important early in development. The mother's behaviour largely determines the physical and social environment of the young, including the social group in which the infant is born and the kind of contacts it has with its members (D'Udine and Alleva 1980). Mice show communal nesting under natural conditions and this means that other individuals, in addition to the (biological) mother take care of the pups under natural conditions (Crowcroft and Rowe 1963, Saylor and Salmon 1969). Thus, the laboratory environment, constrained by the experimental settings, presents the infant with a very limited amount of social experience. In addition to the mother, the littermates are one of the main sources of experience for the mammalian infant. The litter gender composition is an important variable capable of exerting both short and long-term changes in physiological and behavioural parameters, including exploratory patterns, emotionality and pain reactivity (Alleva et al. 1986, Cirulli et al. 1997). Thus, in those cases where there is a strong bias within the litter for one or the other sex, such as in litters with a small number of pups, cross-fostering between females can be adopted as a rearing procedure. When dealing with breeding females and their offspring, special care must be afforded to accommodate the needs of each species. For example precocial rodent species (such as the *Acomys cahirinus*) differ from altricials (*Mus musculus*) in the quality and quantity of maternal care provided to the infants. In designing enrichment procedure one should always take into account the fact that separating mother and infants or manipulating the litter can modify maternal behaviour resulting in long-term consequences on the offspring behaviour and physiology (Cirulli et al. 2003). In those strains or situations (transgenic mice) where maternal behaviour is disrupted, interactions between mother and offspring can be helped by providing the females with cotton or other nesting material. This should also provide the pups with a warmer nest helping them to maintain their body temperature more efficiently.

One important notion in animal welfare is that the animal living in captivity should be able to make daily choices relating its environment. When dealing with very aggressive strains the housing style should be 'strain' and 'sex' specific (Alleva 1993, Alleva et al. 1995). For example, in territorial rodents living in the same cage, such as mice, a shelter should be present to allow those subjects which are repeatedly attacked to protect themselves. Female mice show aggressive behaviour only during pregnancy and lactation to protect their pups (Ostermeyer 1983). If females with offspring cohabit with males, nest-defence should be favoured by providing the cage with a protected nest.

As a final comment, enriching environments with physical, social and sensory stimuli are now established to be beneficial to brain development and ageing. Experiments, conducted in rodents in the early 1960s and influenced by the initial works of Hebb, demonstrated for the first time measurable changes in the brain following training or differential experiences (Rosenzweig et al. 1961, Wiesel and Hubel 1963). Starting from those pioneering studies there are now numerous reports documenting the role played by experience and behaviour in activating brain plasticity mechanisms and remodelling neuronal circuitry in the brain. Exercise, in particular, is a simple and widely practised behaviour that activates molecular and cellular cascades that support and maintain brain plasticity. More specifically, physical exercise, such as voluntary running, induces the expression of genes associated with plasticity, such as that encoding Brain-derived neurotrophic factor (BDNF) in the hippocampus, a highly plastic structure that is normally associated with higher cognitive function, rather than motor activity (Cotman and Berchtold 2002). Thus, enrichment can have both an immediate impact on animal well being, but, most importantly, can exert long-term beneficial effects on an animal's behaviour and physiology leading to successful ageing.

2.2 Non-human primates: The Callitrichidae

The number of non-human primates used in biomedical research varies between species, countries and areas of research. A recent document issued by the European Commission has shown that the countries that are the biggest users of non-human primates as animal models among EU members are, in order, United Kingdom, France and Germany (European Commission 2003b). The areas of research in which non-human primates are more frequently used are biological studies of a fundamental nature, such as neurology and physiology, and toxicological and other safety evaluation (European Commission 2003a).

Within the Old World monkeys (Catarrhine) the macaques are the primate species most used in research and, in particular, the long-tailed macaque (*Macaca fascicularis*). When it comes to New World Monkeys (Platyrrhine), the common marmoset (*Callithrix jacchus*) has replaced the squirrel monkey (*Saimiri sciureus*) in the last decade. Rylands (1997) reports that marmosets became acknowledged laboratory models in biomedical studies in the early 1960' in United States. Before then, these monkeys were just used in anatomical and behavioural studies, or for screening for yellow fever in the wild. The common marmoset presents some advantages as laboratory animal. Among these, there is the relatively small size (an adult male weights approximately 250-300 g), which makes possible to house several marmosets in a smaller space, compared to other species.

Duncan and Fraser (1997) have identified three approaches to the study of animal welfare. The first one proposes the subjective feelings of an individual as the important parameter to be taken into consideration when judging whether an animal is in a good state of welfare. The second point of view considers the good functioning of the biological machinery as essential element for good welfare. Finally, a third view links the level of welfare with the possibility for the animal to perform in the most "natural" way (Duncan and Fraser 1997). We believe that common sense requires an interaction between the three approaches in order to have a more comprehensive view of the welfare of the captive animal. However, as ethologists, we also feel that the possibility to express a good portion of the behavioural repertoire observed in nature can be very important for a captive individual. Therefore, the captive environment should be functional in helping the animal to do so. Although it is impossible to recreate in captivity the African savannah, or the Brazilian forest, a naturalistic laboratory environment should at least give the monkey an adequate level of complexity, variability, and control over it.

In research laboratories it is not always possible give up cages and, for example, to house the monkeys in rooms full of exercise space and elaborate enrichments, or to construct a indoor-outdoor system. However, we review here environmental features that can increase the welfare of laboratory Callitrichids. Before doing that we would like again to emphasise that designing a more naturalistic captive environment for one species can have a limited value for another species. In other words, when providing enrichments, looking at the species-specific needs, both social and physical, is the required exercise. The recent document issued by the European Commission on the welfare of non-human primates used in research, aimed at the revision of the current European legislation on this matter, very clearly stresses this methodological point (European Commission 2003b).

Once the captive environment is adequate enough to completely satisfy the basic physiological needs of the animals, the ecological and ethological characteristics of these monkeys should be considered. Marmosets and tamarins are arboreal monkeys and, therefore, need to develop and exercise their locomotory skills. If there is the possibility to take advantage of a indoor-outdoor facilities, air-conditioning ducting has been proved to be successfully used by tamarins when moving between the two environments (McGrew and McLuckie 1986). This kind of arrangements mimics dispersal movements observed in the wild and the monkeys, especially the young, appear to enjoy running along the ducting. This is also true for common marmosets. At the Istituto Superiore di Sanità in Rome, we house five families of common marmosets. Each family lives in a cage measuring 220x150x80, provisioned with hanging objects, ropes, branches of different sizes and a wooden nest. Each of these cages is connected, by a system of tunnels, to two other cages of the same size and located in an adjacent room. Observational experiments are performed in these latter cages. The animals continuously run back and forth along the tunnels.

In our colony we observed that infants spend a considerable part of their time in trying to balance on branches of different sizes; they also use twigs to exercise their grasping skills. Furthermore, we acquired some adults from a breeding facility in which marmosets were housed in large numbers in small cages. There was no possibility to run over or climb branches. When these adults arrived they were unable to balance themselves correctly on the horizontal branches and, when jumping from the top of the cage to a branch below, they were missed the target, landing on the floor of the cage. These kinds of locomotor difficulties were never observed in adults born in our colony. So, these monkeys should be provided in captivity with wooden perches of various diameter, size and orientation, taking into account species-specific preferences (Buchanan-Smith 1997). Different degrees of firmness could also help the marmosets to exercise their jumping skills. Finally, when furnishing the cage with such substrates one should remember that some Callitrichids species tend to avoid the lower parts of the cage due, perhaps, to an anti-predator strategy.

In our experience, the use of wooden platforms, hung at different heights of the cage (always in the top half) has proved to be beneficial. The monkeys, when visiting the cage provided with such platforms, engaged in more social behaviours, such as grooming and play, using the platforms as substrate (DeRosa, unpublished data).

Feeding enrichments are very important. In nature, monkeys spend a significant part of their time looking for and processing food (Mittermaier et al. 1988); this is an important activity because it has also the function to keep the cognitive abilities exercised (Tomasello and Call 1997). It is possible in captivity, without much effort either of time or of money, to provide the monkeys with such behavioural opportunity. Food can be scattered on the bottom of the cage, and experimental data have shown that, when given the choice, monkeys in captivity prefer to work for their food, rather than obtaining it and consuming it easily (see, for example, Reinhardt 1994). Food can be made more difficult to obtain by using feeding devices or puzzle-feeders. For example, McGrew and collaborators presented common marmosets with a wooden cylinder drilled with holes filled with Arabic gum. The marmosets, who are regular extractors of plant exudates, readily learned to gnaw the device to extract the gum (McGrew et al. 1986). In another case, common marmosets were provided with a foraging tree. This object was a PVC pipe, connected with other PVC at right angles along its length. Holes were drilled on its surface. Food treats were inserted both within the pipes, and in the holes. The results showed that, in the absence of such enrichment the marmosets spent 1-25% of their time budget foraging whereas, in the presence of the food tree, the foraging time increased up to 80%. These data were approaching a more natural division of their time budget. (Byron and Bodri 2001). Puzzle-feeders have been used with marmosets as well. In a study by DeRosa and collaborators marmosets were given the choice between simply collecting pieces of food from food dishes, or working to extract the food from plastic boards, with a series of holes in the front, hung at different heights of the cage. The study showed that, in terms of increasing the relative time spent in feeding activities, the puzzle-feeders appeared to be successful, leading to the expression of a more natural time budget. The exploitation of the enrichment by the family was related to its average age, with families containing younger individuals being more active and interested in the enrichments, than the families composed only of adults (DeRosa et al. 2003). Therefore age composition of the groups can be important in determining the success of certain type of physical enrichment (for a brief review on different types of feeding enrichments, see Reinhardt and Roberts 1997).

A final crucial point is the need to give monkeys the possibility to exercise some degree of control over their captive environment. To cite Buchanan-Smith: "However, many of the devices do not provide ecologically valid stimuli... It is argued that rather than, or in addition to, providing specific enrichment devices, the alternative is to design enclosures, and have routine housing, husbandry and feeding regimes which permit a degree of control. Variability within the environment alone permits an animal some control and exerts preferences, for instance by incorporating unpredictable environmental perturbations which require an adaptive response from the animal." (Buchanan-Smith 1997).

3. ASSESSING AND PREVENTING SOCIAL AND AGGRESSIVE BEHAVIOUR IN LABORATORY MAMMALS

3.1 Rodents

The social ethogram of the experimental species should be the background against which any protocol involving behavioural tests is formulated. Simplified scoring scales used in the laboratory are derived from full ethograms established under naturalistic, or semi-naturalistic, conditions (Alleva et al. 1995). Ethograms using behavioural scores, other than those specifically used in recording aggressive behaviour, can provide useful information in terms of general behavioural arousal and interpretation of an animal's reactions (Table 14-1). When a particular protocol specifies observing the outcome of social interactions, overt fighting should either be excluded entirely from the experiment or at least interrupted at a pre-established early stage.

There are several other methodological strategies which should be considered with the goal of reducing stress and suffering in rodents, for example: i) testing-arena size, subject's familiarity with it and previous provision of safe hiding places should be controlled to minimise excessive stress; ii) repeated, escalating biting episodes should be anticipated to allow for rapid and effective intervention; iii) on-line scoring of exploratory and affiliative phases, eventually culminating in colloquial intraspecific aggression (for example, Markov chains have been used to evaluate those trends) (Bressers et al. 1995). An ethogram for scoring mouse aggressive behaviour should always be available (Table 14-2) (the behavioural elements are described in Maestripieri et al. 1990; Laviola et al. 1991; Bigi et al. 1992). Finally, repeated testing could be necessary to characterise "dominant" and "subordinate" behavioural phenotypes, and an accurate inspection of the "standard opponent" animal confronting a treated mouse can reveal, or help to interpret, behavioural changes.

Table 14-1. An example of the behavioural elements present in an ethogram for mice

Category	Behavioural elements
Nonsocial investigation	Explore, scan, wash, self-groom, scratch, dig, push-dig, kick-dig, shake, jump, eat, drink, sit, turn, flop, displacement, stretched attend posture, substrate, leave, on cage bars, off cage bars
Social investigation	Attend, stretched attend, approach, investigate, nose, groom
Sexual activity	Follow, sniff, attempted mount, mount, genital groom, push-under, crawl-over, push-past
Offence	Threat, attack, bite, chase, aggressive groom, fully aggressive
Offensive ambivalence	Offensive-sideways, offensive-upright, sideways posture
Defensive ambivalence	Oblique, upright posture, defensive-upright, defensive-sideways
Arrested flight	Crouch, freeze, straight legs, kick, on back
Escape	Flag, evade, retreat, flee
Distance ambivalence	Circle, zig-zag, walk round, tail-rattle (Repetitive, shaking movements of the tail)

Different kinds of housing conditions can influence the physiological and behavioural conditions of both male and female rodents (Brain and Nowell 1971ab, Brain and Stanislaw 1988). Individual housing (social isolation) of mice leads to a higher propensity to attack when these subjects are used in an agonistic experimental context. A factor to be taken into account in this case is the possible existence of a sensitive period for the effects of isolation on the level of aggressiveness of certain individual subjects (Cairns et al. 1985). Furthermore, both the size and the group composition of a mouse colony must be monitored and planned in advance. For example, in socially-established mouse colonies, residential mice respond differently to intruders with different fighting experience (Brug and Slotnick 1983, Crawley et al. 1975).

Table 14-2. Ethogram for scoring mouse aggressive behaviour (Luigi De Acetis, personal communication)

A. Fighting behaviour
    Attacks (number, intensity, frequency, duration, latency to first episode)
    Tail-rattling (Repetitive, shaking movements of the tail)
    Aggressive grooming
B. Species-specific displays for aggression-inhibited purposes
    Defensive upright posture
    Submissive upright posture
    Submissive crouched posture
C. Escape responses
    Flee
    Evade
D. Displacement or stereotypic elements for contextual evaluation of agonistic performance
    Self-grooming
    Bar holding
    Digging
    Freezing
    Patrolling
    Defensive burying of the opponent (very rarely observed)

An example of an alternative experimental protocol aimed at reducing the likelihood of physical injuries is the case in which intermale aggressive behaviour can be used as a variable to cause changes in specific mouse central nervous system areas, for example, leading to changes in gene expression (Spillantini et al. 1989). A possible stepwise approach is to consider whether something less dangerous than overt aggression (male aggressive patterns can include violent repetitive biting on the rump and back between the two contestants) could achieve the aims of the study as well or better. For example, male rodents could be simply exposed to social odour cues, such as wood shavings soaked in the urine, secretions from sex glands (which reportedly have a higher communicative value in intraspecific agonistic interactions), or fur debris from another male subject.

Animal models of social stress involve single, intermittent, or chronic exposure of a subject animal to a conspecific. A variant that has been used in the case of studies dealing with intermittent defeat involves housing animals in adjacent areas within visual, auditory, and olfactory, but not tactile, contact. At intervals, the barriers between the two animals are removed such that they can interact directly, establishing a victor and a defeated or submissive member of the pair. When the barrier is replaced, the defeated animal is left in chronic sensory contact with the victor. Thus, in this case, while the exposure to the psychosocial stressor is chronic, the actual fighting experience (and thus the risk of bites or injury) is reduced and intermittent (for a review see Blanchard et al. 2001).

In any case, experimental protocols should always be refined depending upon the specific testing contest and the characteristic of the individual subject being tested. This is especially true in the case of genetic manipulation which can result in hyperaggressive strains. Since significant changes in agonistic patterns may emerge as a by-product of genetic modification, researchers should be able to modify their protocols according to the specific needs of the experimental animal (see also Nelson and Chiavegatto 2001).

3.2 Non-human primates: The case of the common marmoset

The majority of primate species live in very complex social systems (Smuts et al. 1987). Experimental procedures interfere with the social relationships existing between the members of a social group (ensuring every possible effort is made to ensure that the captive population of primates experience an adequate level of sociality). The common marmoset is a good example for aggressive behaviour because its social organisation presents some aspects that can lead to social aggression.

When managing a captive colony of non-human primates there are two situations in which the researcher can potentially be confronted with aggressive behaviours: (i) problems arising within an established group; (ii) problems arising when composing new groups. The latter situation can occur when, for example, new pairs or groups have to be formed to increase the breeding opportunity within the colony, or to solve instances of social intolerance occurring within a previously established group.

The common marmoset usually engages in a monogamous mating system, and the presence of two breeding females in the same group is usually a temporary arrangement (Rothe and Koenig 1991, Rothe and Darms 1993). However, just one female breeds, giving birth to twins, and she inhibits the ovulation of the other females present in the family (Abbott and Hearn 1978, Digby 1999). The female dominance is rarely expressed aggressively. For example, at the Istituto Superiore di Sanità we investigated the dominant role of adult females over other individuals within a family. We presented the monkeys with restricted sources of food. When in pairs, the males were observed just to let the female to go first; when the experiments were carried out in groups, some instances of aggressiveness among the members of the families were observed, but these were never serious enough to warrant interrupting the experiments. Sessions were limited to no more than 10 min, and the cage in which the experiments were carried out had an escape route that could be opened instantaneously in the event of any individual becoming the object of dangerous attacks (Vitale and Scolavino, unpublished data; see also Petto and Devin 1988).

Our colony was founded in 1983 with four breeding pairs, since then the number of animals housed has varied between 12 and 27. One of the reasons for this fluctuation is "social expulsions". This is a phenomenon occurring in colonies of common marmosets with different intensity and frequency. Initially a family member suddenly chases, attack, bites and fights with another member of its own family. The situation can become serious in a couple of days, requiring the removal of one of the two individuals. The other members of the family neither interfere nor show interest in what is happening. We have never been able to fully comprehend and, therefore, to prevent such events. In our colony females as well as males were in turn the expellers and the expelled, without any clear relationship with age. Similar data exist for related species, such as the lion tamarins (*Leontopithecus rosalia*) (Kleinman 1979, Inglett et al. 1989), but it is unreported in common marmosets. Such phenomena mimic aggressive interactions in the wild, which end up with the dispersal of the expelled individual. Obviously, captivity has the effect of exacerbating such events, the attacked monkey cannot escape (violent expulsions have not been observed in the wild) (for a review see Schnaffner and Caine 2000).

It is important to be able to distinguish between day-to-day aggressive interactions, and events leading to expulsion. Promptness in doing so can prevent severe injuries to the target individual. In the first case the events are temporary, the opponents reciprocate, and no obvious consequences on the relationship between the two monkeys are visible. Different behavioural patterns can be a good indication of serious aggression. For example, the target individual rarely reacts to the attack of the opponent, it tries continuously to avoid the presence of the attacker and, every time he sees it, emits vocalisations similar of those uttered by begging infants. This situation, in the case of an expulsion can go on, without solution, day after day. At that point it is advisable to remove one of the two individuals from the family, otherwise severe injuries can occur.

When it comes to the relationships between different families, it has to be remembered that *Callithrix jacchus* is a highly territorial species. Observations carried out on wild populations have shown that when different families meet in the forest, the marmosets engage in a series of facial displays, body postures and vocalizations aimed at scaring and chase away the members of the other family. Physical aggression rarely occurs (Hubrecht 1985). For this reason in captivity, where the animals cannot escape, members of different families must never come in direct physical contact. The constant sight of members of a different family can be a source of stress, therefore, cages housing different groups can be separated by the use of hanging curtains.

The aggressive reactions towards intruders must be taken into account when forming new families within a captive colony. It is good practice, initially, to familiarise the individuals coming from different groups by allowing visual contact, to check for signs of aggressiveness. The sessions can be increased daily. If the animals are calm, then visual contact plus a minimum of physical contact could be allowed by separating the monkeys by a wire mesh. The researcher can decide on the required number of sessions, on the basis of his/her knowledge of the temperament and personal history of the individuals involved. Finally, the monkeys can be housed in the same cage and the staff involved must monitor this final phase very closely.

Relatively few problems are encountered when pairing a male and a female. Usually signs of sexual behaviour can be observed at the first visual contact (e.g., tongue-flicking, Stevenson and Poole 1976) and, although the female will not immediately accept the courtship behaviour of the male, new breeding pairs can be easily formed. The procedure can be more difficult, and sometimes fail, when three individuals are involved. The three phases recommended are essential. For example, Yamamoto (personal communication) when pairing pairs of adult females with a new male, has observed that the females promptly establish a dominance relationship, without obvious signs of physical aggressiveness. However, in some cases one of the two females could decide to expel the potential sexual competitor.

Finally we reiterate that, in order to control the level of aggressiveness within a captive colony and allow the animals to live a relatively satisfying social life, the personnel involved with the animals, at any level, must have a knowledge of the temperament and personal history of each individual under their care.

4. ASSESSING NOCICEPTION IN PAIN EXPERIMENTS IN RODENTS

Pain has been defined as an 'unpleasant sensory and emotional experience associated with actual or potential damage (International Association for the Study of Pain 1979).

Such a definition only emphasises the notion that pain is a subjective experience, and thus difficult to objectively evaluate. Nonetheless, suffering is becoming behaviourally identifiable in animals through progress in applied ethology and animal welfare disciplines. Since a state of suffering or malaise can manifests itself through a deviation from normal behavioural patterns, the diagnosis of pain in animals entails both a clinical and an ethological assessment (Fraser 1984, Viercki and Cooper 1984, Wright et al. 1985, Bateson 1991).

Physical and nociceptive pain in rodents can be measured using a number of tests originally developed to evaluate the efficacy of putative analgesic compounds (Taber 1974). The hot-plate test has been in use for over 50 years as a cost-effective measure of pain sensitivity in both mice and rats. First described in 1944 by Woolfe and MacDonald, it has been repeatedly modified since. Briefly, the test consists of placing the animal on a heated surface and measuring the latencies to various behavioural responses (usually licking of a paw) for a time period usually not exceeding 60 seconds. The plate was initially set at temperatures ranging from 55 to 70 °C but was successively lowered to 55 °C by Eddy and co-workers in 1950 (Eddy et al. 1950) and this temperature level is used in most cases nowadays.

Thermal stimulation is appealing as heat is a natural stimulus that can be easily controlled. Compared to other tests used to assess nociception, the hot-plate test provides a minimum pain and suffering in the experimental animal. This test, in fact, allows the subject to make appropriate behavioural responses in order to avoid the thermal stimulus, such as lifting and licking the paws. The hot-plate temperature commonly used (55 °C) is unlikely to cause any severe pain or damage to the skin of the animal.

Pain sensitivity in the hot-plate test has been measured as the latency to first forepaw licking, latency to first hindpaw licking, or latency to the first licking of any paw. In our experience, the best measure of pain sensitivity is the first licking of the hindpaw because forepaw licking is biased by the occurrence of wall-rearing behaviour. A response-competition between these two measures can be observed: if the subject rears on the walls of the apparatus, the latency to forelimb licking will be longer than expected since the forepaws are not on the heated surface and therefore do not experience the thermal stimulus. This is especially true when the hot-plate is set at low temperatures, such as 50 or 52 °C, since in this case the first behaviour shown by the subjects is wall rearing and not forepaw licking (Figure 14-1). By contrast, when the hot-plate temperature is high, animals lick their forepaws right after they are placed in the apparatus. This explanation might account for some discrepancies found in different studies measuring either forepaw licking or hindpaw licking latencies and suggests that latency to hindpaw licking should be used as the most reliable index of pain sensitivity

(Bigi et al. 1993ab). In addition to paw licking responses, animals can be seen crouched and licking their testicles, behaviour more readily shown at high temperatures. Expression of undesired wall rearing responses may be reduced by choosing an appropriate size (species-, strain- or age-dependent) for the hot-plate testing arena. For example, a standard adult-mouse arena may be inappropriate for developing rats.

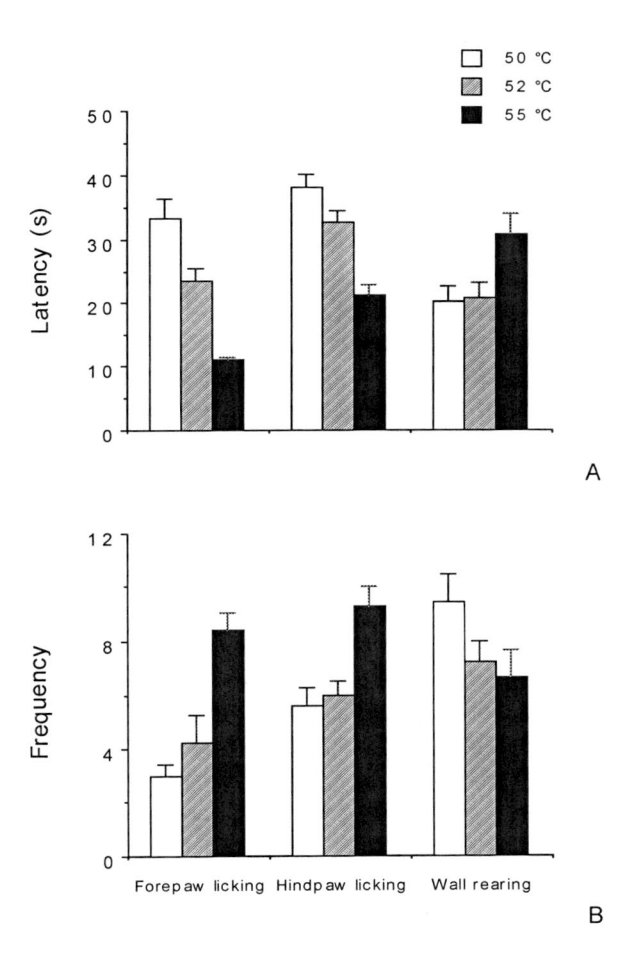

Figure 14-1. Latency and frequency to lick the forepaw, hindpaw and to perform wall rearing in adult CD-1 mice exposed to different temperatures. N = 24 subjects in each group.

In order to argue that a behavioural response to a selected thermal stimulation indicates the presence of pain, the reaction should not only occur preferentially to painful levels of stimulation, but it should also be differentially affected by graded changes in stimulus intensity. We have assessed changes in the latency and frequency of behaviours commonly performed by rats and mice in the hot-plate test by varying the temperature of the plate from 50 to 55 °C. While both forepaw and hindpaw licking latencies show a clear temperature-dependent profile, being lower at high temperatures, the latency to perform wall rearing behaviour shows a very different profile, being lower at 50 and 52 °C, compared to 55 °C (Figure 14-1A). This can be explained observing the behaviour of the subjects at the different temperatures. When the intensity of the thermal stimulus is high, the animals first lick their paws and then perform behaviours which are not pain-related, such as wall rearing. This is also true when measuring frequency: a high frequency of wall rearing can be seen when the hot plate is set at low temperatures, decreasing when the intensity of the thermal stimulus is increased (Figure 14-1B). This profile is opposite to that of forepaw or hindpaw licking and substantiates the notion that wall rearing should not be used as a reliable index of pain.

Another item characterising behavioural response to the hot plate test is the jumping reaction, which in control/untreated (adult) subjects can be scored reliably only following repeated testing in the apparatus (Della Seta et al. 1994). Jumping behaviour has been interpreted as an attempt to avoid actual paw contact with the heated surface, a reduction of exploratory behaviour in a more familiar environment, or both. The peculiar pattern of this response over successive trials (it increases with repeated testing but eventually tends to wane) is suggestive of a learning curve: the subject attempts to avoid the thermal stimulus by jumping, but after a few unsuccessful trials, eventually learns that it cannot escape, thus reducing its responses. Furthermore, it has been hypothesised that in the hot plate test the paw-licking response (short latency) could represent the sensory component of pain, while jumping (long latency) might indicate its affective component (Amir et al. 1979), the latter being a function of stress-induced changes in arousal affecting pain reactivity, such as release of endorphins and enkephalins, which can affect central elaboration of emotions (Amir et al. 1979). In support of this latter hypothesis it has been shown that if the sensory and affective components of pain are selectively modified, for instance by means of changes in endogenous opioid ligands, the jumping response is affected, while the paw lick latency remains unchanged (Amir et al. 1979). Although the hot-plate test is extensively used in biomedical research, very few authors have addressed systematically the issue of which behavioural measure is the most reliable and should be used as an index of pain, while permitting some degree of suffering reduction. This point can be highly relevant also for the quality of research performed. In fact, choosing the most appropriate dependent variable can facilitate the detection of subtle treatment effects (e.g., reducing the risks of ceiling or floor effects) and ultimately result in a reduction in the number of experimental subjects exposed to a highly distressing situation. We propose that an ethological observational scale, composed of three to five items, can allow a single experimenter to discriminate the locomotor, exploratory, and emotional components (this latter being also a function of previous experience of the testing environment or social isolation) underlying the subjects' behaviour and thus to identify the behavioural traits truly indicative of pain. Behaviours shown by rats and mice in the hot-plate test are summarised in Table 14-3. Specific aim of this proposed evaluation scale is to provide an effective methodology, which allows animal suffering to be reduced to a minimal level.

Table 14-3. Mouse and rat ethogram in the hot-plate test

Nociceptive responses
Forepaw licking: Animal standing on hindlimbs and licking forepaws in a washing action
Hindpaw licking: Head tilted towards hindpaw with ventral surface angled upwards
Licking testicles (rare)
Exploratory/escape responses
Wall rearing: Animal standing on hindlimbs and touching the walls of the cage with the forepaws
Rearing: Animal standing on its hindlimbs
Jumping: Animal jumping upwards with both hindpaws away from the hot-plate surface (more readily seen with repeated testing)

In our experience, the temperature at which the hot-plate is set represents a critical variable for discriminating treatment effects. In a study conducted on Swiss albino CD-1 male mice, a significant effect of social isolation on pain threshold was apparent only if the temperature was kept at 52°C, but could not be detected at 55 °C. In addition, previous reports have shown that, while the analgesic effects of morphine are easily identified when the hot plate is set at 55°C, the effects of non-narcotic analgesics producing weak nociceptive effects, can be detected when the temperature is reduced to 50 °C or lower (Hunskaar et al. 1986, Ankier 1974). These same authors have also devised an increasing temperature hot-plate for rats and mice in order to assess the effects of pharmaceutical compounds which cannot be screened with the classical hot-plate (Hunskaar et al. 1986). In other words, social experiences appear important variables to take into account when designing an experiment likely to produce some degree of suffering. Moreover, accurate planning may decrease the painful experience.

Lowering the intensity of the thermal stimulus can be recommended for two main reasons. First, it allows the experimenter to assess subtle treatment effects that are not detectable at high temperatures, thus ameliorating data quality. Second, by reducing the intensity of the painful stimulation, it decreases animal suffering. Both these points are in line with the need for refinement of experimental procedures first expressed by Russel and Burch (Russel and Burch 1959) in their classic test, that is: 'simply to reduce to an absolute minimum the amount of distress imposed on those animals that are still used' (Flecknell 1994).

Using the hot-plate test, it has been possible to demonstrate that a wide range of factors (some associated with animal husbandry or laboratory procedures) can affect pain sensitivity. For example, stressful or aversive events such as electric shock, cold water swimming and restraint can induce analgesia in rodents. Animals might undergo these or similar experimental procedures just before being tested in a hot-plate, thus introducing a bias in the results. For example, using the hot-plate test (Rodgers and Hendrie 1983) it has been shown that, following intraspecific fighting, male mice can show either analgesia or hyperalgesia, depending upon their social status (Alleva 1993).

Early social experiences may exert long-lasting effects on the response to pain as well as on morphine analgesia (Alleva et al. 1986). Male mice raised in all-male litters are less affected by morphine administration, when compared to males reared in litters with a balanced gender composition. Changes in pain sensitivity can also occur in singly-housed rodents ("social isolation"). Rodgers and Hendrie (1983) have shown that male mice kept in isolation present hindpaw licking latencies that are, on average, 80% longer than those of their group-housed counterparts. Other authors, however, using a very similar experimental set-up, failed to show the same effect. Such a discrepancy could be accounted for by strain differences in opiate receptor populations or diurnal variations in opioid activity. Strain differences in pain sensitivity do occur and should be appropriately taken into account. For example, morphine exerts a more powerful analgesic effects on DBA mice, when compared with mice of the C57 strain (Alleva et al. 1980). Unexpected sensitivity is likely to occur in genetically-modified mouse strains, both as a result of unwanted pleiotropic effects or as secondary changes in neural target(s) of genetic disruption. Some of these changes are expected (Lee et al. 1992), while others might be more difficult to foresee. In the above-mentioned case of transgenic animals, it is thus recommendable to start with very low hot-plate temperatures to assess basal pain thresholds before exposing the animals to potential tissue damage or unnecessary pain levels.

In addition to the strain differences mentioned above, species-specific differences in pain sensitivity need to be taken into necessary account.

Standard methodologies should not be transferred from one species to another without a preliminary assessment of responses, which might be unpredictable. For example, different pain thresholds can be found in rodent species when they are exposed to an ecologically-relevant stimulus, such as the odour of a predator (Carere et al. 1999). Differential sensitivity can also occur because of maturational processes occurring during development. Development is a peculiar "transitional" time period during which experimental subjects could be especially vulnerable to pain stimuli. Indeed, we have previously shown that young animals show long latencies in the hot plate test and are more sensitive to the effects of a morphine challenge (Alleva et al. 1987). Furthermore, jumping behaviour is expressed at much higher levels in young CD-1 mice when compared to adult subjects. In young animals, a high frequency of jumping behaviour could be regarded as an index of higher arousability (or enhanced pain perception) although it might also result from an inability of immature subjects to perform adult-like paw-licking responses.

The hot-plate testing is only one among a variety of testing strategies in need of further refinement. Ethological scoring (Alleva et al. 1998), requiring contact between the human experimenter and the animal subject(s) of his/her choice (Visalberghi and Alleva 1998) remains an important tool to standardise scoring methods aimed at minimising animal suffering.

5. CONCLUSIONS

In this contribution we have been able to cover just a few aspects concerning animal welfare and animal experimentation. However, our aim has been to show that an improved level of animal welfare must be pursued before, during and after an experiment. The knowledge of the ethological characteristics and needs of a particular species can play an essential role to this purpose. Only soundly ethical animal experiments will satisfy the general public's demands for humane treatment of laboratory animals. While science does influence the shape of society, the reverse is also true: the communication between science and society must remain open and respectful.

We have focused on the one hand on rodents, these being the most ubiquitous animals in biomedical sciences; on the other on non-human primates, although these latter are the less frequently used animal models. However in most discussions on animal welfare, especially where issues invoking negative subjective emotions are involved, the focus is on non-human primates. Due to the phylogenetic proximity of these animals to men, there is an inclination to believe that non-human primates can suffer in ways similar to humans. Darwinian considerations call for a continuity between the emotions of humans and other animals (Darwin 1872): this point of view carries with it relevant ethical considerations (Rachels 1990).

Hopefully, concerns raised by the use of non-human primates in biomedical sciences will pave the way for more ethical treatment of other vertebrates as well. On this note, it is useful to cite a statement by Marian Dawkins:"Let us not mince words: animal welfare involves the subjective feelings of animals. The growing concern for animals in laboratories, farms, and zoos is not just concern about their physical health, important though that is. Nor is it just to ensure that animals function properly, like well-maintained machines, desirable thought that may be. Rather, it is a concern that some of the ways in which humans treat other animals cause mental suffering and that these animals may experience "pain", "boredom", "frustration", "hunger", and other unpleasant states perhaps not totally unlike those we experience." (Dawkins 1990).

ACKNOWLEDGEMENTS

Work was supported by Italian Ministry of Health project "Biostatistical and ethological approaches for the promotion of welfare of laboratory animals and the quality of experimental data" by intramural ISS research grant to A.V. (Project N 2111/RI-2001/2002) "Quality of social relationship and learning mechanisms: an animal model", and by Intramural ISS research grant to E.A. (Project N 1103/RI-2001/2002) "Neurotrophins and neurobehavioural plasticity: animal models".

REFERENCES

Abbott DH, Hearn JP. Physical, hormonal and behavioural aspects of sexual development in the marmoset monkey (*Callithrix jacchus*). J Reprod Fert 1978; 53:155-166

Alleva E. Assessment of aggressive behavior in rodents. In Methods in Neurosciences: Paradigms for the Study of Behavior. Conn MP ed, Academic Press, New York 1993; vol 14:111-137

Alleva E, Castellano C, Oliverio A. Effects of L- and D-amino acids on analgesia and locomotor activity of mice: their interaction with morphine. Brain Res 1980; 198:249-252

Alleva E, Caprioli A, Laviola G. Postnatal social environment affects morphine analgesia in male mice. Physiol Behav 1986; 36:779-781

Alleva E, Laviola G, Bignami G. Morphine effects on activity and pain reactivity of developing mice with or without late prenatal oxazepam exposure. Psychopharmacology 1987; 92:438-40

Alleva E, Petruzzi S, Ricceri L. Evaluating the social behaviour of rodents: Laboratory, seminaturalistic and naturalistic approaches. In Behavioural Brain Research in Naturalistic and

Semi-naturalistic Settings. Alleva E, Fasolo A, Lipp HP, Nadel L, Ricceri L eds, Kluwer Academic, Dordrecht, The Netherlands 1995; 359-374

Alleva E, Cirulli F, Bianchi M, Bondiolotti GP, Chiarotti F, De Acetis L, Panerai AE. Behavioural characterization of interleukin-6 overexpressing or deficient mice during agonistic encounters. Eur J Neurosci 1998; 12:3664-3672

Amir S, Brown ZW, Amit Z. The role of endorphins in stress: evidence and speculations. Neurosci Biobehav Rev 1979; 4:77-86

Ankier SI. New hot plate tests to quantify antinociceptive and narcotic antagonist activities. Eur J Pharmacol 1974; 27:1-4

Bateson P. Assessment of pain in animals. Anim Behav 1991; 42:827-839

Bigi S, Maestripieri D, Aloe L, Alleva E. NGF decreases isolation-induced aggressive behavior, while increasing adrenal volume in adult male mice. Physiol Behav 1992; 51:337-343

Bigi S, De Acetis L, Chiarotti F, Alleva E. Substance P effects on intraspecific aggressive behaviour of isolated male mice: an ethopharmacological analysis. Behav Pharmacol 1993a; 4:495-500

Bigi S, De Acetis L, De Simone R, Aloe L, Alleva E. Neonatal capsaicin exposure affects isolation-induced aggressive behavior and hypothalamic substance P levels in adult male mice. Behav Neurosci 1993b; 107:363-369

Blanchard RJ, McKittrick CR, Blanchard DC. Animal models of social stress: effects on behavior and brain neurochemical systems. Physiol Behav 2001; 73:261-271

Brain PF, Nowell NW. Isolation versus grouping effects on adrenal and gonadal functions in albino mice: I. The male. Gen Comp Endocrinol 1971a; 16:149-154

Brain PF, Nowell NW. Isolation versus grouping effects on adrenal and gonadal functions in albino mice: II. The female. Gen Comp Endocrinol 1971b; 16:155-159

Brain PF, Stanislaw H. A reevaluation of the effects of differential housing on physiology and behaviour in male and female mice. Aggress Behav 1988; 8:130-132

Bressers WMA, Haccou P, Kruk MR, Meelis M, Van Erp AMM, Willekens-Bramer DC. A time-structured analysis of hypothalamically induced increases in self-grooming and activity in rat. Behav Neurosci 1995; 109:1158-1171

Brug DR, Slotnick BM. Response of colony mice to intruders with different fighting experience. Aggress Behav 1983; 9:49-58

Buchanan-Smith HM. Environmental control: An important feature of good captive Callitrichids environments. In Marmosets and Tamarins in Biological and Biomedical Research. Pryce C, Scott L, Schnell C eds, DSSD Imagery, Salisbury 1997; 47-53

Byron JK, Bodri MS. Environmental enrichment for laboratory marmosets. Lab Anim Europe 2001; 1:34-37

Cairns RB, Hood KE, Midlam J. On fighting in mice: Is there a sensitive period for isolation effects? Anim Behav 1985; 33:166-180

Carere C, Casetti R, De Acetis L, Perretta G, Cirulli F, Alleva E. Behavioural and nociceptive response in male and female spiny mice (Acomys cahirinus) upon exposure to snake odour. Behav Proc 1999; 47:1-10

Caston J, Devulder B, Jouen F, Lalonde R, Delhaye-Bouchaud N, Mariani J. Role of an enriched environment on the restoration of behavioral deficits in Lurcher mutant mice. Dev Psychobiol 1999; 35:291-303

Cirulli F, Adriani W, Laviola G. Sexual segregation in infant mice: Behavioural and neuroendocrine responses to d-amphetamine administration. Psychopharmacology 1997; 134:140-152

Cirulli F, Berry A, Alleva E. Early disruption of the mother-infant relationship: effects on brain plasticity and implications for psychopathology. Neurosci Biobehav Rev 2003; 27:73-82

Cotman C, Berchtold NC. Exercise: a behavioural intervention to enhance brain health and plasticity. Trends Neurosci 2002; 25:295-300

Crowcroft P, Rowe FP. Social organization and territorial behaviour in the wild house (Mus musculus L.). Proc Zool Soc Lond 1963; 140:517-531

Crawley JN, Schleidt WM, Contrera JF. Does social environment decrease propensity to fight in male mice? Behav Biol 1975; 73-83

Darwin C. The Expression of the Emotions in Man and Animals. Murray, London 1872.

Dawkins MS. Behavioural deprivation: a central problem in animal welfare. Appl Anim Behav Sci 1988; 20:209-225

Dawkins MS. From an animal's point of view: motivation, fitness, and animal welfare. Behav Brain Sci 1990; 13:1-25

Della Seta D, De Acetis L, Aloe L, Alleva E. NGF effects on hot plate behaviors in mice. Pharmacol Biochem Behav 1994; 49:701-705

DeRosa C, Vitale A, Puopolo M. The puzzle-feeder as feeling enrichment for common marmosets (*Callithrix jacchus*): a pilot study. Lab Anim 2003; 37:100-107

Digby LJ. Sexual behaviour and extragroup copulation in a wild population of common marmosets (*Callithrix jacchus*). Folia primatol 1999, 70:136-145

D'Udine B, Alleva E. On the teleonomic study of maternal behaviour. In The Dialetics of Biology Group. Muir A, Rose S eds, Allison and Busby, London, UK 1980; 50-61

Duncan IJH, Fraser D. Understanding animal welfare. In Animal Welfare. Appleby MC, Hughes BO eds, CAB International, Wallingford 1997; 19-31

Eddy NB, Touchberry CF, Lieberman JE. Synthetic analgesics. I. Methadone isomers and derivatives. J Pharmacol Exp Ther 1950; 98:121-37

Estep DQ, Lanier DL, Dewsbury DA. Copulatory behaviour and nesting-building behaviour of wild house mice (Mus musculus L.). Anim Learn Behav 1975; 3:329-336

European Commission. Third report from the Commission to the Council and the European Parliament on the statistics on the number of animals used for experimental and other scientific purposes in the member States of the European Union 2003a.

European Commission. The welfare of non-human primates used in research. Report of the scientific committee on animal health and animal welfare 2003b; http://europa.eu.int/comm/food/fs/sc/scah/out83_en.pdf

Flecknell PA. Refinement of animal use - assessment and alleviation of pain and distress. Lab Anim 1994; 28:222-231

Fraser AF. The behaviour of suffering in animals. Appl Anim Behav Sci 1984; 13:1

Galef BG, Sorge RE. Use of PVC conduits by rats of various strains and ages housed singly and in pairs. J Appl Anim Welf Sci 2000; 3:279-292

Hubrecht RC. Home-range size, use and territorial behavior in the common marmoset (Callithrix jacchus jacchus), at the Tapacurà Field Station, Recife, Brazil. Int J Primatol 1985; 6:553-560

Hunskaar S, Berge OG, Hole KA. Modified hot-plate test sensitive to mild analgesics. Behav Brain Res 1986; 21:101-108

Inglett BJ, French JA, Simmons LG, Vires KW. Dynamics of intrafamily aggression and social reintegration in lion tamarins. Zoo Biol 1989; 8:67-78

International Association for the Study of Pain. Report of subcommittee on taxonomy. Pain 1979; 6: 249-252

Kleinman DG. Characteristics of reitroduction and sociosexual interactions in pairs of lion tamarins (*Leontopithecus rosalia*) during the reproductive cycle. In Biology and Conservation of the Callitrichidae. Kleinman DG ed, Smithsonian Institution Press, Washington DC 1979; 181-190

Laviola G, De Acetis L, Bignami G, Alleva E. Prenatal oxazepam enhances mouse maternal aggression in the offspring, without modifying acute chlordiazepoxide effects. Neurotoxicol Teratol, 1991; 13:75-81

Lee KF, Li E, Huber LJ, Landis SC, Sharpe AH, Chao MV, Jaenisch R. Targeted mutation of the gene encoding the low affinity NGF receptor p75 leads to deficits in the peripheral sensory nervous system. Cell 1992; 69:737-749

Maestripieri D, De Simone R, Aloe L, Alleva E. Social status and nerve growth factor serum levels after agonistic encounters in mice. Physiol Behav 1990; 47:161-164

McGrew WC, McLuckie EC. Phylopatry and dispersion in the cotton-top tamarin, Saguinus o. oedipus: An attempted laboratory simulation. Int J Primatol 1986; 7:401-422

McGrew WC, Brennan JA, Russell J. An artifical "gum-tree" for marmosets (*Callithrix j. jacchus*). Zoo Biol 1986; 5:45-50

Mittermaier RA, Rylands AB, Coimbra-Fihlo AF, Fonseca GAB eds. Ecology and Behaviour of Neotropical Primates. World Wildlife Fund, Washington DC 1988

Nelson RJ, Chiavegatto S. Molecular basis of aggression. Trends Neurosci 2001; 24:713-719

Nevison CM, Hurst JL, Barnard CJ. Strain-specific effects of cage enrichment in male laboratory mice (*Mus musculus*). Anim Welf 1999; 8:361-379

Newberry RC. Environmental enrichment: increasing the biological relevance of captive environments. Appl Anim Behav Sci 1995; 44:229-243

Ostermeyer NC. Maternal aggression. In Parental Behaviour of Rodents. Elwood RW ed, Wiley, Chichester 1983: 151-179

Petto AJ, Devin M. Food choices in captive common marmoset (*Callithrix jacchus*). Lab Primate Newsl 1988; 27:7-9

Pietropaolo S, Branchi I, Chiarotti F, Alleva E. Utilization of a physically-enriched environment by laboratory mice: age and gender differences. Submitted.

Rachels J. Created from Animals: The Moral Implications of Darwinism. Oxford University Press, Oxford 1990

Reinhardt V. Caged rhesus monkeys voluntary work for ordinary food. Primates 1994; 35:95-98

Reinhardt V, Roberts LR. Effective feeding enrichment for non-human primates: a brief review. Anim Welf 1997; 6:265-272

Rodgers RJ, Hendrie CA. Social conflict activates status-dependent endogenous analgesic or hyperalgesic mechanisms in male mice: effects of naloxone on nociception and behaviour. Physiol Behav 1983; 30:775-780

Rosenzweig MR, Krech D, Bennett EL. Heredity environment, brain biochemistry, and learning. In Current Trends in Psychological Theory. Pittsburg PA, University of Pittsburgh Press 1961; 87-110

Rothe H, Darms K. The social organisation of marmosets: a critical evaluation of recent concepts. In Marmosets and Tamarins: Systematics, Behaviour and Ecology. Rylands AB ed, Oxford University Press, Oxford 1993; 176-199

Rothe H, Koenig A. Variability of social organisation in captive common marmoset (*Callithrix jacchus*). Folia primatol 1991; 57:28-33

Russell WMS, Burch RL. The Principles of Humane Experimental Technique. University Federation for Animal Welfare, South Mimms, England 1959

Rylands AB. The Callitrichidae: a biological review. In Marmosets and Tamarins in Biological Research. Pryce C, Scott L, Schnell C eds, DSSD Imagery, Salisbury 1997; 1-9

Saylor A, Salmon M. Communal nursing in mice: Influence of multiple mothers and growth of the young. Science 1969; 164:1309-1310

Schnaffer CM, Caine NG. The peacefulness of cooperatively breeding primates. In Natural Conflict Resolution. Aureli F, De Waal FBM eds, University of California Press, Berkeley 2000; 155-169

Sherwin CM. Preferences of individually housed TO strain laboratory mice for loose substrate or tubes for sleeping. Lab Anim 1996; 30:245-251

Sherwin CM. Observations on the prevalence of nest building in non-breeding TO strain mice and their use of two nesting materials. Lab Anim 1997; 31:125-132

Smuts BB, Cheney DL, Seyfarth RM, Wrangham RW, Struhsaker TT. Primate Societies. Chicago University Press, Chicago 1987

Spillantini MG, Aloe L, Alleva E, De Simone R, Goedert M, Levi-Montalcini R. Nerve growth factor mRNA and protein increase in hypothalamus. Proc Natl Acad Sci USA 1989; 86:8555-8559

Stevenson MF, Poole TP. An ethogram of the common marmoset (*Callithrix jacchus jacchus*): general behavioural repertoire. Anim Behav 1976; 24:428-451

Taber RI. Predictive value of analgesic assays in mice and rats. In Narcotic Antagonists. Advances in Biochemical Psychopharmacology. Braude MC, Harris LS, May EL, Smith JP, Villarreal JE eds, Raven Press, New York 1974; 191-212

Tomasello M, Call J. Animal Cognition. Oxford University Press, New York 1997

Viercki CJ Jr, Cooper BY. Guidelines for assessing pain modulation in laboratory animal subjects. In Advances in Pain Research and Therapy. Kruger L, Liebeskind JC eds, Raven Press, New York 1984; 305-22

Visalberghi E, Alleva E. Book review of: The Inevitable Bond: Examining Scientist-Animal Interactions (Davis D and Balfour D, eds. Cambridge UP 1992). Quart J Exptl Psychol 1998; 46B:222-224

Vitale A, Alleva E. Ethological and welfare considerations in the study of aggression in rodents and nonhuman primates. In Animal Models of Human Emotion and Cognition. Haug M, Whalen RE eds, American Psychological Association, Washington DC 1999; 283-295

Wiesel TN, Hubel DH. Single-cell responses in striate cortex of kittens deprived of vision in one eye. J Neurophysiol 1963; 26:1003-1017

Woolfe G, MacDonald AD. The evaluation of the analgesic action of Pethidine hydrochloride (Demerol). J Pharmacol Exp Ther 1944; 80:300-307

Wright EM Jr, Marcella KL, Woodso JF. Animal pain: evaluation and control. Lab Anim 1985; 20

Index

GENERAL PRINCIPLES

3R's 15, 19, 37, 81
AAALAC 21, 46, 47
AALAS 22
accreditation 21
ad libitum 53, 61, 64, 68, 69, 76
anaesthesia 86, 87
    alpha-chloralose 100
    alphaxolone-alphadolone 98
    amphibia 103
    apnoea 89
    artificial ventilation 91, 100
    atipamezole 99, 101
    body temperature 90
    chloral hydrate 100
    choloroform 93
    circuits 95, 96, 97
    dehydration 101
    depth 88
    emergencies 102
    EMLA 103
    ether 93
    fentany/fluonisone/midazolam 99
    fish 103
    fluorinated hydrocarbons 93, 94
    gaseous agents 93
    halothane 93
    injectable agents 97, 98
    intra-operative care 101
    intubation 95, 96
    isoflurane 93
    ketamine 99
    local 102
    mean alveolar concentration 94, 95
    medetomidine 99
    minute volume 94
    neuromuscular blocking agents 91
    nitrous oxide 95
    pentobarbitone 99
    premedication 90
    propofol 98
    recovery 102
    reversal 101
    sedatives 91
    stages 89
    starvation 92
    tranquillisers 91
    tribomoethanol 99
    urethane 100
    zylazine 99
analgesia 103
    buprenorphine 105
    non-steroidal anti-inflammatory drugs
        NSAIDS 105, 106
    opioids 105
    pain treatment 103, 105
    pre-emptive 105
    signs of pain 104
animal care and housing 16
animal welfare 3, 4, 12, 15, 16, 24, 42, 51, 52, 62, 66, 73, 76, 81
Appendix A 44, 45, 46, 47

INDEX

barrier 25, 27, 33
bedding 40, 41, 43, 45
biological functioning 4, 10, 11, 51, 53, 57, 60
cage cleaning 40
cage height 45, 46
CCAC 18
choice test 6
circadian clock 65
circadian rhythm 51, 67, 68, 77
competency 20
contamination 24, 26
conventional 26, 27
coping 9, 10, 42, 61
coprophagy 55
cost-benefit analysis 15, 20
Council of Europe 19, 42, 44, 45, 46
density 42, 43
diagnostic laboratories 30
dietary composition 58
diets 55
    natural-ingredient 55, 56
    purified 56
diseases 23
education 21
embryonic stem (ES) cells 26
energy expenditure 66
energy requirements 53
enrichment 16, 41, 42, 45, 74
environment 8, 37, 42, 45
Environmental Protection Agency 58
ethical evaluation 15, 19
European Convention .. 16, 20, 42, 44, 45, 47
European Directive 16, 20, 44
European Parliament 20, 21
euthanasia 106, 107
    physical methods 107
    anesthetic overdose 107
    carbon dioxide 108
    human endpoints 108, 109
    signs for euthanasia 110
experimental procedures 83-86
    administration 83
    blood sampling 84
    faeces sampling 86
    good practices 83, 85
    saliva sampling 86
    urine sampling 86

fasting 72, 73
feeding 57, 68
feeding restricted 64, 67, 69, 76
feeding schedules 60, 76
feelings 5, 6
FELASA 20, 21, 28, 29, 30, 31
five freedoms 12
floor area 42, 45, 46
floor solid 43, 45
floor grid 43, 45
floor perforated 43
food deprivation 72
food restriction 61
food-anticipatory behaviour 67
frustration 12
gavage 63
genetically modified animals 25, 33
gnawing 43, 57
gnotobiotic 26
group housing 16, 42, 45, 70, 72, 76
group size 42
Guide for the Care and Use 17, 18, 46, 47
GV-SOLAS 28
handling 43, 82
harmonisation 16
health monitoring ... 24, 27, 29, 30, 31, 32
health reports 25, 33
health status 25, 27
hedonism 5, 11
homeostasis 38, 51
homeothermic zone 39
housing and care 37, 40, 42, 44, 45
housing individual/single ... 40, 42, 70, 72
human contact 43
human endpoints 108, 109
hybrid view 11, 12
IACUC 18, 21, 46, 47
immunodeficient animals 25, 28
individually ventilated cage IVC ... 26, 39, 41
infections 23, 24, 25, 33
ingestive behaviour 67
isolator 26, 33
legislation 44
long-term effects 8
macroclimate 39
MAP/RAP test 26
meal size 66
meal training 63

mental states 5
microbiological quality 24, 26
microclimate 39
micro-organisms 24, 27, 28
murine pathogen-free MPF 27
narrow-hedonism 5
National Research Council 57, 58, 62
natural behaviour 4, 10, 12, 63, 64, 73
natural environment 10
natural functioning 12
neophobic reaction 69
nephrocalcinosis 56
non-barrier 33
nutritional requirements 53
    fats 54
    National Research Council 53, 54
    proteins 54
    vitamins 54
operant test 6
Paigen's diet 59
pair-feeding 62
palatability 57
pathogen-free 27
pathogens 24, 25, 28, 31
PCR 26
perfectionism 9, 11, 12
perforated floor 43
poor welfare 10, 11, 12
power 38
preferences 6, 7, 8, 9, 11, 12, 52, 57, 74, 76
preference theory 4
preference-hedonism 5
Public Health Service PHS 17, 21
reduction 15, 18, 19, 37
refinement 15, 18, 19, 37, 82
replacement 15, 18, 19, 37
scientific integrity 19
sentinels 32
social status 70
space requirements 16, 42
species-specific behaviours 11
specific pathogen free SPF 26, 27
standardisation 37, 41, 52, 60, 62
stereotypes 69, 73
stress signs 82
stomach-tubing 63
subclinical infections 23
suffering 11

temperature 39
thermogenic effect 51
thermoregulation 70
toxic levels 58
training 21, 43
U.S. Animal Welfare Act 17, 21, 46
value-laden presumptions 3
variation 23, 24, 37, 38, 39, 41, 55, 56, 77
virus-antibody-free VAF 27
western-type diet 59, 60
working for food 74, 75, 76
xenotransplantation 26

SPECIES SPECIFIC
Mouse
acclimatisation 136
ad libitum 135
aggression 123
animal welfare 120, 121
Appendix A 131
barbering 134
bedding 130, 131, 132
cages 131
cannibalism 126
circadian rhythm 122
coping 121, 147
Council of Europe 131
diseases 137
embryonic stem (ES) cells 120
enrichment 123, 127, 128, 130, 135
environment physical . 127, 128, 129, 130
environment social 128, 129
estrous cycle 125
European Convention 131
European Directive 131, 133
experimental techniques 138
    administration 143, 144, 145
    anaesthesia, analgesia 145
    body temperature 140
    euthanasia 146
    handling 139
    identification 138
    post-operative care 145
    sampling blood 141, 142
    sampling faeces, urine 143
    sampling milk, sperm 143
    sexing 139
fasting 135
feeding 135

genetic monitoring 120
genetically modified 120, 121
Guide for the care and use 131
health 136
health monitoring 137
homeostasis 137
housing 129
    group 123, 134
    individual 123, 134
humidity 133
inbred 119, 120, 123
individually ventilated cages IVC 132
life span 125
lighting 133
macroenvironment 132
microbiological status 136, 137
microenvironment 132
nest building 126
nesting material 128, 130, 131, 133
noises 133
pheromones 125
preferences 133, 147
reproduction 125
senses hearing, olfaction, taste, touch,
    vision 124
sexual maturity 125
social behaviour 123
social housing 134
space recommendations 131, 132
standardisation 120, 128
stereotypies 137
temperature 132, 133
transgenic 120, 121
transport 135
water 135
welfare assessment 147

Rat

ad libitum 170
Appendix A 160
bedding 162, 163
behavioural needs 158
cage grid floor 161, 162
cage material 160
cage size 159, 160
cage solid bottom 161, 162
circadian rhythm 154
coprophagy 155
Council of Europe 160

development 155
disease models 173
diseases 172
enrichment 164, 165, 166
environment 158, 164, 165, 166
environment natural 154
experimental techniques 170, 171, 172
    best practices 170
    handling 169, 171
feeding 170
group size 167
health 172
health monitoring 173
Home Office 160
housing group 166, 167
husbandry procedurs 168, 169
hypothalamic-pituitary-adrenal 156
individually ventilated cages IVC 169
infections 172, 173
lifespan 155
macroenvironment 159
maternal aggression 168
microenvironment 159
National Research Council 160, 170
nesting material 164,165
nutritional requirements 155
post-natal period 156
preferences 161
restricted feeding 170
senses smell, touch, hear, vision 155
sexual maturity 155
signs for poor welfare 172
social play 157
social stress 167
standardisation 166
stress hyporesponsive period SHRP ... 157
stress responses 169
variation 166

Guinea pig

ad libitum 201
aggression 195
bedding 202
behaviour 184, 185
behavioural masculinisation 197, 198
cage size 202
care and handling 203
circadian rhythm 187
development 187

INDEX

domestication 184
dominance 191, 192, 195
environment 202, 204
experimental techniques 203
feeding 201
housing general 202
natural behaviour 183
origin 181, 182, 183
pituitary-adrenocortical 186, 189, 192, 193, 195, 196
senses auditory, olfactory, vision 187
sexual maturity 187
social bonding 193, 194
social housing 198, 199, 200, 201
social interactions 185
social organisation 190, 191, 192
social prenatal 197
social support 193
stress response 186
sympathetic-adrenomedullary ... 186, 189, 192, 193, 196
welfare assessment 188

Rabbit
aggressiveness 216, 221, 227
ammonia 222
behaviour 217, 218
behaviour abnormal 218, 219
behaviour foraging 217
behaviour natural 215
behaviour social 216
behavioural needs 220
breeding group 215
Appendix A 224, 225
Council of Europe 224, 225
cage floor area 225
cage height 224
cage type 224, 226
care and handling 231
castration 228
circadian rhythm 217, 221
Convention of Europe 223
development 214
diet, natural 213
dietary enrichment 230
diseases 231, 232
domestication 212
exercise arena 228
experimental techniques 235

euthanasia 236
immunisation 235
post operative care 236
feeding 230, 235
FELASA 232
Guide for the care and use 223
handling 231
health 231, 232
Health monitoring 232
Housing floor area 229
Housing groups 226, 227, 228, 229
Housing in cages 224
Housing in floor pens 226
Housing individually 228
humidity 221
hygiene 222
hnfections baterial, viral, parasites 233
injuries 234
origin 212
ovulation 214
reproduction 213, 214
scent glands 216
senses scent, taste, hearing, vision 213
sexual maturity 213
social organisation 216
sounds 221
Specific Pathogen Free SPF 232
stereotypies 218
temperature 213, 221
transportation 222, 223, 224
water 231
welfare research 237
ventilation 222
WHHL 232
World Rabbit Science Association 224, 225
Virus Antibody Free VAF 232

Dog
ad libitum 264
aggression 252, 257
Appendix A 245, 253
barking 248
beagle 249
behaviour abnormal 249
breeding 268
care 264, 265
clinical health examination 267
conspecifics 251

coping 249
Council of Europe 251, 254, 258
development 250
diet 249, 264
diseases 266, 267
domestication 246
enclosure 256
enrichment 260, 261, 263
environment 249, 250, 253
European Convention 245, 253
experimental procedures 250, 269
feeding 263
group size 258
handling 250, 269
health 266
halth conditions 268
health screening 268
housing in groups/social 251, 257, 258
housing single 251, 258, 269
housing 256, 257, 260, 261, 270
infections 266
maturation 247
National Research Council 254
natural behaviour 247
natural environment 258
noises 255, 263
nutrition 263
origin 246, 247
outdoor area 259
pain 265
pheromones 248
senses hearing 248, 254, 255
senses olfaction 248, 254, 263
senses taste, vision 248
separation 251
sleeping area 253
social contacts 247, 252
social needs 251, 252
socialisation 250
space requirements 256, 258, 259
stereotypies 249, 256
stress behavioural indications 265
temperament 268
temperature 253
vaccination 266
vocalisations 248

Pigs and minipigs
abnormal behaviours 280
acclimatisation 284
ad libitum 283
aggression 277, 282
anatomy 278
bedding 281
behaviour 277
behavioural measurements 279
breeds 276
circadian rhythm 277
diseases 279, 280
domestication 276, 277
enrichment 281
European Union 275
feeding *ad libitum*/restricted 283
flooring 281
genetically engineered 284, 285
housing individual 281
humidity 282
injuries 279, 280
lighting 282
miniature pigs 277
natural behaviour 277
noise 282, 283
nutrient requirements 283
origin 276, 277
physiology 278
research 284
social grouping 281, 282
space allowances 283
stereotypies 280, 281
temperature 282
transgenic 284, 285, 286
welfare assessment 279
xenotransplantation 275, 285, 286

Non-human primates
abnormal behaviour 294, 302
adaptation 301
aggression 294, 295
anthropomorphism 293
anxiety 307
Appendix A 296
breeding 305
cage size 296
coping 301
Council of Europe 296
diet 299
discomfort 300
diseases 300, 303

distress 302
enrichment 302, 303
environment 302, 303
European Commission 292
experimental procedures 298
fear 301
five freedom 299
food 299
handling 301, 304
health hazards 304
housing group 295, 300
housing in groups 294
housing pair- 295, 296
housing singly 292, 295, 296
injuries 300
natural behaviour 302, 303
pain 300, 307
research 291, 292
social needs 294
stereotypies 302
stressors 307
time budget 295
transportation 306
water 299
weaning 305
welfare 296, 293
welfare assessment 306, 307, 308
well-being 292, 293

Ethological perspectives
non-human primates
    aggressiveness 325, 326, 327
    colony management 325, 326, 327
    enrichment 320, 321, 322
    European Commission 319, 320
    feeding enrichment 321
    housing 320
    natural behaviour 320
    New World Monkeys 319
    Old World Monkeys 319
    social systems 325
rodents
    ageing 318
    aggressiveness 318, 323, 324
    development 318
    enrichment physical/social ... 316, 317
    ethogram 322, 323, 324, 331
    hot-plate test ....
    .... 328, 329, 330, 331, 332, 333

housing individual 323
maternal behaviour 318
nociception 327, 328
pain 327, 329, 333
pain sensitivity 328, 329, 330, 331
periadolescent 317
social experience 318, 331, 332
social stress 324

Printed in the United Kingdom
by Lightning Source UK Ltd.
128786UK00001B/5/A